J. Paulo Davim

Editor

Machining

Fundamentals and Recent Advances

 Springer

J. Paulo Davim
Department of Mechanical Engineering
University of Aveiro
Campus Universitário de Santiago
3810-193 Aveiro
Portugal

ISBN 978-1-84800-212-8   e-ISBN 978-1-84800-213-5

DOI 10.1007/978-1-84800-213-5

British Library Cataloguing in Publication Data
Machining : fundamentals and recent advances
  1. Machining
  I. Davim, J. Paulo
  671.3'5
ISBN-13: 9781848002128

Library of Congress Control Number: 2008931573

Cover design: eStudio Calamar S.L., Girona, Spain

Printed on acid-free paper

9 8 7 6 5 4 3 2 1

springer.com

Preface

Machining is the broad term used to describe the removal of material from a workpiece and is one of the most important manufacturing processes. Parts manufactured by other processes often require further operations before the product is ready for application. Machining operations can be applied to work metallic and non-metallic materials such as polymers, wood, ceramics, composites and exotic materials.

M. E. Merchant has written *"Today in industrialized countries, the cost of machining amounts to more than 15% of the value of all manufactured products in those countries."* For this reason and others, machining as part of manufacturing science and technology is very important for modern manufacturing industries.

This book aims to provide the fundamentals and the recent advances in machining for modern manufacturing engineering.

The first three chapters of the book provide the fundamentals of machining with special emphasis on three important aspects: metal cutting mechanics (finite element modelling), tools (geometry and material) and workpiece surface integrity.

The remaining chapters of the book are dedicated to recent advances in machining, namely, machining of hard material, machining of particulate-reinforced metal matrix composites, drilling polymeric matrix composites, ecological machining (near-dry machining), sculptured surface machining, grinding technology and new grinding wheels, micro- and nanomachining, advanced (non-traditional) machining processes and intelligent machining (computational methods and optimization).

The present book can be used as a textbook for a final undergraduate engineering course or for a course on machining at the postgraduate level. However, in general, this textbook can be used for teaching modern manufacturing engineering. It can also serve as a useful reference for academics, manufacturing and materials researchers, manufacturing and mechanical engineers, and professionals in machining and related industries. The scientific interest of this book is evident from the many important centres of research, laboratories and universities in the world working in this area. Therefore, it is hoped that this book will inspire and enthuse further research in this field of science and technology.

I would like to thank Springer for the opportunity to publish this book and for their competent and professional support. Finally, I would like to thank all the chapter authors for making their work availability for this book.

University of Aveiro, Portugal,   *J. Paulo Davim*
December 2007

Contents

1

Metal Cutting Mechanics, Finite Element Modelling

Viktor P. Astakhov[1] and José C. Outeiro[2]

[1]General Motors Business Unit of PSMi, 1255 Beach Ct., Saline MI 48176, USA.
Email: astvik@gmail.com
[2]Portuguese Catholic University, Faculty of Engineering, Estrada Octávio Pato,
2635-631 Rio de Mouro, Lisboa, Portugal.
Email: jcouteiro@crb.ucp.pt

This chapter presents a short analysis of the basics of traditional metal cutting mechanics, outlining its components and the basics of finite element modelling (FEM) of the metal cutting process. Based on a previously proposed definition of metal cutting, advanced metal cutting mechanics considers the power spent in metal cutting as the summation of four components: the power spent on the plastic deformation of the layer being removed, the power spent on the tool–chip interface, the power spent on the tool–workpiece interface, and the power spent in the formation of new surfaces (cohesive energy). Energy partition in the cutting system and the relative impact of the parameters of the machining regime are discussed. Analyzing the basics of FEM and presenting examples, this chapter considers the errors in such modelling and their major sources. It points out the importance of the selection, verification and validation of the physically justifiable model.

1.1 Advanced Metal Cutting Mechanics

1.1.1 Objective of Metal Cutting Mechanics

The major objectives of metal cutting mechanics are to determine the cutting force and the cutting power through analyzing the thermomechanical processes involved in the cutting process.

1.1.2 State of the Art

Although these objectives can reportedly be achieved by analyzing a relevant model of the cutting process and all available experimental information, this is not the case in practice. The old trial-and-error experimental method, originally developed in the middle of the 19th century (well summarized in [1]) is still in wide use in metal cutting research and development activities. Its modern form, known as the unified or generalized mechanics approach, has been pursued by Armarego and co-workers for years [1], then spreading as the mechanistic approach in metal cutting [2]. It was developed as an alternative to metal cutting theory because the latter did not prove its ability to solve even the simplest of practical problems.

One of the most important yet least understood operation parameters of a machining operation is the cutting force. In general, this force is thought of as a three-dimensional (3D) vector which is represented by three components, namely, the power component, the radial component and the axial component in the tool coordinate system as shown in Figure 1.1(a) [3]. Out of these three components, the largest is normally the power component which is often called the cutting force. This simplification will be used through the body of this Section. As this force is of high importance, one might assume that theoretical and experimental methods for its determination have been developed and thus be available in the literature. Unfortunately, this is not the case.

When it comes to the possibility of theoretical determination, the foundation of the force and energy calculations in metal cutting is based on the oversimplified orthogonal force model known as Merchant's force circle diagram or a condensed force diagram [4, 5] shown in Figure 1.1(b). In this figure, the total cutting force R is resolved into the tool face–chip friction force F and the normal force N. The angle μ between F and N is thus the friction angle. The force R is also resolved along the shear plane into the shear(ing) force, F_s which, in Merchant's opinion, is responsible for the work expended in shearing the metal, and the normal force F_n, which exerts a compressive stress on the shear plane. The force R is also resolved along the direction of tool motion into F_c, termed by Merchant the cutting force, and F_T, the thrust force.

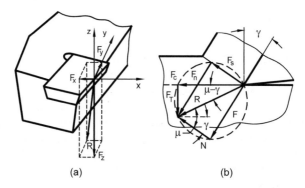

(a)   (b)

Figure 1.1. The force system in cutting: (a) turning and (b) the orthogonal force model proposed by Merchant

The determination of the cutting force is based upon the calculation of the shearing force, F_s. In 1941 Ernst and Merchant [6] proposed the following equation to determine this force:

$$F_s = \frac{\tau_y A_c}{sin\,\varphi} \qquad (1.1)$$

where τ_y is the shear strength of the work material, φ is the shear angle and A_c is the area of shearing (the uncut chip area, equal to the product of the uncut chip thickness and the uncut chip width).

According to Ernst and Merchant, the work material deforms when the stress on the shear plane reaches its shear strength. Later researchers published a great number of papers showing that τ_y should be thought of as the shear flow stress, which is somewhat higher than the yield strength of the work material depending upon particular cutting conditions. Still, this stress remains today the only relevant characteristic of the work material characterizing its resistance to cutting.

It follows from Figure 1.1(b) that

$$F_c = \frac{F_s \cos(\mu - \gamma)}{\cos(\varphi + \mu - \gamma)} \qquad (1.2)$$

and combining Equations (1.1) and (1.2), one can obtain

$$F_c = \frac{\tau_y A_c \cos(\mu - \gamma)}{sin\,\varphi \cos(\varphi + \mu - \gamma)} \qquad (1.3)$$

The cutting power P_c is then calculated as

$$P_c = F_c v \qquad (1.4)$$

This power defines the energy required for cutting, cutting temperatures, plastic deformation of the work material, machining residual stress and other parameters.

However, everyday practice of machining shows that these considerations do not much reflect reality even to the first approximation. For example, machining of medium carbon steel AISI 1045 (tensile strength, ultimate $\sigma_R = 655$ MPa, tensile strength, yield $\sigma_{y0.2} = 375$ MPa) results in much lower total cutting forces (Figure 1.8 in [7]), greater tool life, lower required power, cutting temperature, machining residual stresses than those obtained in the machining of stainless steel AISI 316L ($\sigma_R = 517$ MPa; $\sigma_{y0.2} = 218$ MPa) [8]. The prime reason for that is that any kind of strength of the work material in terms of its characteristic stresses cannot be considered alone without corresponding strains, which determine the energy spent in deformation of the work material [9, 10]. Only when one knows the stress and corresponding strain, can one calculate other parameters and outcomes of the metal cutting process [9].

When it comes to experimental determination of the cutting force, there are at least two problems:

- The first and foremost is that the cutting force cannot be measured with reasonable accuracy. Even if extreme care is taken, a 50% variation is found [11].
- Second, many tool and cutting inserts manufacturers (not to mention manufacturing companies), do not have adequate dynamometric equipment to measure the cutting force. Many dynamometers used in the field are not properly calibrated because the known literature sources did not provide proper experimental methodology for cutting force measurements using piezoelectric dynamometers [12].

Therefore, to make practical calculations of the cutting force and thus energy spent in machining another approach has to be found. This Section is to present such an advanced approach.

1.1.3 Advanced Methodology

The proposed methodology is based on the definition of the metal cutting process proposed by Astakhov [10] and on the model of energy partition in the metal cutting system developed using this definition (Figure 2.1 in [7]). According to this model, the power balance in the cutting system can be written as

$$P_c = F_c v = P_{pd} + P_{fR} + P_{fF} + P_{ch} \tag{1.5}$$

from which the cutting force is calculated as

$$F_c = \frac{P_{pd} + P_{fR} + P_{fF} + P_{ch}}{v} \tag{1.6}$$

where P_{pd} is the power spent on the plastic deformation of the layer being removed, P_{fR} is the power spent at the tool–chip interface, P_{fF} is the power spent at the tool–workpiece interface and P_{ch} is the power spent in the formation of new surfaces.

The power spent on the plastic deformation of the layer being removed, P_{pd}, can be calculated knowing the chip compression ratio and parameters of the deformation curve of the work material as follows [7, 9]

$$P_{pd} = \frac{K\left(1.15 ln\,\zeta\right)^{n+1}}{n+1} v A_w \tag{1.7}$$

where K is the strength coefficient (N/m^2) and n is the hardening exponent of the work material, ζ is the chip compression ratio [7, 9, 10] and A_w is uncut chip cross-sectional area (m^2)

$$A_w = d_w f \tag{1.8}$$

where d_w is the depth of cut (m) and f is the cutting feed per revolution (m/rev).

The practical methods of experimental determination of the chip compression ratio ζ are discussed for various machining operation by Astakhov [7, 9].

Power spent due to friction at the tool–chip interface is calculated as

$$P_{fR} = \tau_c l_c b_{1T} \frac{v}{\zeta} \qquad (1.9)$$

where $\tau_c = 0.28\sigma_{UTS}$ is the average shear stress at the tool–chip contact (N/m^2) [7], σ_{uts} is the ultimate tensile strength of the work material (N/m^2), l_c is the tool–chip contact length (m) and b_{1T} is the true chip width (m).

The tool–chip contact length is calculated as [7]

$$l_c = t_{1T} \zeta^{1.5} \qquad (1.10)$$

where t_{1T} is the true uncut chip thickness (m) [7].

The true uncut chip thickness and the true chip width depend on the configuration of the projection of the cutting edge into the main reference plain. Formulae to calculate t_{1T} and b_{1T} for various configuration presented by Astakhov [7, 10]. The most common case of machining is when the cutting insert, with tool cutting edge angle κ_r and tool minor cutting edge angle κ_{r1}, is made with a nose radius r_n and set so that the depth of cut d_w is greater than the nose radius, as shown Figure 1.2. If the following relationships are justified

$$d_w \geq r_n \left(1 - \cos \kappa_r\right), \quad f \leq 2r_n \sin \kappa_{r1} \qquad (1.11)$$

then the formulas for calculation of t_{1T} and b_{1T} are as follows

$$t_{1T} = \frac{f}{c_1} \sin \arctan \frac{c_1}{\left[1 - e_1(1 - \cos \kappa_r)\right] \cot \kappa_r + e_1(\sin \kappa_r + g_1)} \qquad (1.12)$$

and

$$b_{1T} = \frac{c_1 d_w}{\sin \arctan \dfrac{c_1}{\left[1 - e_1(1 - \cos \kappa_r)\right] \cot \kappa_r + e_1(\sin \kappa_r + g_1)}} \qquad (1.13)$$

where

$$g_1 = \frac{f}{2r_n}, \quad e_1 = \frac{r_n}{d_w}, \quad c_1 = 1 - e_1\left(1 - \sqrt{1 - g_1}\right) \qquad (1.14)$$

Power spent due to friction at the tool–workpiece interface is calculated as

$$P_{fF} = F_{fF} v \qquad (1.15)$$

where F_{fF} is the friction force on the tool–workpiece interface

$$F_{fF} = 0.625 \tau_y \rho_{ce} l_{ac} \sqrt{\frac{Br}{\sin \alpha^*}} \qquad (1.16)$$

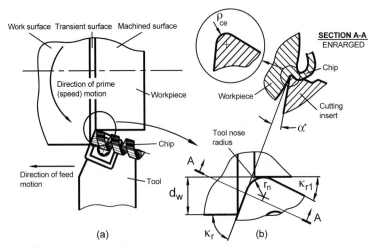

Figure 1.2. Visualization of terms used in the considered case of machining

where τ_y is the shear strength of the tool material (N/m^2), ρ_{ce} is the radius of the cutting edge (m), α^* is the normal flank angle (deg) and l_{ac} is the length of the active part of the cutting edge (the length of the cutting edge engaged in cutting) (m). In the considered case (Figure 1.2)

$$l_{ac} = r_n\left(0.018\kappa + \frac{r_n + \cos\kappa}{\sin\kappa}\right) \tag{1.17}$$

Br is the Briks similarity criterion [7, 10],

$$Br = \frac{\cos\gamma}{\zeta - \sin\gamma} \tag{1.18}$$

where γ is the normal rake angle (deg).

The power spent in the formation of new surfaces P_{ch} is calculated as the product of the energy required for the formation of one shear plane and the number of shear planes formed per second, *i.e.*,

$$P_{ch} = E_{fr} \cdot f_{cf} \tag{1.19}$$

where f_{cf} is the frequency of chip formation, *i.e.*, the number of shear planes formed per second and E_{fr} is the energy of fracture per shear plane.

The frequency of chip formation determines how many shear planes form per second of machining time. This frequency depends primarily on the work material and on the cutting speed, as discussed by Astakhov [10]. Figure 1.3 provides some data for common work materials.

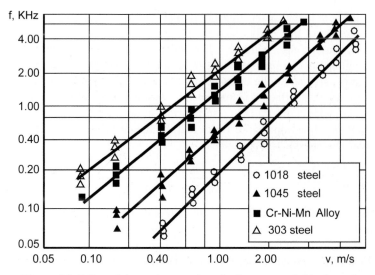

Figure 1.3. Effect of the cutting speed on the frequency of chip formation

The work of fracture per a shear plane is

$$E_{fr} = E_{fr-u} \cdot A_{fr} \tag{1.20}$$

where E_{ch} is the cohesive energy (J/m^2) [13] and A_{ch} is the area of fracture (m^2). The area of fracture is the area of the shear plane determined as

$$A_{fr} = L_{sh} \cdot b_{1T} \tag{1.21}$$

where the length of the shear plane L_{sh} is calculated as

$$L_{sh} = \frac{t_{1T}}{\sin \arctan Br} \tag{1.22}$$

1.1.4 Combined Influence of the Minor Cutting Edge

The influence of the minor cutting edge (Figure 1.2(b)) on the cutting force and power consumption is seldom considered in the literature on metal cutting. At best, the influence of the tool minor cutting edge angle κ_{r1} is mentioned in the consideration of the theoretical roughness of the machined surface or the geometric component of roughness [14, 15]. In the authors' opinion, the term *minor* probably misled many researchers in the field, causing a common perception that this cutting edge does not affect the cutting process to any noticeable degree.

Everyday practice of machining and even simple observation of the chip formed in a common machining operation show that the chip side formed by the minor cutting edge is always more deformed and has a darker colour. Zorev [3] provided a detailed analysis of the chip formation by the minor cutting edge. Zorev studied the velocity hodograph, associated plastic deformation and flows in this region. Using the results of this study, one can visualize the chip cross-sectional area cut by the minor cutting edge with the help of Figure 1.4. Figure 1.4(a) shows a hypothetical single-point cutting tool having $\kappa_{r1} = 90°$, *i.e.*, practically no minor cutting edge. Figure 1.4(b) show the cross-sectional area ABC of a *tooth* of the surface profile left after this surface was machined by this tool. Real cutting tools have the minor cutting edge with $\kappa_{r1} = 90°$ so that the surface profile left by the cutting tool is ADC as shown in Figure 1.4(c) and the height h_m of this surface profile (theoretical roughness) is calculated as

$$h_m = \frac{f}{\cot \kappa_r + \cot \kappa_{r1}} \tag{1.23}$$

Then, the part ABC shown in Figure 1.4(c) is cut by the minor cutting edge.

According to Zorev [3], the contribution of the cutting and deformation processes on the minor cutting edge to the overall power spent in cutting depends on the tool minor cutting edge angle κ_{r1} and on the cutting feed. When the feed becomes significant, the minor cutting edge takes the role of the major cutting edge so that thread cutting is the case. In real cutting tools, the tool nose radius is always made to connect the major and minor cutting edges. At moderated cutting feeds, the crater tool wear, commonly found when machining wide variety of steels occurs while when the feed rate becomes greater, wear of tool nose takes

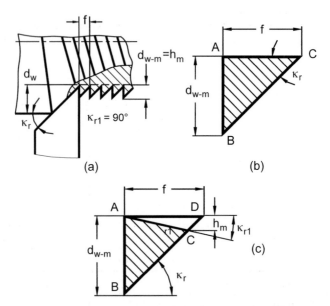

Figure 1.4. The cross-sectional area of the chip cut by the minor cutting edge: (a) hypothetic tool having a 90° tool cutting edge angle of the minor cutting edge, (b) cross-sectional area ABC of a *tooth* of the surface profile, and (c) the cross-section of the chip cut by the minor cutting edge

place [7]. This is because the energy spend due to cutting by the minor cutting edge becomes great so that the prime mode of tool wear changes from crater to nose wear.

An analysis of a great body of the experimental results and the results obtained by Zorev [3] and Astakhov [7, 10] showed that, when the tool minor cutting edge angle $30° \leq \kappa_{r1} \leq 45°$, the total power should be increased by 14%, when $15° \leq \kappa_{r1} < 30°$ by 17%, when $10° \leq \kappa_{r1} < 15°$ by 20%, and when $\kappa_{r1} < 10°$ by 23%.

1.1.5 Influence of the Cutting Speed, Depth of Cut and Cutting Feed on Power Partition

Tables 1.1 and 1.2 show the comparison of the calculated and experimental results for E52100 steel and aluminium alloy. Fairly good agreement between the calculated and the experimental results confirms the adequacy of the proposed methodology. The major advantage of the proposed methodology is that it allows not only calculating the total power and thus the cutting force, but also provides a valuable possibility to analyze the energy partition in the cutting system.

The results presented in Tables 1.1 and 1.2 are valid for new tools (a fresh cutting edge of a cutting insert). Tool wear significantly increases the cutting force. For steel E52100, $VB_B = 0.45$ mm causes a 2.0–2.5 times increase in the cutting force when no plastic lowering of the cutting edge [16] occurs (for cutting speeds 1 and 1.5 m/s) and a 3.0–3.5 increase when plastic lowering is the case (for cutting speeds 3 and 4 m/s).

The proposed methodology allows accessing the absolute and relative impacts of various variables of a metal cutting operation on the power required and thus on the cutting force according to strictures of Equations (1.5) and (1.6). Figures 1.4–1.6 present some results for steel E52100.

Table 1.1. Comparison of the experimental and calculated results for AISI steel E52100

Cutting Speed (m/s)	Feed (mm/rev)	Depth of cut (mm)	CCR	Frequency (kHz)	Cutting force Exp. (N)	Cutting force Calc. (N)
1	0.20	3	3.12	1.0	1580	1608
1.5	0.20	3	2.54	1.6	1348	1389
3	0.20	3	2.03	3.2	1076	1104
4	0.20	3	1.67	4.7	873	945
1.5	0.30	3	2.08	1.6	1562	1606
1.5	0.40	3	1.76	1.6	1640	1678
1.5	0.20	2	2.64	1.6	940	998
1.5	0.20	5	2.52	1.6	2202	2256

Table 1.2. Comparison of the experimental and calculated results for aluminium 2024 T6

Cutting Speed (m/s)	Feed (mm/rev)	Depth of cut (mm)	CCR	Frequency (kHz)	Cutting force Exp.(N)	Cutting force Calc.(N)
1	0.45	4	4.96	1.0	1223	1256
3	0.45	4	3.84	2.6	1038	1076
5	0.45	4	2.65	4.2	794	854
7	0.45	4	1.92	5.8	601	625
3	0.75	4	2.82	2.6	1393	1476
3	0.50	3	3.75	2.6	906	932
3	0.50	2	3.82	2.6	632	658
3	0.30	4	3.94	2.6	787	834

The relative impact of the cutting speed on the energy partition is shown in Figure 1.5. As seen, the power required for the plastic deformation of the layer being removed in its transformation into the chip is the greatest. However, the greater the cutting speed, the greater powers spent on the rake and flank faces of the cutting tool. When the cutting speed is 1 m/s, the power of the plastic deformation, P_{pd} is 67% while the power spent at the tool–chip interface, P_{fR}, is 18% and the power spent at the tool–workpiece interface, P_{fF}, is 9%. When the cutting speed is 4 m/s, P_{pd} is 45%, P_{fR} is 25, and P_{fF} is 22%, *i.e.*, the sum of the powers spent on the tool–chip and tool–workpiece interfaces (P_{fR} and P_{fF}) is greater than the power spent on the plastic deformation P_{pd}. This result signifies the role of tribology in high-speed machining [7]. The power spent in the formation of new surfaces P_{ch} is 6% in both considered case, although the frequency of chip formation is much greater when $v = 4$ m/s.

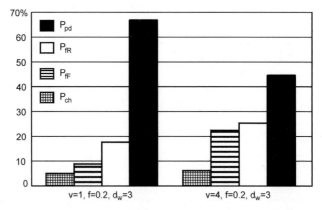

Figure 1.5. Relative impact of the cutting speed on the energy partition

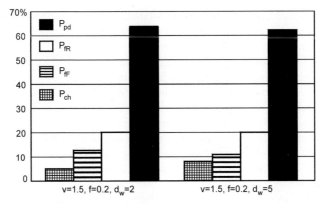

Figure 1.6. Relative impact of the depth of cut on the energy partition.

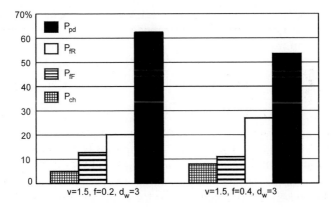

Figure 1.7. Relative impact of the cutting feed on the energy partition

The relative impacts of the depth of cut and the cutting feed are shown in Figures 1.6 and 1.7. As seen in Figure 1.6, a 2.5-fold increase in the depth of cut does not affect the energy partition. A twofold increase in the cutting feed reduces P_{pd} from 62% to 54% while P_{fR} increases from 20% to 27%.

Practically the same results were obtained for aluminium. When the cutting speed is 1 m/s, the power of the plastic deformation, P_{pd} is 67% while the power spent at the tool–chip interface, P_{fR} is 20% and the power spent at the tool–workpiece interface, P_{fF}, is 6% and P_{ch} is 7%. When the cutting speed is 7 m/s, P_{pd} is 50%, P_{fR} is 25, P_{fF} is 25% and P_{ch} is 6%.

1.1.6 Concluding Remarks

The proposed methodology uses the major parameters of the cutting process and the chip compression ratio as two of the most important process outputs (in terms of process evaluation and optimization). The apparent simplicity of the proposed methodology is based upon a great body of theoretical and experimental studies that establish the correlations among the parameters in metal cutting. This simplicity allows the use of this methodology even on the shop floor for practical evaluations and optimization of machining operations.

The results of calculations indicate that the power required for the deformation of the layer being removed is greatest in the metal cutting system within the practical cutting speed limits. When the cutting speed increases, the relative impact of this power decreases while the powers spend at the tool–chip and tool–workpiece interfaces increase. At high cutting speeds, the sum of these latter powers may exceed that required for the plastic deformation of the layer being removed. This result signifies the role of metal cutting tribology at high cutting speed.

The effects of cutting feed and the depth of cut on the energy partition seem to be insignificant.

Although it is conclusively proven that metal cutting is the purposeful fracture of the layer being removed [10, 17, 18], the notions and theory of traditional fracture mechanics are not applicable in metal cutting studies as this analysis presupposes the existence of an infinitely sharp crack leading to singular crack tip fields. In real materials, however, neither the sharpness of the crack nor the stress levels near the crack tip region can be infinite. As an alternative approach to this singularity-driven fracture approach, Barrenblatt [19] and Dugdale [20] proposed the concept of the cohesive zone model. This model has evolved as a preferred method to analyze fracture problems in monolithic and composite materials as discussed by Shet and Chandra [21]. This is due to the fact that this method not only avoids the singularity but also can be easily implemented in analytical and numerical methods of analysis.

Although a particular cohesive zone model for metal cutting is yet to be selected and justified among the many available models [21], the simplest practical way to account for the fracture (and thus for the energy associated with the formation of new surfaces) in metal cutting is the use of the so-called cohesive energy J (J/m^2), which can be determined experimentally for any work material using a relatively simple test [21]. Then, this energy multiplied by the area of fracture in metal cutting, which is the area of the shear plane, defines the mechanical work involved in the fracture and formation of new surfaces. The problem then arise of what to do with the result obtained, i.e., how to incorporate this result in the metal cutting model to calculate the cutting force, power and other characteristics of a practical machining operation.

For many years, Atkins [17, 18] has argued that fracture is the case in metal cutting even of ductile materials and that the energy associated with this fracture is significant so it has to be accounted for in metal cutting models and calculations. Atkins [17] and Rosa et al. [22] proposed a method of experimental determination of the cohesive energy and incorporation this energy in the metal cutting model to calculate the cutting force. In the authors' opinion, however, this attempt to combine the improper chip formation and thus force model [7] with the concept of cohesive energy does not account for the real discrete metal cutting process, i.e., for the number of shear planes formed per unit time.

Although it is well known and depictured in any book on metal cutting that the chip formation is discrete, i.e., at some point, a transition from one shear plane to the next has to happen, this simple fact has never been accounted for in the known models of chip formation as discussed by Astakhov [7]. As the cohesive energy is associated with a single surface of fracture, the number of surfaces of fracture that occur per unit time is essential for the determination of the power needed for such fracture process. In the proposed methodology, it is accounted for through the frequency of chip formation.

Although increasing attention is played to the role of the so-called cohesive energy in metal cutting, the results obtained show that, when accounted for properly, the relative impact of this factor is insignificant. This can be readily explained by the very small area of fracture in metal cutting.

1.2   Finite Element Analysis (FEA)

Experimental studies in metal cutting are expensive and time consuming. Moreover, their results are valid only for the experimental conditions used and depend greatly on the accuracy of calibration of the experimental equipment and apparatus used. An alternative approach is numerical methods. Several numerical methods have been used in metal cutting studies, for instance, the finite difference method, the finite element method (FEM), the boundary element method *etc.* Amongst the numerical methods, FEM is the most frequently used in metal cutting studies. The goal of finite element analysis (FEA) is to predict the various outputs and characteristics of the metal cutting process as the cutting force, stresses, temperatures, chip geometry *etc.*

In the last three decades, FEM has been progressively applied to metal cutting simulations. Starting with two-dimension simulations of the orthogonal cutting more than two decades ago, researches progressed to three-dimensional FEM models of the oblique cutting, capable of simulating metal cutting processes such as turning and milling [23–25]. Increased computation power and the development of robust calculation algorithms (thus widely availability of FEM programs) are two major contributors to this progress. Unfortunately, this progress was not accompanied by new developments in metal cutting theory so the age-old problems such as the chip formation mechanism and tribology of the contact surfaces are not modelled properly. Therefore, although these FEM simulations can provide detailed information about the distribution of stress, deformations, temperatures and residual tensions, in the deformation zone the above referred problems raise questions about the validity of such information.

Applying FEM, one should clearly realize that the results will not contain more physics than the inputs. In other words, if the model and its boundary conditions are not represented adequately with physically justified assumptions and simplifications, then one should not expect meaningful results.

This section aims to present a brief analysis of some basic aspects of FEM used in metal cutting simulations. A bibliographical review of the FEM applied to the simulation of metal cutting is presented in [26].

1.2.1 Numerical Formulations

Two major numerical formulations are used in finite element (FE) simulations: *Lagrangian* and *Eulerian*. In the *Lagrangian* formulation, broadly used in problems related to mechanics of solids, the FE mesh is constituted by elements that cover exactly the whole of the region of the body under analysis. These elements are attached to the body and thus they follow its deformation. This formulation is particularly convenient when unconstrained flow of material is involved, *i.e.*, when its boundaries are in frequent mutation. In this case, the FE mesh covers the real contour of the body with sufficient accuracy. On the other hand, the Eulerian formulation is more suitable for fluid-flow problems involving a control volume. In this method, the mesh is constituted of elements that are fixed in the space and cover the control volume. The variables under analysis are calculated at fixed spatial location as the material flows through the mesh. This formulation is more suitable for applications where the boundaries of the region of the body under analysis are known *a priori*, such as in metal forming.

Although both of these formulations have been used in modelling metal cutting processes, the Lagrangian formulation is more attractive due to the ever-mutating of the model used. The Eulerian formulation can only be used to simulate steady-state cutting. As a result, when the Lagrangian formulation is used, the chip is formed with thickness and shape determined by the cutting conditions. However, when one uses the Eulerian formulation, an initial assumption about the shaped of the chip is needed. This initial chip shape is used for a matter of convenience, because it considerably facilitates the calculations in an incipient stage, where frequent problems of divergence of algorithm are found.

The Lagrangian formulation, however, also has shortcomings. First, as metal cutting involves severe plastic deformation of the layer being removed, the elements are extremely distorted so the mesh regeneration is needed. Second, the node separation is not well defined, particularly when chamfered and/or negative-rake or heavy-radiused cutting edge tools are involved in the simulation [27]. Although the severity of these problems can be reduced to a certain extent by a denser mesh and by frequent re-meshing, frequent mesh regeneration causes a lot of other problems [24].

These problems do not exist in the Eulerian formulation as the mesh is spatially fixed. This eliminates the problems associated to high distortion of the elements, and consequently no re-meshing is required. The mesh density is determined by the expected gradients of stress and strain. Therefore, the Eulerian formulation is more computationally efficient and suitable for modelling the zone around the tool cutting edge, particularly for ductile work materials [27]. The major drawback of this formulation, however, is that the chip thickness should be assumed and kept constant during the analysis, as well as the tool–chip contact length and contact conditions at the tool–chip and tool–workpiece interfaces [28–31]. As discussed by Astakhov [7], the chip thickness is the major outcome of the cutting process that defines all other parameters of this process so it cannot be assumed physically. Consequently, the Eulerian formulation does not correspond to the real deformation process developed during a real metal cutting process.

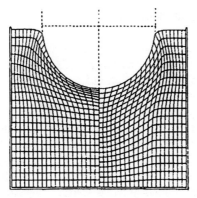

Figure 1.8. Comparison between a mesh produced by the Lagrangian formulation (left side) and a mesh produced by the ALE formulation (right side)

To address the problems associated with modelling metal forming processes using the Lagrangian and the Eulerian formulations, the arbitrary Lagrangian–Eulerian (ALE) formulation was introduced. In this formulation, the mesh is neither attached to the material nor fixed in space. This mesh moves, which allows the advantages of the Lagrangian and the Eulerian formulations to be combined. In the ALE formulation, material displacement is described as the sum of mesh displacement and relative displacement. The former is the Eulerian displacement by which the mesh is controlled to reduce the numerical errors resulted from mesh distortions during the deformation process. The latter represents the Lagrangian displacement associated with deformation. Figure 1.8 compares the mesh produced using the Lagrangian formulation (left side) with the mesh produced by the ALE formulation (right side) when modelling metal forging. The Lagrangian mesh presents a high distortion, while the ALE is more regular and presents low distortion.

The ALE formulation has two major drawbacks. Often the ALE formulation cannot prevent the need for a complete re-meshing. The re-mapping of state variables is also a drawback compared to the updated Lagrangian formulation. When the re-map is performed inaccurately, the history of the material is not taken into account properly.

1.2.2   Modelling Chip Separation from the Workpiece and Chip Segmentation

There are a number of numerical techniques to model chip separation from the rest of the work material. The *node-splitting* technique is the oldest, where chip separation is modelled by the separation of nodes of the mesh ahead of the tool cutting edge along the predefined cutting line. This technique is usually used with the Lagrangian formulation to simulate steady-state cutting. A number of separation criteria grouped as *geometrical* and *physical* have been developed [28, 32–40]. According to the geometrical criteria, the separation of two nodes occurs when the distance D between the tool cutting edge (point O, in Figure 1.9) and the node immediately ahead (node a) becomes less than a predefined critical value. According to the physical criteria, the

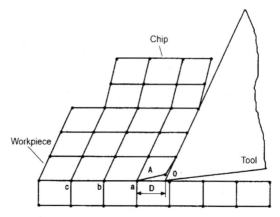

Figure 1.9. Separation of nodes based in the distance **D** between the tool cutting edge (point O) and the node immediately ahead (node a)

separation of two nodes occurs when the value of a predefined physical parameter, such as stress, strain or strain energy density, at node a or element A (Figure 1.9) achieves a predefined critical value, selected depending upon the work material properties and the cutting regime.

The physical criteria seem to be more adequate for modelling node splitting as they are based on physical measurable properties of the work material. The problem, however, is in the selection of suitable representations of these properties. For example, Chen and Black [38] argued that the critical strain energy density commonly used as a separation criterion is determined from a uniaxial tensile test and thus cannot be considered as relevant in metal cutting.

1.2.3 Mesh Design

The first step in any finite element or finite boundary analysis includes dividing the continuum or solution region into finite elementary regions (lines, areas or volumes) called elements. This procedure to converting a continuous region into a discrete region is referred to as *discretization*, included in a large topic named as *mesh design*.

The problems related to mesh design are not restricting to the initial discretization procedure. In metal cutting modelling, the common problem is related to the element distortion during the simulations due to severe plastic deformation. The distortion can cause a deterioration of the FE simulation in terms of convergence rate and numerical errors, or cause the Jacobian determinant to become negative, which makes further analysis impossible. It is often necessary to redefine the mesh after some stages of deformation.

Several techniques are used to reduce the element distortion: re-meshing, smoothing and refinement. These techniques include the generation of a completely new finite element mesh out of the existing mesh, increasing the local element density by reducing the local element size (Figure 1.10) and/or reallocating the individual nodes to improve the local quality of the elements (Figure 1.11).

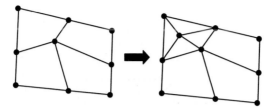

Figure 1.10. Increase local mesh density (refinement)

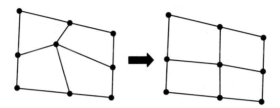

Figure 1.11. Reallocation of the nodes (smoothing)

The discussed techniques are used in the so-called *adaptive mesh* procedure. Adaptive mesh refers to a scheme for finite difference and finite element codes wherein the size and distribution of the mesh is changed dynamically during the simulations. In the regions of strong gradients of variables involved, a higher mesh density is needed in order to decrease the solution errors. As these gradients are not known *a priori*, the adaptive mesh generation procedure starts with a relatively coarse primary mesh and, after obtaining the solution on this primary mesh, the mesh density is increased for the strong gradients.

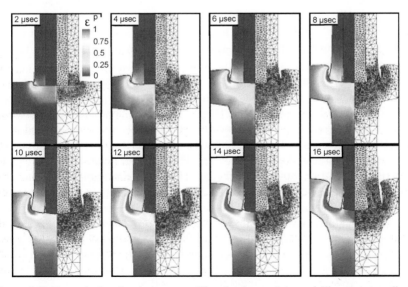

Figure 1.12. Example showing the concept of the adaptive mesh in modelling strong gradients

The adaptive mesh procedure can improve the accuracy and efficiency of the calculation/simulation in some cases, for example those with complicated geometry and large gradients [41]. For this reason, the adaptive mesh procedure is increasingly being used together with the ALE formulation to simulate the metal forming and metal cutting processes. An example of this procedire is shown in Figure 1.12. As seen, the solution begins with a relatively coarse primary mesh and, after obtaining the solution on this primary mesh, the mesh is modified in some regions in order to increase the mesh density in the areas of strong gradients.

1.2.4 Work Material Modelling

Several material constitutive models are used in FEM of metal cutting, including rigid-plastic, elasto-plastic, viscoplastic, elasto-viscoplastic, *etc*. These models take into account the high strains and temperatures reportedly found in metal cutting. Among others, the most widely used is the Johnson and Cook model [42]. This is a thermo-elasto-visco-plastic material constitutive model represented as

$$\sigma_{eq} = \left(A + B\varepsilon_{eq}^n\right)\left(1 + C\ln\left(\frac{\dot{\varepsilon}_{eq}}{\dot{\varepsilon}_{eq}^0}\right)\right)\left[1 - \left(\frac{T - T_0}{T_F - T_0}\right)^m\right] \tag{1.24}$$

where ε_{eq}^n is the equivalent plastic strain, $\dot{\varepsilon}_{eq}$ is the equivalent plastic strain rate, $\dot{\varepsilon}_{eq}^0$ is the reference equivalent plastic strain rate (normally $\dot{\varepsilon}_{eq}^0 = 1s^{-1}$), T is the temperature, T_0 is the room temperature, T_F is the melting temperature and A, B, C, n and m are constants that depend on the material determined through material tests.

Because the strain rates in conventional material tests (tensile, compressive or torsion) are in the range of $10^{-3} - 10^{-1}$ s^{-1}, non-conventional material tests, referred to as dynamic tests, are usually preferred by many researchers. Split Hopkinson pressure bar (SHPB) impact testing is the most common. This test is used to determine the material behaviour at strain rates up to 10^5 s^{-1} [42]. This test can be conducted at an elevated temperature (500°C or even more).

There are two major problems with the use of the discussed model and its method of the determination of its constants. First, only few laboratories and specialist in the world can conduct SHPB testing properly, assuring the condition of dynamic equilibrium. None of the known tests in metal cutting was carried out in these laboratories. Second, the high strain rate in metal cutting is rather a myth than reality, as discussed by Astakhov [7]. Third, the temperature in the so-called primarily deformation zone where the complete plastic deformation of the work materials takes place can hardly exceed 250°C. It is understood that the mechanical properties of the work material obtained at room temperature are not affected by this temperature so metal cutting is a cold working process [10], although the chip appearance can be cherry-red. Fourth, it is completely unclear how to correlate the properties of the work materials obtained in SHPB uniaxial impact testing with those in metal cutting with a strong degree of stress triaxiality.

1.2.5   Modelling of Contact Conditions

Although the contact conditions at the tool–chip and tool–workpiece interfaces are complicated [7], many known FE metal cutting simulations use simple Coulomb friction condition. Moreover, the friction coefficient is frequently assumed to be constant over these interfaces.

As discussed by Astakhov [7], if the friction coefficient $\mu_f \geq 0.577$, no relative motion can occur at the tool–chip interface, as follows from the fundamental properties of work materials. The experiment data are in direct contradiction with this limiting value: Zorev [3] obtained $\mu_f = 0.6 - 1.8$, Kronenberg [43] 0.77–1.46, Armarego and Brown [44] 0.8–2.0, Finnie and Shaw [45] 0.88–1.85, Usui and Takeyama [46] 0.4–2.0 *etc.* In FEM of metal cutting, Stenkowsky and Moon [29] used $\mu_f = 0.2$, Komvopoulos and Erpenbeck [16] 0.0–0.5, Lin *et al.* [47] 0.074, Lin and Lin [22] 0.001, Carroll and Stenkowski [28] 0.3, Olovsson *et al.* [48] 0.1 *etc.* As seen, the reported values of μ_f obtained in metal cutting tests are well above 0.577. On the other hand, the values of μ_f used in FEM modelling are always below the limiting value to suit the sliding condition at the interface. Interestingly, the results of FEM modelling were always found to be in a good agreement with the experimental results regardless of the particular value of the friction coefficient selected for such a modelling.

It is true that, in general, the coefficient of friction for sliding surfaces remains constant within wide ranges of the relative velocity, apparent contact area and normal load. In contrast, in metal cutting the coefficient of friction varies with respect to the normal load, the relative velocity and the apparent contact area. The coefficient of friction in metal cutting was found to be so variable that Hahn [49] doubted whether this term served any useful purpose. Moreover, Finnie and Shaw [45] concluded that the concept of the coefficient of friction is inadequate to characterize the sliding between chip and tool and thus recommended to discontinue using the concept of the coefficient of friction in metal cutting. An extensive analysis of the inadequacy of the concept of the friction coefficient in metal cutting was presented by Kronenberg (pp. 18–25 in [43]) who stated "I do not agree with the commonly accepted concept of coefficient of friction in metal cutting and I am using the term 'apparent coefficient of friction' wherever feasible until this problem has been resolved." Unfortunately, it has never been resolved, although more than 40 years passed since Kronenberg made this statement.

The proper modelling of contact conditions can be achieved if the relevant tribological parameters at the tool–chip and tool–workpiece interfaces are considered and the proper boundary conditions that adequately describe these conditions are part of the FEM [7].

1.2.6   Numerical Integration

Because the metal cutting process is highly non-linear and dynamic, its simulation requires a robust time integration method. Direct methods of time integration, which includes the *implicit* and *explicit* direct methods of time integration, are used for non-linear problems [50]. The explicit method presents some advantages when applied to dynamic problems. For both the explicit and the implicit time integration method, equilibrium is defined in terms of external applied forces, P, the internal forces, I, and the nodal accelerations

$$M \cdot \ddot{u} = P - I \qquad (1.25)$$

where M is the mass matrix and $\ddot{u}$ is the acceleration. Both methods use the same element calculations to determine the internal element forces. The significant difference between the two methods lies in the manner in which the nodal accelerations are computed [51]. The implicit method solves a set of linear equations using a direct solution method (such as the Newton–Raphson method), performing iterations per increment of time unttil a converged solution is reached within a given tolerance. The explicit method determines the solution without evoking iterations, but uses a central difference rule to integrate the equations of motion explicitly through time.

The explicit method has a number of advantages when used for highly dynamic problems, especially for transient problems involving complex models. It is particularly suitable for problems involving complex contact interactions between many independent bodies, as it happens in metal cutting simulations. For such contact conditions, the convergence of the implicit method becomes questionable as it needs a large number of iterations [51].

1.2.7 Errors

Modelling of cutting process using sophisticated FEM software has become quite popular and many researchers try to refer to the results of such modelling as something conclusively proven. There are a thousand points of doubt in every complex computer program. It should be very clear that FEM software incorporates many assumptions that cannot be easily detected by its users but that affect the validity of the results. There are several sources of errors which can lead to improper results [41]: (1) poor input data due to the lack of information about the process, (2) unreasonable simplifications, idealization and assumptions of metal cutting process, (3) improper modelling of the boundary conditions, (4) numerical round-off (in solving the simultaneous equations), (5) discretisation error and (6) errors associated with re-mapping.

1.2.8 Example

Figure 1.12 shows an example of the results of FEM modelling. The following conditions were considered in this modelling:

- Tool: normal rake angle $\gamma_n = 0°$, normal flank angle $\alpha_n = 7°$, inclination angle $\lambda_s = 0°$, radius of the cutting edge $\rho_{ce} = 0.005$ mm, tool material, K15.
- Work material: AISI steel 316L, $\sigma_{UTS} = 517$ MPa and $\sigma_{YT} = 218$ MPa.
- Cutting regime: cutting speed $v = 75$ m/min, uncut chip thickness $t_1 = 0.2$ mm, width of cut $d_w = 2$ mm.

Figures 1.13(a) and (b) show the state of the deformation zone for two distinctive phases of chip formation of the saw-toothed chip. Figures 1.13(c) and (d) show the temperature distribution in the deformation zone for these two stages. As can be

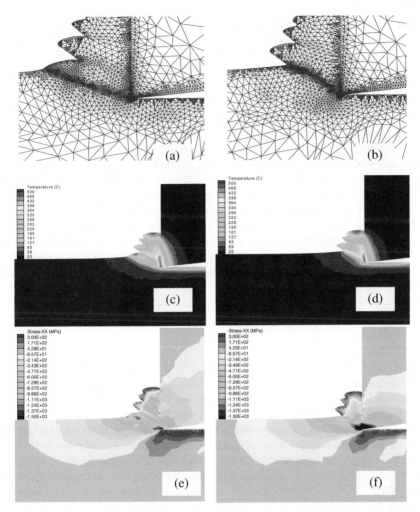

Figure 1.13. FEM modelling of the formation of saw-toothed continuous fragmentary chip: (a,b) shape of the chip, (b,c) temperature distribution, (e,f) stress distribution

seen, the temperatures in the deformation zone vary within a chip formation cycle [10]. Figures 1.13(e) and (f) present the stress distributing in the deformation zone for the discussed phases.

1.2.9 Advanced Numerical Modelling

Standard FEM works well when it is applied to problems which do not involve severe plastic deformations of the work material. In metal cutting, however, the maximum deformation equal to the strain at fracture has to be achieved to accomplish the process. This presents a challenging problem for computational methods as it is necessary to deal with extremely large deformations and thus distortion of the mesh. Moreover, because metal cutting is the purposeful fracture of the layer being removed [7, 10], it is necessary to model the propagation of cracks with arbitrary and complex path and track the growth and phase boundaries and extensive microcracking.

These models are not well suited for conventional computational methods such as FE, finite volume or finite difference methods. The underlying structure of these methods, which originates from their reliance on a mesh, is not well suited to the treatment of discontinuities which do not coincide with the original mesh lines. To deal with moving discontinuities in methods based on meshes is to re-mesh in each step of the evolution so that the mesh lines remain coincident with the discontinuities throughout evolution of the problem. Being the most viable, this strategy introduces numerous difficulties such as the need to project between meshes in successive stages of the evolution, which leads to degradation of accuracy of the solution in unpredictable manner and greatly increases the complexity of the computer programs [52–54].

An emerging new technique is the element-free Galerkin (EFG) method, also called the meshless or meshfree method. The objective of EFG methods is to eliminate at least part of this structure by construction the approximation function entirely in terms of nodes. Thus, it becomes possible to solve large classes of problems which are very awkward with mesh-based methods.

1.2.10 Model Validation

FEM analysis has proved to be a powerful investigative tool capable of encompassing various aspects of the metal cutting process. However, despite its great potential, there are a number of limitations and approximations in the commercially available FE programs. Although many papers have been published on FEA of the metal cutting process, the issue of accuracy and adequacy of the modelled results remains widely open due to the inherent drawbacks of the traditional FEA, highly idealized models involved and inadequate representation of boundary conditions.

The highly structured nature of FE approximations imposes severe penalties in the solution of metal cutting problems, and it is the decrease in structure that makes meshless methods appealing. Meshless methods, however, require considerable more central processing unit (CPU) time [52].

The results obtained using FEA strongly depend on the experience and judgment of the engineers involved in the analysis of the problem and definition of a simulation model. It should be clear that FEM makes sense if and only if a physically sound and adequate model of the metal cutting process is included at the very foundation of this modelling. Therefore, the construction of such a model, careful selection and justification of the proper boundary conditions and then model validation should be the cornerstone of FEM. The market is overpopulated with FE codes and tools for metal cutting simulations, but the issue of model validity is silently set aside. How can one seriously propose to eliminate a physical test with a digital simulation if he cannot say that his model has, say, a 95% level of confidence, or credibility, attached to it?

The best way to validate the model used in FE simulation is through relevant testing conducted under the same conditions assumed in the model construction. However, this is rarely done because there are a number of problems with adequate metal cutting tests. The existence of scatter in metal cutting experiments, it appears, has now been widely accepted and is no longer under discussion. The uncertainty behind work and tool materials properties, working loads, boundary conditions at the tool–chip and tool–workpiece interfaces and other particularities of the machining system is the main reason why two supposedly identical machining operations will yield different results in terms of tool life, quality of the machined part and achievable production rate. This fact is well known and yet, surprisingly, engineers are involved in building very detailed models with the intention of capturing the results of one particular test. It is difficult to understand what exactly is detailed in these models, given the fact that the real world is neither detailed, nor accurate, nor precise. Now, since tests are, for very good physical reasons, nonrepeatable, this race for accuracy is lost from the very beginning.

The simplest yet most accurate way to validate a FEM in metal cutting is through the chip compression ratio ζ [7, 9, 10], defined as

$$\zeta = \frac{t_2}{t_1} = \frac{v}{v_1} \qquad (1.26)$$

where v is the cutting velocity (speed), v_1 is the chip velocity, t_1 is the thickness of the layer being removed (uncut chip thickness) and t_2 is the chip thickness.

Astakhov showed that the chip compression ratio is the one of the most important process output (in terms of process evaluation and optimization). Besides the cutting force and energy considered in this chapter, it directly correlates with the tool–chip contact length, contact stresses and temperatures [7, 9, 10]. Using this parameter, one can easily validate a FE model.

Such a validation can be as simple as that: (a) adjust the test setup to the same conditions assumed in the construction of the model including boundary conditions, *i.e.*, use the corresponding work material, cutting tool and machining regime, (b) collect a typical chip and determine the chip compression ratio using the simple known methods [10], (c) using the determined chip compression ratio, calculate the tool–chip contact area and (d) compare the chip shape, chip compression ratio and the tool–chip contact length obtained using the test with those obtained in FEM. Figure 1.14 presents an example of such a comparison. Although the shape of the chips in both cases are rather similar, the experimentally obtained value is $\zeta = 3.42$ while, as follows from Figure 1.14(b) $\zeta = 1.65$. The chip–tool contact length, calculated using the experimentally obtained chip compression ratio is $l_c = 1.58$ mm while that in Figure 1.13(b) is $l_c = 0.65$ mm. Therefore, the FE model cannot be considered as valid.

The FEM of thermal process in metal cutting may need more sophisticated validation. Commercially available and specially designed infrared imaging equipment, which includes both hardware and software, are available for temperature distribution measurements in the deformation zone in three-dimensional cutting process. The high flexibility and low cost of this equipment make it suitable to evaluate high temperature gradients, such as those produced by metal cutting

[8, 55]. Figure 1.15 shows an example of such measurements. These results should be compared with those obtained using FEM. Then the model used in the modelling should be fine tuned to mach the experimentally obtained result.

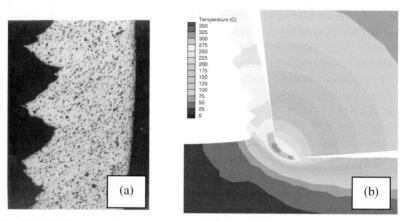

Figure 1.14. Validation of a FE model by comparison the chips obtained experimentally (a) and through FEM (b). Orthogonal cutting of the AISI 316L; tool material ISO M10 tungsten carbide; tool geometry: normal rake angle, $\gamma_n = -4.29°$; normal relief (flank) angle, $\alpha_o = 4.29°$; normal wedge angle, $\beta_o = 90°$; inclination angle of the cutting edge, $\lambda_s = -14°$; tool cutting edge angle, $\kappa_r = 72°$; nose radius, $r_\varepsilon = 0.8$ mm. The cutting speed was 145 m/min; the cutting feed was 0.25 mm/rev; the depth of cut was 2.5 mm. No cutting fluid was used in the tests

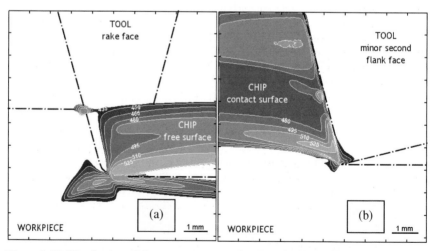

Figure 1.15. Distribution of the temperatures on the: (a) chip free surface (b) chip contact surface and minor second tool flank face. Cutting conditions: work material AISI 1045 steel; uncoated tool; $V_C = 125$ m/min; $f = 0.05$ mm/rev; $p = 2.5$ mm (b) coated tool; $V_C = 125$ m/min; $f = 0.05$ mm/rev; $p = 5$ mm

The foregoing considerations show that the model used in FEM of metal cutting process should be adequate to the process. Moreover, the concept of FE model must be broadened in order to embrace important facets of physics including uncertainty, which unfortunately has been axiomatized out of modern metal cutting research. Model validation should be considered as the mandatory step in FEM. Breakthroughs in these directions will have considerable impact by making metal cutting numerical simulation useful for practical optimization of various metalworking operations including the cutting and machine tools, the metal working fluids and fixtures, adaptive CNC controllers and programs.

References

[1]   Armarego EJA (1995) Predictive modelling of machining operations – a means of bridging the gap between the theory and practice – A keynote paper at the 13th Symposium on Engineering Applications of Mechanics. Hamilton, ON, Canada: CMSE

[2]   Endres WJ, Devor RE, Kapoor SG (1995) A dual-mechanism approach to the prediction of machining forces, Part 1: model development. ASME J Eng Ind 117: 526–533

[3]   Zorev NN (1966) Metal Cutting Mechanics. Pergamon: Oxford

[4]   Komanduri R (1993) Machining and grinding: A historical review of the classical papers. Appl Mech Rev 46: 80–132

[5]   Merchant E (2003) An Interpretive Review of 20th Century US Machining and Grinding Research. TechSolve Inc, Cincinnati (OH) USA

[6]   Ernst H, Merchant ME (1941) Chip formation, friction and high quality machined surfaces. Surface Treatment Metals ASM 29: 299–378

[7]   Astakhov VP (2006) Tribology of Metal Cutting. Elsevier, London

[8]   Outeiro JC (2003) Application of Recent Metal Cutting Approaches to the Study of the Machining Residual Stresses. PhD Thesis, Department of Mechanical Engineering. 2003, University of Coimbra: Coimbra. p. 340

[9]   Astakhov VP, Shvets S (2004) The assessment of plastic deformation in metal cutting. J Mater Process Technol 146: 193–202

[10]   Astakhov VP (1998) Metal Cutting Mechanics. CRC, Boca Raton, USA

[11]   Ivester RW (2004) Comparison of machining simulations for 1045 steel to experimental measurements. SME Paper TPO4PUB336: 1–15

[12]   Astakhov VP, Shvets SV (2001) A novel approach to operating force evaluation in high strain rate metal-deforming technological processes. J Mater Process Technol 117: 226–237

[13]   Shet C, Chandra N (2002) Analysis of energy balance when using cohesive zone models to simulate fracture process. ASME J Eng Mater Technol 124: 440–450

[14]   Shaw MC (1984) Metal Cutting Principles. Oxford Science, Oxford

[15]   Stephenson DA, Agapiou JS (1996) Metal Cutting Theory and Practice. Marcel Dekker, New York

[16]   Astakhov VP (2004) The assessment of cutting tool wear. Int J Machine Tools Manuf 44: 637–647

[17]   Atkins AG (2003) Modelling metal cutting using modern ductile fracture mechanics: quantitative explanations for some longstanding problems. Int J Mech Sci 43: 373–396

[18]   Atkins AG, Mai YW (1985) Elastic and Plastic Fracture: Metals, Polymers, Ceramics, Composites, Biological Materials. Wiley, New York

[19] Barrenblatt GI (1962). Mechanical theory of equilibrium cracks. Advances in Applied Mechanics. Academic, New York, pp. 55–125
[20] Dugdale DS (1960) Yielding of steel sheets containing slits. J Mech Phys Solids 8: 100–104
[21] Shet C, Chandra N (2002) Analysis of energy balance when using cohesive zone models to simulate fracture process. ASME J Eng Mater Technol 124(4): 440–450
[22] Rosa PAR, Martins PAF, Atkins AG (2007) Revising the fundamentals of metal cutting by means of finite elements and ductile fracture mechanics. Int J Mach Tools Manuf 47: 607–617
[23] Ceretti E, Lazzaroni C, Menegardo L, Altan T (2000) Turning simulations using a three-dimensional FEM code. J Mater Process Technol 98(1): 99–103
[24] Marusich TD, Ortiz M (1995) Modelling and simulation of high-speed machining. Int J Numer Methods Eng 38: 3675–369
[25] Guo YB, Liu CR (2002) 3D FEA modelling of hard turning. J Manuf Sci Eng 124(2): 189–199
[26] Mackerle J (1999) Finite-element analysis and simulation of machining: a bibliography (1976–1996). J Mater Process Technol 86(1–3): 17–44
[27] Movahhedy M, Gadala MS, Altintas Y (2000) Simulation of the orthogonal metal cutting process using an arbitrary Lagrangian–Eulerian finite-element method. J Mater Process Technol 103: 267–275
[28] Carroll JT, Strenkowski JS (1988) Finite element models of orthogonal cutting with application to single point diamond turning. Int J Mech Sci 30: 899–920
[29] Strenkowski JS, Moon K-J (1990) Finite element prediction of chip geometry and tool/workpiece temperature distributions in orthogonal metal cutting. J Eng Ind 112(4): 313–318
[30] Kim KW, Sin H-C (1996) Development of a thermo-viscoplastic cutting model using finite element method. Int J Mach Tools Manuf 36(3): 379–397
[31] Wu J-S, Dillon JR (1996) Thermo-viscoplastic modelling of machining process using a mixed finite element method. J Manuf Sci Eng 118: 470–482
[32] Komvopoulos K, Erpenbeck SA (1991) Finite element modelling of orthogonal metal cutting. J Eng Ind 113(3): 253–267
[33] Zhang B, Bagchi A (1994) Finite element simulation of chip formation and comparison with machining experiment. J Eng Ind 116(3): 289–297
[34] Shih AJ (1995) Finite element simulation of orthogonal metal cutting. J Eng Ind 117(1): 84–93
[35] Hashemi J, Tseng AA, Chou PC (1994) Finite element modelling of segmental chip formation in high-speed orthogonal cutting. J Mater Eng Performance 3(5): 712–721
[36] Iwata K, Osakada K, Terasaka T (1984) Process modelling of orthogonal cutting by the rigid-plastic finite element method. J Eng Mater Technol 106: 132–138
[37] Lin ZC, Lin SY (1992) A coupled finite element model of thermo-elastic-plastic large deformation for orthogonal cutting. J Eng Mater Technol 114: 218–226
[38] Chen AG, Black TJ (1994) FEM modelling in metal cutting. Manuf Rev 7: 120–133
[39] Ceretti E, Fallböhmer P, Altan T (1996) Application of 2D FEM to chip formation in orthogonal cutting. J Mater Process Technol 59(1–2): 160–180
[40] Obikawa T, Usui E (1996) Computational machining of titanium alloy – finite element modelling and a few results. J Manuf Sci Eng 118: 208–215
[41] Mackerle J (2001) 2D and 3D finite element meshing and remeshing: A bibliography (1990–2001). Engineering Computations. Int J Comput-Aided Eng 18(8): 1108–1197
[42] Johnson GR, Cook WH (1983) A constitutive model and data for metals subjected to large strain, high strain rates and high temperature. In: Proceedings of seventh international symposium on ballistic, The Hague, The Netherlands

[43] Kronenberg M (1966) Machining Science and Application. Theory and Practice for Operation and Development of Machining Processes. Pergamon, London

[44] Armarego EJ, Brown RH (1969) The Machining of Metals. Prentice-Hall, New Jersey, USA

[45] Finnie I, Shaw MC (1956) The friction process in metal cutting. Trans ASME 77: 1649–1657

[46] Usui E, Takeyma H (1960) A photoelastic analysis of machining stresses. ASME J Eng Ind 81: 303–308

[47] Lin ZC, Pan WC, Lo SP (1995) A study of orthogonal cutting with tool flank wear and sticking behaviour on the chip-tool interface. J Mater Process Technol 52: 524–538

[48] Olovsson L, Nilsson L, Simonsson K (1998) An ALE formulation for the solution of two-dimensional metal cutting problems. Comput Struct 72: 497–507

[49] Hahn RS (1952) On the temperature development at the shear plane in the metal cutting process. In Proceedings of the First US Nat. Appl. Mech. 1952: ASME, New York, 112–118

[50] Sun, J.S., Lee KH, Lee HP (2000) Comparison of implicit and explicit finite element methods for dynamic problems. J Mater Process Technol 105: 110–118

[51] Hibbitt, Karlsson, and Sorenson, Inc. (2001) ABAQUS Theory and Users' Manuals, Version 6.2-1, Providence, RI

[52] Belytshko T, Krongauz Y, Organ D, Fleming M, Krysl P (1996) Meshless methods: An overview and resent developments. Comput Methods Appl Mech Eng 139: 3–47

[53] Chen Y, James Lee J, Eskandarian A (2006) Meshless Methods in Solid Mechanics. Springer Science, NY

[54] Liu GR (2002) Mesh Free Methods Moving beyond finite element method. CRC, Boca Raton, USA

[55] Outeiro JC, Dias AM, Lebrun JL (2004) Experimental assessment of temperature distribution in three-dimensional cutting process. Mach Sci Technol 8/3: 357–376

Tools (Geometry and Material) and Tool Wear

Viktor P. Astakhov[1] and J. Paulo Davim[2]

[1]General Motors Business Unit of PSMi, 1255 Beach Ct., Saline MI 48176, USA.
E-mail: astvik@gmail.com
[2] Department of Mechanical Engineering, University of Aveiro, Campus Santiago
3810-193 Aveiro, Portugal.
E-mail: pdavim@ua.pt

This chapter presents the basic definitions and visualisations of the major components of the cutting tool geometry important in the consideration of the machining process. The types and properties of modern tool materials are considered as well, as a closely related topic, as these properties define to a great extent the limitations on tool geometry. The basic mechanisms of tool wear are discussed. Criteria and measures of tool life are also considered in terms of Taylor's tool life models as well as in terms of modern tool life assessments for cutting tools used on computer numerical control (CNC) machines, manufacturing cells and production lines.

2.1 Essentials of Tool Geometry

2.1.1 Importance of the Cutting Tool Geometry

The lack of information on cutting tool geometry and its influence on the outcomes of machining operation can be explained as follows. Many great findings on the tool geometry were published a long time ago when CNC grinding machines capable of reproducing any kind of tool geometry were not available and computers to calculate parameters of such geometry were not common; it was therefore extremely difficult to reproduce proper tool geometries using manual machines. As a result, once-mighty chapters on tool geometry in metal cutting and tool design books were reduced to a few pages, in which no correlation between tool geometry and performance was normally considered. What is left is a general perception that the so-called *positive geometry* is somehow better than the *negative geometry*. As such, there is no quantitative translation of the word "better" into the language of technical data, although a great number of articles written in many professional magazines discuss the qualitative advantages of *positive geometry*.

During recent decades, the metalworking industry underwent several important changes that should bring the cutting tool geometry to the forefront of tool design and implementation:

- For decades, the measurement of the actual tool geometry of real cutting tools was a cumbersome and time-consuming process as no special equipment besides toolmakers' microscopes was available. Today, automated tool geometry inspection systems as the ZOLLER Genius 3, Helicheck®, Heli-Toolcheck® etc. are available on the market.
- A modern tool grinder is typically a CNC machine tool, usually with four, five, or six axes. Extremely hard and exotic materials are generally no problem for today's grinding systems and multi-axis machines are capable of generating very complex geometries.
- Advanced cutting-insert manufacturing companies have perfected the technology of insert pressing (for example, spray drying) so practically any desired shape of cutting insert can be produced with a very tight tolerance.
- Many manufacturing companied have updated their machines, fixtures and tool holders. Modern machines used today have powerful rigid high-speed spindles, high-precision feed drives and shrinkfit tool holders.
- Many manufacturing companies have established tight controls and maintenance of their coolant units. Control of the coolant concentration, temperature, chemical composition, pH, particle count, contaminations as tramp oil, bacteria etc. is becoming common.

All this pushed tool design, including primarily tool materials and geometry, to the forefront as none of the traditional excuses for poor performance of cutting tools can be accepted.

The cutting tool geometry is of prime importance because it directly affects:

1. **Chip control**. The tool geometry defines the direction of chip flow. This direction is important to control chip breakage and evacuation.
2. **Productivity of machining**. The cutting feed per revolution is considered the major resource in increasing productivity. This feed can be significantly increased by adjusting the tool cutting edge angle. For example, the most common use of this feature is found in milling, where increasing the lead angle to 45° allows the feed rate to be increased 1.4-fold. As such, a wiper insert is introduced to reduce the feed marks left on the machined surface due to the increased feed.
3. **Tool life**. The geometry of the cutting tool directly affects tool life as this geometry defines the magnitude and direction of the cutting force and its components, the sliding velocity at the tool–chip interface, the distribution of the thermal energy released in machining, the temperature distribution in the cutting wedge etc.
4. **The direction and magnitude of the cutting force and thus its components**. Four components of the cutting tool geometry, namely, the rake an-

gle, the tool cutting edge angle, the tool minor cutting edge angle and the inclination angle, define the magnitudes of the orthogonal components of the cutting force.

5. **Quality (surface integrity and machining residual stress) of machining**. The correlation between tool geometry and the theoretical topography of the machined surface is common knowledge. The influence of the cutting geometry on the machining residual stress is easily realized if one recalls that this geometry defines to a great extent the state of stress in the deformation zone, *i.e.*, around the tool.

2.1.2 Basic Terms and Definitions

The geometry and nomenclature of cutting tools, even single-point cutting tools, are surprisingly complicated subjects [1–4]. It is difficult, for example, to determine the appropriate planes in which the various angles of a single-point cutting tool should be measured; it is especially difficult to determine the slope of the tool face. The simplest cutting operation is one in which a straight-edged tool moves with constant velocity in a direction perpendicular to the cutting edge of the tool. This is known as the two-dimensional or orthogonal cutting process, illustrated in Figure 2.1. The cutting operation can best be understood in terms of orthogonal cutting parameters. Figure 2.2 shows the application of a single-point cutting tool in a turning operation. It helps to correlate the orthogonal and oblique non-free cutting.

In orthogonal cutting (Figure 2.1), the two basic surfaces of the workpiece are considered:

- The work surface: the surface of the workpiece to be removed by machining.
- The machined surface: the surface produced after the cutting tool passes.

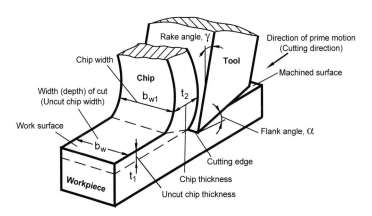

Figure 2.1. Terminology in orthogonal cutting

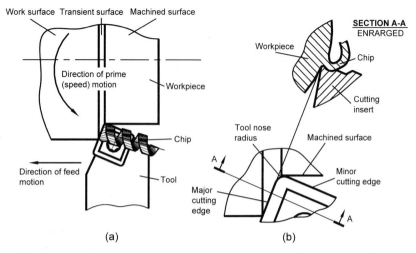

Figure 2.2. Turning terminology

In many practical machining operations an additional surface is considered:

- The transient surface: the surface being cut by the major cutting edge (Figure 2.2). Note that this surface is always located between the work surface and machined surface.

Its presence distinguishes orthogonal cutting and other machining operations from simple shaping, planning and broaching where the cutting edge is perpendicular to the cutting speed. One should clearly understand that, in most real machining operations, the cutting edge does not form the machined surface. As clearly seen in Figure 2.2, the machined surface is formed by the tool nose and minor cutting edge. Unfortunately, not much attention is paid to these two important components of tool geometry, although their parameters directly affect the integrity of the machined surface including the surface finish and machining residual stresses.

2.1.3   System of Considerations

As pointed out by Astakhov [4], there are three basic systems in which the tool geometry should be considered depending upon the objectives, namely, the tool-in-hand, tool-in-machine (holder) and tool-in-use systems. One should appreciate the neccessity of such consideration and the need for transformation matrixes if one considers a simple cutting insert used in an indexable turning, milling or drilling tool. The insert has its own geometry, assigned by the insert drawing and shown in the catalogues of the tool manufacturers. This geometry, however, may be considerebly altered through a wide range depending upon the tool holder used. In turn, the resultant geometry can be considerably altered depending upon the tool location in the machine with respect to the workpiece. Finally, the tool-in-use system becomes relevant when the directions of the cutting speed and cutting feed(s) become known. Naturally, the tool geometry in the tool-in-use system should be of prime concern. It should be used in any kind of modelling of the machining operation and in assuring tool-free penetration into the workpiece without interference. Knowing the tool-in-use geometry, the tool geometry in the other two systems can be obtained using transfomation matrices.

2.1.4 Basic Tool Geometry Components

There are two basic standards for utting tool geometry: (a) the American National Standard B94.50–1975 "Basic Nomenclature and Definitions for Single-Point Cutting Tools 1", reaffirmed date 2003, (b) ISO 3002/1 "Basic quantities in cutting and grinding – Part 1: Geometry of the active part of cutting tools – General terms, reference systems, tool and working angles, chip breakers", second edition 1982-08-01. Both standards deal with the tool-in-hand tool geometry. Although both standards are outdated and thus do not account for significant changes in the metal machining industries and for the advances of metal cutting theory and practice, they can be used to represent the basic cutting tool geometry.

The cutting tool geometry includes a number of angles measured in different planes [3]. Figure 2.3 visualizes the definition of the main reference plane P_r as perpendicular to the assumed direction of primary motion (the z-direction in Figure 2.3). This figure also sets the tool-in-hand coordinate system. In this figure, v_f is the assumed direction of the cutting feed, line 12 is the major cutting edge and 13 is the minor cutting edge. The tool-in-hand system consists of five basic planes defined relative to the reference plane P_r [4], some of which are illustrated in Figure 2.3:

- Perpendicular to the reference plane P_r and containing the assumed direction of feed motion is the assumed working plane P_f
- The tool cutting edge plane P_s is perpendicular to P_r, and contains the major cutting edge (1–2 in Figure 2.3)

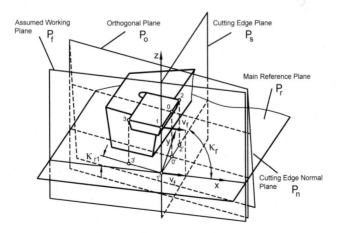

Figure 2.3. Reference planes

- The tool back plane P_p (not shown) is conicident with the zy-plane and thus is perpendicular to P_r and P_f
- Perpendicular to the projection of the cutting edge into the reference plane is the orthogonal plane P_o (in Figure 2.3 shown as passing thorugh the point 0' selected on the projection of the cutting edge)
- The cutting edge normal plane P_n is perpendicular to the cutting edge

The geometry of the cutting tool is defined by a set of the basic tool angles in the corresponding reference planes shown in Figure 2.3. The definitions of basic tool angles in the tool-in-hand system are as follows:

- Ψ is the tool approach angle; it is the acute angle that P_s makes with P_p and is measured in the reference plane as shown in Figure 2.4;
- The rake angle is the angle between the reference plane (the trace of which in the considered plane of measurement appears as the normal to the direction of primary motion) and the intersection line formed by the considered plane of measurement and the tool rake plane. The rake angle is defined as always being acute and positive when looking across the rake face from the selected point and along the line of intersection of the

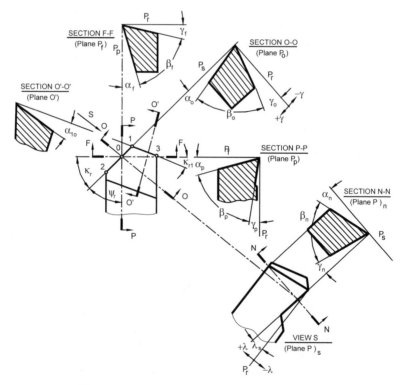

Figure 2.4. Tool angles in the tool-in-hand system

face and plane of measurement. The viewed line of intersection lies on the opposite side of the tool reference plane from the direction of primary motion in the measurement plane for γ_f, γ_p, γ_o or a major component of it appears in the normal plane for γ_n. The sign of the rake angles is well defined (Figure 2.4).

- The flank angles are defined in a way similar to the rake angles, although here, if the viewed line of intersection lies on the opposite side of the cutting edge plane P_s from the direction of feed motion (assumed or actual as the case may be) then the flank angle is positive. Angles α_f, α_p, α_o, and α_n are clearly defined in the corresponding planes as seen in Figure 2.4. The flank (clearance) angle is the angle between the tool cutting edge plane P_s and the intersection line formed by the tool flank plane and the considered plane of measurement, as shown in Figure 2.4.
- The wedge angles β_f, β_p, β_o and β_n are defined in the planes of measurements. The wedge angle is the angle between the two intersection lines formed as the corresponding plane of measurement intersects with the rake and flank planes.
- The orientation and inclination of the cutting edge are specified in the tool cutting edge plane P_s. In this plane, the cutting edge inclination angle λ_s is the angle between the cutting edge and the reference plane.
- The definition of the tool cutting edge angle, κ_r, is shown in Figure 2.2. It is defined as the acute angle that the tool cutting edge plane makes with the assumed working plane and is measured in the reference plane P_r. Similarly, the tool minor (end) cutting edge angle, κ_{r1}, is the acute angle that the minor cutting edge plane makes with the assumed working plane and is measured in the reference plane P_r.

2.1.5   Influence of the Tool Angles

The tool cutting edge angle significantly affects the cutting process because, for a given feed and cutting depth, it defines the uncut chip thickness, width of cut, and thus tool life. The physical background of this phenomenon can be explained as follows: when κ_r decreases, the chip width increases correspondingly because the active part of the cutting edge increases. This results in improved heat removal from the tool and hence tool life increases. For example, if the tool life of a high-speed steel (HSS) face milling tool having $\kappa_r = 60°$ is taken to be 100% then when $\kappa_r = 30°$ its tool life is 190%, and when $\kappa_r = 10°$ its tool life is 650%. An even more profound effect of κ_r is observed in the machining with single-point cutting tools. For example, in rough turning of carbon steels, the change of κ_r from 45° to 30° sometimes leads to a fivefold increase in tool life. The reduction of κ_r, however, has its drawbacks. One of these is the corresponding increase of the radial component of the cutting force, which reduces the accuracy and stability of machining particularly when the machine, tool holder and workpiece fixture are not sufficiently rigid.

Rake angles come in three varieties: positive, zero (sometimes referred to as neutral) and negative, as indicated in Figure 2.4. It is generally accepted that an increase in the rake angle reduces horsepower consumption per unit volume of the layer being removed at the rate of 1% per degree starting from $\gamma = -20°$. As a result, the cutting force and tool–chip contact temperature change in approximately the same way. So, it seems to be reasonable to select a high positive rake angle for practical cutting operations. Everyday machining practice, however, shows that there are number of drawbacks of increasing the rake angle.

The main drawback is that the strength of the cutting wedge decreases when the rake angle increases. When cutting with a positive rake, the normal force on the tool–chip interfaces causes bending of the tip of the cutting wedge. The presence of the bending significantly reduces the strength of the cutting wedge, causing its chipping. Moreover, the tool–chip contact area reduces with the rake angle so the point of application of the normal force shifts closer to the cutting edge. On the contrary, when cutting with a tool having a negative rake angle, the mentioned normal force causes the compression of the tool material. Because tool materials have very high compressive strength, the strength of the cutting edge in this case is much higher, although the normal force is greater than that for tools with positive rake angles. Another essential drawback is that the region of the maximum contact temperature at the tool–chip interface shifts toward the cutting edge when the rake angle is increased, which lowers tool life as discussed by Astakhov [5].

Realistically, the rake angle is not an independent variable in the process of tool geometry selection because the effect of the rake angle depends upon other parameters of the cutting tool geometry and the cutting process. Moreover, the necessity of applying chip breakers of different shapes often dictates the resulting rake angle rather than other parameters of the cutting process such as tool life, power consumption and cutting force.

Flank angle. If the flank angle $\alpha = 0°$ then the flank surface of the cutting tool is in full contact with the workpiece. As such, due to spring-back of the workpiece material, there is a significant friction force in such a contact that usually leads to tool breakage. The flank angle affects the performance of the cutting tool mainly by decreasing the rubbing on the tool's flank surfaces. When the uncut chip thickness is small (less than 0.02 mm), this angle should be in the range 30–35° to achieve maximum tool life.

The flank angle directly affects tool life. When the angle α increases, the wedge angle β decreases, as seen in Figure 2.4. As such, the strength of the region adjacent to the cutting edge decreases as well as the heat dissipation through the tool. These factors lower tool life. On the other hand, the following advantages may be gained by increasing the flank angle: (a) the cutting edge radius decreases with the flank angle, which leads to corresponding decreases in the frictional and deformation components of the flank force. This effect becomes noticeable in cutting with small feeds. As a result, less heat is generated, which leads to an increase in tool life, (b) as the flank angle becomes larger, more tool material has to be removed (worn out) to reach the same flank wear VB, increasing tool life. As a result of such contrary effects, the influence of the flank angle on tool life always has a well-defined maximum. In other word, there is always an optimal flank angle that should be found for a given machining operation.

Inclination angle. The sense and sign of the inclination angle λ_s is clearly shown in Figure 2.4 and is defined earlier as the angle between the cutting edge and the reference plane; experience shows that there are certain difficulties and confusions in understanding this angle. When the angle λ_s is positive, the chip flows to the right and when it is negative the chip flows to the left. The direction of chip flow, however, is defined not only by the angle λ_s but also by the cutting edge angle κ_r.

2.2   Tool Materials

Many types of tool materials, ranging from high-carbon steels to ceramics and diamonds, are used as cutting tool materials in today's metalworking industry. It is important to be aware that differences exist among tool materials, what these differences are and the correct application for each type of material [6].

The three prime properties of a tool material are:

- **Hardness:** defined as the resistance to indenter penetration. It is directly correlates with the strength of the cutting tool material [7]. The ability to maintain high hardness at elevated temperatures is called hot hardness. Figure 2.5 shows the hardness of typical tool materials as a function of temperature.
- **Toughness:** defined as the ability of a material to absorb energy before fracture. The greater the fracture toughness of a tool material, the better it resists shock load, chipping and fracturing, vibration, misalignments, runouts and other imperfections in the machining system. Figure 2.6 shows

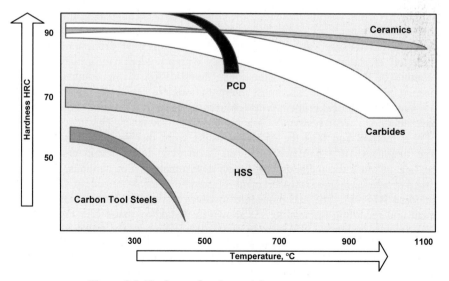

Figure 2.5. Hardness of tool materials versus temperature

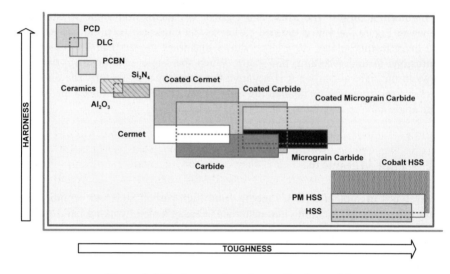

Figure 2.6. Hardness and toughness of tool materials

that, for tool materials, hardness and toughness change in opposite directions. A major trend in the development of tool materials is to increase their toughness while maintaining hardness.

- **Wear resistance:** In general, wear resistance is defined as the attainment of acceptable tool life before tools need to be replaced. Although seemingly very simple, this characteristic is the least understood.

Wear resistance is not a defined characteristic of the tool material and the methodology of its measurement. The nature of tool wear, unfortunately, is not yet sufficiently clear despite numerous theoretical and experimental studies. Cutting tool wear is a result of complicated physical, chemical, and thermo-mechanical phenomena. Because various simple mechanisms of wear (adhesion, abrasion, diffusion, oxidation *etc.*) act simultaneously with a predominant influence of one or more of them in different situations, identification of the dominant mechanism is far from simple, and most interpretations are subject to controversy. As the most common experimental device used by hard tool material manufacturers to characterize wear resistance is a pin-on-disk tribometer. The unacceptability of this method and thus the obtained results were discussed by Astakhov [5]. The toughness of a hard tool material is an even less relevant characteristic bearing in mind the methods used in its determination. For carbides, the *short-rod fracture toughness* measurement is common, as described in the ASTM standard B771-87. The test procedure involves testing of chevron-slotted specimens and recording the loading. As shown by Astakhov (page 150, Figure 4.8 in [4]), fracture toughness can vary by 300% depending on the loading conditions (stress state, strain rate and temperature). Therefore, the toughness of the tool materials should be determined using loading conditions similar to that occurred in machining.

Although a number of different tool materials are available today, five most important groups will be outlined in this section: carbides, ceramics, polycrystalline cubic boron nitrides (PCBNs), polycrystalline diamonds (PCDs) and solid or thick film diamond (SFDs or TFDs).

2.2.1 Carbides

Carbide as a tool material was discovered in the search for a replacement for expensive diamond dies used in the wire drawing of tungsten filaments. Initiated by a shortage of industrial diamonds at the beginning of World War I, researchers in Germany had to look for alternatives. On 10 June 1926 the name WIDIA (from the German term *Wie Diamant, i.e.*, like diamond) was entered into the register of trademarks and an arduous period of work started to transform laboratory-scale experiments into industrial production. The first product (Widia N – WC-6Co) was presented at the Leipzig Spring Fair in 1927.

2.2.1.1 Composition

Today, carbide tool materials include silicon and titanium carbides (called cerments) and tungsten carbides and titanium carbides as well as other compounds of a metal (Ti, W, Cr, Zr) or metalloid (B, Si) and carbon. Carbides have excellent wear resistance and high hot hardness. The terms *tungsten carbide* and *sintered carbide* for a tool material describe a comprehensive family of hard carbide composits used for metal cutting tools, dies of various types and wear parts [8]. A carbide tool material consists of carbide particales (carbides of tungsten, titanium, tantalum or some combination of these) bound together in a cobalt matrix by sintering. Normally, the size of the carbide particles is less than 0.8 μm for micrograins, 0.8–1.0 μm for fine grains, 1–4 μm for medium grains, and more than 4 μm for coarse-grain cutting inserts. The amount of cobal significantly affects the properties of carbide inserts. Normally, the cobalt content is 3–20%, depending upon the desired combination of toughness and hardness. As the cobalt content increases, the toughness of a cuting insert increases while its hardness and strength decrease. However, the correct combination of carbide insert composition (grade), coating materials, layer sequence and the selection of the appropriate coating technology makes it possible to increase metal cutting productivity substantially without sacrificing insert wear resistance.

2.2.1.2 Selection

The selection of the most advantageous carbide grade has become as sophisticated a factor as the design of the tooling itself. A wide variety of new carbide grades and coatings available today continue to complicate the manufacturing engineer's task of selecting the optimum grade as it relates to work material machinability, hardness and desired productivity, efficiency and quality. Coupled with newer, high-speed, powerful machines and coolant brands and supply techniques, this selection have created a real cutting tool insert selection dilemma for many specialists in the field. Because many manufacturing facilities do not have the luxury

Table 2.1. Guides to select carbide grade for a given application

Cutting conditions	Code	Colour
Finishing steels, high cutting speeds, light cutting feeds, favourable work conditions	P01	Blue
Finishing and light roughing of steels and castings with no coolant	P10	
Medium roughing of steels, less favourable conditions. Moderate cutting speeds and feeds.	P20	
General-purpose turning of steels and castings, medium roughing	P30	
Heavy roughing of steels and castings, intermittent cutting, low cutting speeds and feeds	P40	
Difficult conditions, heavy roughing/intermittent cutting, low cutting speeds and feeds	P50	
Finishing stainless steels at high cutting speeds	M10	Yellow
Finishing and medium roughing of alloy steels	M20	
Light to heavy roughing of stainless steel and difficult-to-cut materials	M30	
Roughing tough skinned materials at low cutting speeds	M40	
Finishing plastics and cast irons	K01	Red
Finishing brass and bronze at high cutting speeds and feeds	K10	
Roughing cast irons, intermittent cutting, low speeds and high feeds	K20	
Roughing and finishing cast irons and non-ferrous materials. Favourable conditions	K30	

of a machining laboratory or even the time to carry out machining evaluations for different cutting parameters, cutting tool manufacturers offer a guide for the initial selection, as shown in Table 2.1.

2.2.1.3 Coating

One of the most revolutionary changes in the metal cutting industry over the last 30 years has been thin-film hard coatings and thermal diffusion processes. These methods find ever-increasing applications and brought significant advantages to their users. Today, 50% of HSS, 85% of carbide and 40% of super-hard tools used in industry are coated [5]. A great number of coating materials, methods and regimes of application on substrates or whole tools and multi-layer coating combinations are used.

Carbides are excellent substrates for all coatings such as TiN, TiAlN, TiCN, solid lubricant coatings and multilayer coatings. Coatings considerably improve tool life and boost the performance of carbide tools in high-productivity, high-speed and high-feed cutting or in dry machining, and when machining of difficult-to-machine materials. Coatings: (a) provide increased surface hardness, for greater wear, (b) increase resistance (abrasive and adhesive wear, flank or crater wear), (c) reduce friction coefficients to ease chip sliding, reduce cutting forces, prevent adhesion to the contact surfaces, reduce heat generated due to chip sliding *etc.*, (d) reduce the portion of the thermal energy that flows into the tool, (e) increase corrosion and oxidation resistance, (f) improve crater wear resistance and (g) improved the surface quality of finished parts.

Common coatings for carbides applied in single- or multi-layers are shown in Figure 2.7. They are:

- TiN: general-purpose coating for improved abrasion resistance. Colour – gold, hardness HV (0.05) – 2300, friction coeficient – 0.3, thermal stability – 600°C.
- TiCN: multi-purpose coating intended for steel machining. Higher wear resistance than TiN. Available in mono- and multi-layer. Colour – grey-violet, hardness HV (0.05) – 3000, friction coeficient – 0.4, thermal stability – 750°C.
- TiAlN and TiAlCN – High-performance coating for increased cutting parameters and higher tool life; also suitable for dry machining. Reduces heating of the tool. Multi-layered, nanostructured or alloyed versions offer even better performance. Colour – black-violet, hardness HV (0.05) – 3000–3500, friction coeficient – 0.45, thermal stability – 800–900°C.

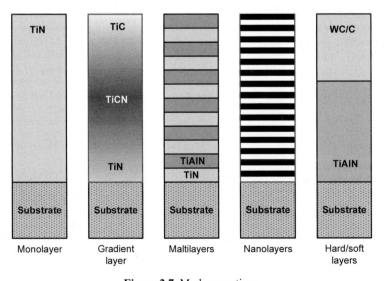

Figure 2.7. Modern coatings

- WC-C and MoS_2 – Provides solid lubrication at the tool–chip interface that significantly reduces heat due to friction. Has limited temperaure resistance. Recommended for high-adhesive work materials such as aluminium and copper alloys and also for non-metallic materials. Colour – gray-black, hardness HV (0.05) – 1000–3000, friction coeficient – 0.1, thermal stability – 300°C.
- CrN – Intended for copper alloys such as brass, bronze *etc.* Colour – metallic.

Coating fracture toughness is as important as coating hardness in crack retardation. Balance between high compressive stress (poor adhesion) and low residual stress (no crack retardation) is necessary.

A great attempt to correlate the counting materials and their performance was made by Klocke and Krieg [9]. It was pointed out that there are basically four major groups of coating materials on the market. The most popular group is titanium-based coating materials as TiN, TiC and Ti(C,N). The metallic phase is often supplemented by other metals such as Al and Cr, which are added to improve particular properties such as hardness or oxidation resistance. The second group represents ceramic-type coatings as Al_2O_3 (alumina oxide). The third group includes super-

Table 2.2. Basic PVD coatings

Coating	Characteristics
Titanium nitride, TiN	This gold-coloured coating offers excellent wear resistance with a wide range of materials, and allows the use of higher feeds and speeds. Forming operations can expect a decrease in galling and welding of workpiece material with a corresponding improvement in the surface finish of the formed part. A conservative estimate of tool life increase is 200–300%, although some applications see as high as 800%.
Titanium carbonitride, TiN(C,N)	Bronze-coloured Ti(C,N) offers improved wear resistance with abrasive, adhesive or difficult-to-machine materials such as cast iron, alloys, tool steels, copper and its alloys, Inconel and titanium alloys. As with TiN, feeds and speeds can be increased and tool life can improve by as much as 800%. Forming operations with abrasive materials should see improvements beyond those experienced with TiN.
Titanium aluminium nitride, (Ti,Al)N	Purple/black in colour, (Ti,Al)N is a high-performance coating which excels at machining of abrasive and difficult-to-machine materials such as cast iron, aluminium alloys, tool steels and nickel alloys. (Ti,Al)N's improved ductility makes it an excellent choice for interrupted operations, while its superior oxidation resistance provides unparalleled performance in high-temperature machining.
Chromium nitride, CrN	Silver in colour, CrN offers high thermal stability, which in turn helps in the aluminium die casting and deep-draw applications. It can also reduce edge build-up commonly associated with machining titanium alloys with Ti-based coatings.

hard coatings, such as chemical vapor deposition (CVD) diamond. The fourth group includes solid lubricant coating such as amorphous metal-carbon. Additionally, to reduce extensive tool wear during cut-in periods, some soft coatings as MoS_2 or pure graphite are deposited on top of these hard coatings. The basic physical vapor deposition (PVD) coatings are listed in Table 2.2. The effectiveness of various coatings on cutting tools is discussed by Bushman and Gupta [10].

2.2.2   Ceramics

Introduced in the earlier 1950s, ceramic tool materials consist primarily of fine-grained aluminium oxide, cold-pressed into insert shapes and sintered under high pressure and temperature. Pure alumimum oxide ceramics are called white ceramics while the addition of titanium carbide and zirconiou oxide results in black cermets (not to be confuse with the carbide cermets discussed earlier).

The prime benefit of ceramics is high hardness (and thus abrasive wear resistance) at elevated temperatures, as seen in Figure 2.6. All tool materials soften as they become hotter, but ceramics do so at a much slower rate because they are not metal limited. Among the major advantages of ceramic cutting tools is also chemical stability. In practical terms this means that the ceramic does not react with the material it is cutting, *i.e.*, there is no diffusion wear, which is the weakest spot of carbides in high-speed machining applications.

Ceramics are suitable for machining the majority of ferrous materials, including superalloys. It should not be used, however, for copper, brass and aluminium due to the formation of an excessive built-up edge. There are indications that aluminium-oxide-based ceramics are being replaced by PCBN. PCBN is taking over much of the ceramic work because it works better for softer materials.

The downside to these ceramic materials is a slightly higher cost and brittleness. To protect their cutting edges, ceramics are typically made with a heavy edge preparation such as a T-land or honed edge or with modern edge preparation features.

There are two basic kinds of ceramics. The first is aluminium oxide. It is wear-resistant but brittle, and used chiefly on hardened steel. The other major type is silicon nitride, which is relatively soft and tough and is used on cast irons. Between aluminium oxide and silicon nitride fall a whole host of ceramic materials called Si-AlONs that combine the two. The greater the proportion of aluminium oxide, the harder the material. The more silicon nitride included, the tougher the material.

In the leap-frog race between work materials and tools, the laurels still go to the tools. The cutting ability of the tool is still slightly ahead of the applications (including available machine tools and their relevant characteristics) because there is a reluctance to apply the available cutting tool technology. However, when one moves to high-speed machining, one has to make major changes to the existing machining operation, including fixturing, chucks, guards, programming, coolants and a lot of other housekeeping issues. Not everyone wants to take the trouble, or spend the money, to do this. In making the ultimate decision, lot size determines to a large extent whether high speed, and therefore ceramics, is practical.

It was a little disappointed that ceramic-reinforcement technology has not moved ahead as quickly as initially supposed. Reinforcements offer a lot of strength advantages. They are available, but not in widespread use. It now seems that a new area that will offer a lot of new advantages in ceramic tools is nanotechnology. The most advanced ceramics today are micrograin materials, while the latest developments aim to move to nanograins or particle sizes of less than a micron. This technology is coming along well. The main advantage it offers is that the smaller particle size increases strength because more grain area is exposed to bonding. This strength increase translates into greater impact resistance and improved wear properties.

Coatings are rarely used with ceramic inserts. On ceramics, coatings do some good but the cost is high and usually does not justify the end result, because of weak adhesion between the coating materials and ceramic substrate.

For ceramics, the future is bright because of the push for high-speed machining. Modern machines now typically operate in the 600–900 m/min range while speeds of 1500 m/min are being tested. Only advanced cutting-tool materials can handle this speed. There has been a lot of improvement in wear, chiefly through the adoption of small grain sizes. For example, in hard turning applications with ceramics, tool life is improved up to 20-fold with modern grades. Cermets are a slow-growth product in many countries except Japan, where part manufacturing starts with a blank that is near net shape, which in many cases requires just finishing with cermets.

2.2.3   Cubic Boron Nitride (CBN)

Polycrystalline CBN blanks are manufactured from cubic boron nitride crystals utilizing an advanced high-temperature, high-pressure process. The cubic boron nitride crystals are sintered together with a binder phase and integrally bonded to a tungsten carbide substrate. The binder phase, usually either a metallic or ceramic matrix, provides chemical stability, enabling the PCBN qualities to be utilized in high-speed machining environments. The tungsten carbide substrate in the PCBN blanks provides the high impact resistance necessary for the depths of cuts and high speeds associated with machining of hardened ferrous materials. PCBN cutting tools offer excellent heat dissipation and wear resistance. Cutting tool geometries can be prepared to withstand interrupted cuts with a T-land and/or honed to stabilize the cutting edge and prolong tool life.

PCBN tools offer the following benefits: (a) machine-hardened and heat-treated steels, (b) an excellent surface finish that allows eliminate grinding, (c) high productivity rate that can be more than four times higher than that in grinding, (d) great resistance to abrasion which is twice that of ceramics and ten times than that of carbide. PCBN tools are recommended for machining cast irons including compacted graphite iron (CGI), sintered iron and superalloys hardened steels. Typical machining regimes are shown in Table 2.3.

To promote good tool life, the cutting edge of the PCBN insert must be reinforced with proper edge preparation. This can range from a small hone for finish-machining cast irons, to a T-land measuring 0.2 mm wide by 15° for heavy roughing of white iron. Combined lands and hones may also be used. Table 2.4 shows typical parameters of the T-land applied for PCBN inserts.

Users of superhard materials such as PCBN and PCD in cutting tools commonly believe that chamfering, also known as applying a T-land or K-land, is

Table 2.3. Typical machining regimes for CBN

Work material	Cutting speed (m/min)	Feed (mm/rev)
Hardened steel	120–150	0.10–0.2
Gray cast iron		
<240 HBN	450–1000	0.25–0.50
>240 HBN	300–600	0.25–0.50
Supealloys	150–300	0.10–0.25
Powder metals	90–300	0.08–0.20
Thermal spray	150–300	0.08–0.20
Bearing steel	110–150	0.05–0.20

Table 2.4. Edge preparation parameters for PCBN inserts

Work material	Roughning	Finishing
Hardened steel	$f_{ch}=0.2–0.5$, $\gamma_{ch}=20°$	$f_{ch}=0.2$, $\gamma_{ch}=20°$
Gray cast iron	$f_{ch}=0.2$, $\gamma_{ch}=15–20°$	Hone $r=0.2$
Hard cast iron	$f_{ch}=0.2–0.5$, $\gamma_{ch}=15–20°$	$f_{ch}=0.2$, $\gamma_{ch}=20°$
Powder metal	$f_{ch}=0.2$, $\gamma_{ch}=20°$	$f_{ch}=0.2$, $\gamma_{ch}=20°$
Superalloys	$f_{ch}=0.2–0.5$, $\gamma_{ch}=20°$	

necessary for extending tool life. This practice is so widely accepted that many industry professionals have never even seen a CBN cutting tool without a chamfer; and they assume that it is a necessary feature of the tool. In fact, chamfering, in most cutting tool applications, has been proven to be a sub-optimal solution that limits tool life and diminishes cutting performance. With the advent of advanced edge preparation technology and edge preparation machines, alternatives to chamfering now exist so chamfering is no longer a necessity for CBN and PCD cutting tools.

2.2.4 Polycrystalline Diamond (PCD) and Solid Film Diamond (SFD)

Being the most southt-after gemstone in the world, diamond may well also be the world's most versatile engineering material. Diamond is the strongest and hardest known substrate, it has the highest thermal conductivity of any material at room temperature, low-friction surface and optical transparancy. This unique combination of properties cannot be matched by any other material [11].

To produce PCD used in cutting tools, a layer of diamond crystals, made out of a mixture of graphite and a catalyst (typically nickel) under a pressure of approximately 7000 MPa and temperature of 1800°C, is placed on a carbide substrate and subjected to a high-temperature high-pressure process (6000 MPa, 1400°C). During this process, cobalt from the tungsten substrates becomes the binder of the diamond crystals giving polycrystalline diamond the required toughness.

PCD tool materials typically provide abrasion resistance up to 500 times that of tungsten carbide and high thermal conductivity. PCD tools have replaced tungsten carbide, ceramics and natural diamond in a range of high-performance applications including the turning, boring, milling, slotting and chamfering of materials such as high-silicon aluminium, metal matrix composites (MMC), ceramics, reinforced epoxies, plastics, carbon-fibre-reinforced plastics (CFRP) and engineered wood products. The extended tool life and increased productivity provided by PCD tools often offset the higher initial cost by lowering the unit cost of parts produced. Useful tool life may be further extended through multiple re-sharpenings. Table 2.5 shows typical machining regimes for PCD tools.

Selecting the optimum grade of PCD tooling for a specific application is generally a function of surface finish requirements and tool life expectations. Material removal rates, tool geometries and material characteristics also affect the relationship between machining productivity, tool life and surface finish. Coarse-grade PCD is designed with a larger diamond particle size than a fine-grade PCD. Generally, PCD with larger diamond particles exhibits greater abrasion resistance, but results in a rougher cutting edge. Conversely, smaller diamond particle will result in a sharper cutting edge, producing a superior workpiece surface finish, but tool life is reduced.

Having high abrasion resistance and great hardness, PCDs suffer from relatively low toughness. To overcome this shortcoming, the development of new prime grades of PCDs relies on structural changes that enhance toughness. One of the most promising directions is to combine diamond particle of different sizes (for example, 30 and 2 μm, as proposed by Element 6 Co.) in the mixture to increase the diamond packing density, as shown in Figure 2.8. The improved packing density results in a higher degree of contiguity between diamond grains, thereby enhancing resistance to chipping of the cutting edge. An added advantage of the increased packing density is the quality of the ground cutting edge as the filling of the area between the coarse diamond grains with fine diamond yields a continuous as opposed to the micro-serrated irregular cutting edge obtained with usual PCD grades.

Thick-film diamond (TFd) tools constitute a major breakthrough in the science of cutting tools. The company SP3 has been developing thick-film diamond technology for several years, and now offers a new product line of TFd cutting tools. A stand-alone sheet of thick-film diamond is grown in a chemical vapour deposition reactor. Typical films are 500 μm thick and come in flat sheets. These sheets are than laser cut into tips, which are secured into tool bodies using a specially developed brazing process. Axial end tools such as as drills, reamers, boring tips, cartridges for boring bars and milling tools are produced. Application-specific tool

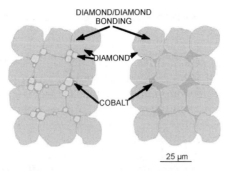

Figure 2.8. Improving packing density by combining diamonds of considerably different sizes

design with TFd is now under extensive development at the most advanced auto-motive manufacturing power-train facilities.

TFd provides three distinct advantages over PCD tools: (a) it is intrinsically harder and more wear resistant than PCD because it is solid diamond with no binder material; (b) when machining abrasive metals with TFd, the tool wears primarily on the flank. This causes the cutting edge to remain sharper than PCD as the tool wears. This is particularly imeortant in applications where burr control is crucial to producing good parts. The life of TFd tools is depend-ent on edge recession and is not limited by premature failure related to edge sharpness; and (c) There is no possibility of chemical interaction with the cool-ant or by-products of the workpiece material because there is no binder in TFd. As a result, the tool life of TFd tools is substantially longer than that of PCD tools (Figure 2.9).

Thick-film diamond tools have demonstrated tool life two to three times that of PCD tools in tests conducted by an independent test laboratory. Thick-film diamond tools are the first to evidence performance exceeding that of PCD in 25 years.

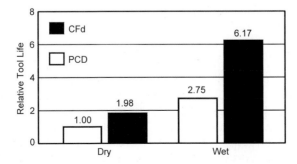

Figure 2.9. Relative performance of TFd versus PCD in turning of high-silicon aluminium alloy 390 (up to 18% Si)

Table 2.5. Recommended machining regimes

Work material	Cutting speed (m/min)	Cutting feed (mm/rev)	Depth of cut (mm)
Aluminium alloys			
<12% Si	1000–3000	0.1–0.4	5
>12% Si	200–600	0.1–0.4	1
Metal matrix composites (MMC)	150–600	0.1–0.4	0.5
Brass	600–2000	0.1–0.4	1.5
Hard plastics	1000–7000	0.1–0.7	2.5
Carbon-fibre-reinforced plastics (CFRP)	500–2000	0.05–0.4	4
Sintered tungsten carbide 18% Co	40–60	0.05–0.2	0.5
Precious metals	100–500	0.05–0.4	1.5

2.3   Tool Wear

Tool wear leads to tool failure. According to many authors, the failure of cutting tool occurs as premature tool failure (*i.e.*, tool breakage) and progressive tool wear. Figure 2.10 shows some types of failures and wear on cutting tools.

Generally, wear of cutting tools depends on tool material and geometry, work-piece materials, cutting parameters (cutting speed, feed rate and depth of cut), cutting fluids and machine-tool characteristics.

2.3.1 Tool Wear Types

Normally, tool wear is a gradual process. There are two basics zones of wear in cutting tools: flank wear and crater wear.

Flank and crater wear are the most important measured forms of tool wear. Flank wear is most commonly used for wear monitoring. According to the standard ISO 3685:1993 for wear measurements, the major cutting edge is considered to be divided in to four regions, as shown in Figure 2.11:

- *Region C* is the curved part of the cutting edge at the tool corner;
- *Region B* is the remaining straight part of the cutting edge in zone C;
- *Region A* is the quarter of the worn cutting edge length b farthest away from the tool corner;

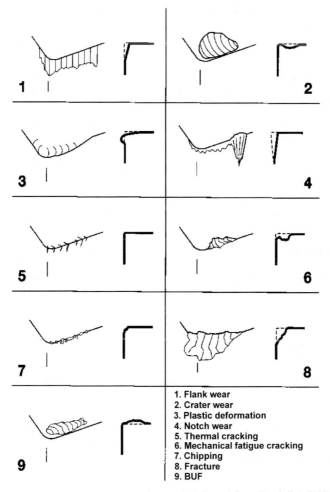

Figure 2.10. Types wear on cutting tools (adapted from Sandvik® [12])

1. Flank wear
2. Crater wear
3. Plastic deformation
4. Notch wear
5. Thermal cracking
6. Mechanical fatigue cracking
7. Chipping
8. Fracture
9. BUF

- *Region N* extends beyond the area of mutual contact between the tool workpiece for approximately 1–2 mm along the major cutting edge. The wear is of notch type.

The width of the flank wear land, VB_B, is measured within zone B in the cutting edge plane P_s (Figures 2.3 and 2.11) perpendicular to the major cutting edge. The width of the flank wear land is measured from the position of the original major cutting edge.

The crater depth, KT, is measured as the maximum distance between the crater bottom and the original face in region B.

Tool wear is most commonly measured using a toolmaker's microscope (with video imaging systems and a resolution of less than 0.001 mm) or stylus instrument similar to a profilometer (with ground diamond styluses).

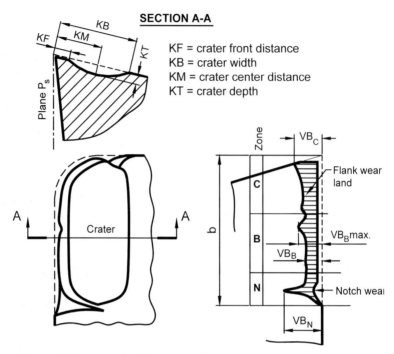

KF = crater front distance
KB = crater width
KM = crater center distance
KT = crater depth

Figure 2.11. Types of tool wear according to standard ISO 3685:1993 [13]

2.3.2   Tool Wear Evolution

Tool wear curves illustrate the relationship between the amount of flank (rake) wear and the cutting time, τ_m, or the overall length of the cutting path, L. Figure 2.12(a) shows the evolution of flank wear VB_B max, as measured after a certain length of cutting path. Normally, there are three distinctive regions that can be observed in such curves. The first region (region I in Figure 2.12(a)) is the region of primary or initial wear. The relatively high wear rate (an increase of tool wear per unit time or length of the cutting path) in this region is explained by accelerated wear of the tool layers damaged during manufacturing or re-sharpening. The second region (region II in Figure 2.12(a)) is the region of steady-state wear. This is the normal operating region for the cutting tool. The third region (region III in Figure 2.12(a)) is known as the tertiary or accelerated wear region. Accelerated tool wear in this region is usually accompanied by high cutting forces, temperatures and severe tool vibrations. Normally, the tool should not be used in this region.

In practice, the cutting speed is of prime concern in the consideration of tool wear. As such, tool wear curves are constructed for different cutting speeds keeping other machining parameters constant. In Figure 2.12(b), three characteristic tool wear curves (mean values) are shown for three different cutting speeds, v_1,

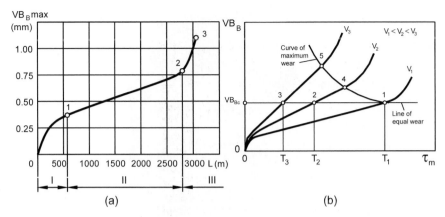

Figure 2.12. Wear curves: (a) normal wear curve, (b) evolution of flank wear land VBB as a function of cutting time for different cutting speeds

v_2, and v_3. Because v_3 is greater than the other two, it corresponds to the fastest wear rate. When the amount of wear reaches the permissible tool wear VB_{Bc}, the tool is said to be worn out.

Typically VB_{Bc} is selected from the range 0.15–1.00 mm depending upon the type of machining operation, the condition of the machine tool and the quality requirements of the operation. It is often selected on the grounds of process efficiency and often called the criterion of tool life. In Figure 2.12(b), T_1 is the tool life when the cutting speed v_1 is used, T_2 – when v_2, and T_3 – when v_3 is the case. When the integrity of the machined surface permits, the curve of maximum wear instead of the line of equal wear should be used (Figure 2.12(b)). As such, the spread in tool life between lower and higher cutting speeds becomes less significant. As a result, a higher productivity rate can be achieved, which is particularly important when high-speed CNC machines are used.

Figure 2.13 shows an example of typical flank wear of a CVD diamond tool, which can be observed during machining of a high-silicon aluminium alloy (MMC).

The criteria recommended by ISO3685:1993 [13] to define the effective tool life for cemented carbides tools, high-speed steels (HSS) and ceramics are:

Cemented carbides:

1. $VB_B = 0.3$ mm, or
2. $VB_{B,max} = 0.6$ mm, if the flank is irregularly worn, or;
3. $KT = 0.06 + 0.3\,f$, where f is the feed.

HSS and ceramics:

1. Catastrophic failure, or;
2. $VB_B = 0.3$ mm, if the flank is regularly in region B; or
3. $VB_{B,max} = 0.6$mm, if the flank is irregularly in region B.

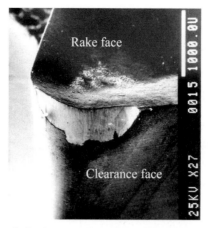

Figure 2.13. Example of flank wear for CVD diamond tool in machining MMC (V_c = 50 m/min, f = 0.2 mm/rev, d_{oc} = 1mm and cutting time 4.5 min) [14]

Table 2.6. Recommendations used in industrial practice for limit of flank wear VB_B for several cutting materials

Tool material		HSS	Cemented carbides	Carbides coateds	Ceramics	
Operation	(mm)				Al_2O_3	Si_3N_4
Roughing	VB_B	0.35–1.0	0.3–0.5	0.3–0.5	0.25–0.3	0.25–0.5
Finishing	VB_B	0.2–0.3	0.1–0.25	0.1–0.25	0.1–0.2	0.1–0.2

General recommendations used in industrial practice for the limit of flank wear VB_B for several cutting materials are given in Table 2.6.

2.3.4   Mechanisms of Tool Wear

The general mechanisms that cause tool wear, summarized in Figure 2.14, are: (1) abrasion, (2) diffusion, (3) oxidation, (4) fatigue and (5) adhesion. The fundaments of there tool wear mechanisms are explained for several authors, for example, Shaw [15] and Trent and Wright [16]. Most of these mechanisms are accelerated at higher cutting speeds and consequently cutting temperatures.

2.4   Tool Life

Tool life is important in machining since considerable time is lost whenever a tool is replaced and reset. Tool life is the time a tool will cut satisfactorily and is expressed as the minutes between changes of the cutting tool. The process of wear

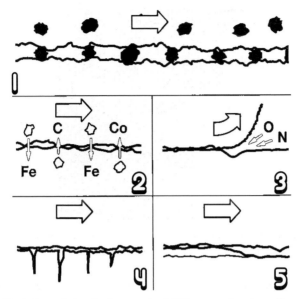

Figure 2.14. Evolution of the flank wear land VB_B as a function of cutting time for different cutting speeds [12]

and failures of cutting tools increases the surface roughness, and the accuracy of workpieces deteriorates.

2.4.1 Taylor's Tool Life Formula

Tool wear is almost always used as a lifetime criterion because it is easy to determine quantitatively. The flank wear land VB_B is often used as the criterion because of its influence on workpiece surface roughness and accuracy. Figure 2.15 shows the wear curves (VB_B versus cutting time) for several cutting velocities (1, 2 and 3) and the construction of the life curve (cutting velocity versus tool life).

Taylor [17] presented the following equation:

$$V_c T^n = C \tag{2.1}$$

where V_c is the cutting speed (m/min), T is the tool life (min) taken to develop a certain flank wear (VB_B), n is an exponent that depends on the cutting parameters and C is a constant. Note that C is equal to the cutting speed at $T=1$ min.

Therefore, each combination of tool material and workpiece and each cutting parameter has it is own n and C values, to be determined experimentally. For example, choosing two extreme points (Figure 2.15(a)), points 1 and 3, $V_c=200$ m/min, $T=40$ min and $V_c=400$ m/min and $T=10$ min, respectively, we have:

$$200 \times 40^n = C \tag{2.2}$$

$$400 \times 10^n = C \tag{2.3}$$

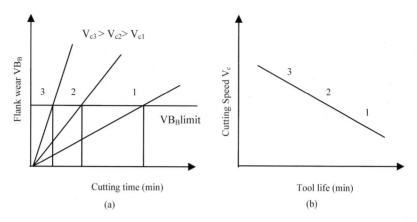

Figure 2.15. Wear curves for several cutting speeds (1, 2 and 3) (a) and life curve (b)

Taking natural logarithms of each term gives

$$ln\,200 + n\,ln\,40 = ln\,400 + n\,ln\,10 \tag{2.4}$$

$$5.298 + n \times 3.689 = 5.991 + n \times 2.303 \tag{2.5}$$

$$n = 0.5 \tag{2.6}$$

Substituting this value of n into Equations (2.2) and (2.3), one can calculate the corresponding values of C

$$C = 200 \times 40^{0.5} = 804 \text{ or } C = 400 \times 10^{0.5} = 1264.9 \tag{2.7}$$

The Taylor equation for the data show in Figure 2.15 is:

$$V_c T^{0.5} = 1264.9 \tag{2.8}$$

Table 2.7 presents the range of n values determined in practice for some tool materials.

Table 2.7. Values of n observed in practice for several cutting tool materials

Tool material	HSS	Cemented carbides	Ceramics
n	0.1–0.2	0.2–0.5	0.5–0.7

2.4.2   Expanded Taylor's Tool Life Formula

According to the original Taylor tool life formula, the cutting speed is the only parameter that affects tool life. This is because this formula was obtained using high-carbon and high-speed steels as tool materials. With the further development of carbides and other tool materilas, it was found that the cutting feed and the depth of cut are also significant. As a result, the Taylor's tool life formula was modified to accommodate these changes as:

$$V_c T^n f^a d^b = C \qquad (2.9)$$

where d is the depth of cut (mm) and f is the feed (mm/rev). The exponents a and b are to be determined experimentally for each combination of the cutting conditions. In practice, typical values for HSS tools are $n=0.17$, $a=0.77$ and $b=0.37$ [18]. According to this information, the order of importance of the parameters is: cutting speed, then feed, then depth of cut. Using this parameters, Equation (2.9) for the expanded Taylor tool life formula model can be rewritten as:

$$T = C^{\frac{1}{n}} V_c^{\frac{-1}{n}} f^{\frac{-a}{n}} d^{\frac{-b}{n}} \text{ or } T = C^{5.88} V^{-5.88} f^{-4.53} d^{-2.18} \qquad (2.10)$$

Although cutting speed is the most important cutting parameter in the tool life equation, the cutting feed and the depth of cut can also be the significant factors.

Finally, the tool life depends on the tool (material and geometry); the cutting parameters (cutting speed, feed, depth of cut); the brand and conditions of the cutting fluid used; the work material (chemical composition, hardness, strength, toughness, homogenity and inclusions); the machining operation (turning, drilling, milling), the machine tool (for example, stiffness, runout and maintanace) and other machining parameters. As a result, it is nearly impossible to develop a universal tool life criterion.

2.4.3   Recent Trends in Tool Life Evaluation

Although Taylor's tool life formula is still in wide use today and lies at the very core of many studies on metal cutting, including at the level of national and international standards, one should remember that it was introduced in 1907 as a generalization of many years of experimental studies conducted in the 19th century using work and tool materials and experimental technique available at that time. Since then, each of these three components has undergone dramatic charges. Unfortunately, the validity of the formula has never been verified for these new conditions. Nobody has yet proved that it continues to be valid for cutting tool materials other than carbon steels and high-speed steels.

Moreover, one should clearly realize that tool life is not an absolute concept but depends on what is selected as the tool life criteria. In finishing operations, surface integrity and dimensional accuracy are of primary concern, while in roughing operations the excessive cutting force and chatter are limiting factors. In both applications, material removal rate and chip breaking could be critical factors. These criteria, while important from the operational point of view, have little to do with the physical conditions of the cutting tool.

To analyze the performance of cutting tools on CNC machines, production cells and manufacturing lines, the dimension tool life is understood to be the time period within which the cutting tool assures the required dimensional accuracy and required surface integrity of the machined parts.

Although there are a number of representations of the dimension tool life, three of them are the most adequate [5]:

- The dimension wear rate is the rate of shortening of the cutting tip in the direction perpendicular to the machined surface taken within the normal wear period (region II in Figure 2.12(a)), i.e.,

$$v_h = \frac{dv_r}{dT} = \frac{h_r - h_{r-i}}{T - T_i} = \frac{vh_{l-r}}{1000} = \frac{vfh_s}{100} \quad (\mu m/min) \qquad (2.11)$$

  where h_r and h_{r-i} are the current and initial radial wear, respectively, T and T_i are the total and initial operating time, respectively, and h_s is the surface wear rate. It follows from Equation (2.11) that the dimension wear rate is inversely proportional to the tool life but does not depend on the selected wear criterion (a particular width of the flank wear land, for example).

- *The surface wear rate* is the radial wear per 1000 cm^2 of the machined area (S)

$$h_s = \frac{dh_r}{dS} = \frac{(h_r - h_{r-i})100}{(l - l_i)f} \quad (\mu m/10^3 cm^2) \qquad (2.12)$$

  where h_{r-i} and l_i are the initial radial wear and initial length of the tool path, respectively, and l is the total length of the tool path. It follows from Equation (2.12) that the surface wear rate is reverse proportional to the overall machined area and, in contrast, does not depend on the selected wear criterion.

- *The specific dimension tool life* is the area of the workpiece machined by the tool per micron of radial wear

$$T_{UD} = \frac{dS}{dh_r} = \frac{1}{h_s} = \frac{(l - l_i)f}{(h_r - h_{r-i})100} \quad (10^3 cm^2/\mu m) \qquad (2.13)$$

The surface wear rate and the specific dimension tool life are versatile tool wear characteristics because they allow the comparison of different tool materials for different combinations of the cutting speeds and feeds using different criteria selected for the assessment of tool life.

References

[1]   Rodin PR (1972) The Basics of Shape Formation by Cutting (in Russian). Visha Skola, Kyev (Ukraine)
[2]   Granovsky GE, Granovsky VG (1985) Metal Cutting (in Russian), Vishaya Shkola, Moscow
[3]   Oxley PLB (1989) Mechanics of Machining: An Analytical Approach to Assessing Machinability. Wiley, New York, NY
[4]   Astakhov VP (1998) Metal Cutting Mechanics. CRC, Boca Raton, USA
[5]   Astakhov VP (2006) Tribology of Metal Cutting. Elsevier: London
[6]   Davis JR (Editor) (2005) Tool Materials (ASM Specialty Handbook), ASM International, Materials Park, OH, USA
[7]   Isakov E (2004) Engineering Formulas for Metalcutting. Industrial, New York, NY
[8]   Upadbyaya GS (1998) Cemented Tungsten Carbides. Production, Properties, and Testing. Noyes, Westwood, NJ
[9]   Klocke F, Krieg T (1999) Coated tools for metal cutting – features and applications. Ann CIRP 48: 515–525
[10] Bushman B, Gupta BK (1991) Handbook of Tribology-Materials, Coatings, and Surface Treatments. McGraw–Hill, New York, NY
[11] Whitney ED (1994) Ceramic Cutting Tools. Materials, Development, and performance. Noyes, Westwood, NJ
[12] Modern Metal Cutting, A practical Handbook, Sandvik Coromant
[13] Tool-life testing with single-point turning tools, ISO 3685:1993
[14] Davim JP (2002) Diamond tool performance in machining metal–matrix composites. J Mater Process Technol 128: 100–105
[15] Shaw MC (1984) Metal cutting principles. Oxford Science, Oxford, UK
[16] Trent EM, Wright PK (2000) Metal cutting. Butterworth–Heinemann, Boston, MA
[17] Taylor FW (1907) On the art of cutting metals. Trans ASME 28: 31–58
[18] Boston OW (1941) Metal processing. Wiley, New York, NY

3

Workpiece Surface Integrity

Joël Rech[1], Hédi Hamdi[1] and Stéphane Valette[2]

[1]Ecole Nationale d'Ingénieurs de Saint-Etienne, LTDS, 58 Rue Jean Parot, 42023 Saint-Etienne cedex 2, France.
E-mail: joel.rech@enise.fr, hédi.hamdi@enise.fr.
[2]Ecole Centrale de Lyon, LTDS, 36 avenue Guy de Collongue, 69134 Ecully cedex, France.
E-mail: stephane.valette@ec-lyon.fr.

This chapter presents an analysis of workpiece surface integrity. The definition and material and mechanical aspects of surface integrity are discussed.

3.1 What Does Surface Integrity Mean?

The choice of manufacturing processes is based on cost, time and precision. The precision of a surface is usually based on two criteria: dimensional accuracy and surface roughness. However, another criterion has become increasingly important: the performance of the surface. The term *performance* has different meanings depending on the context but is mostly linked to fatigue, corrosion, wear and strength. It is usually assumed that performance is directly related to surface texture. The irregularities of the surface, especially valleys or grooves, induce stress concentrations that enable the plastification of the material and crack propagation. As a consequence, a smooth surface limits the risk of crack initiation. An example provided by [1] illustrates the influence of the surface texture generated by grinding on the fatigue strength (Figure 3.1). Curves 1 and 2 correspond to a standard grinding operation with conditions leading to so-called *gentle* operation. On the contrary, curve 3 corresponds to high-productivity operation with extreme cutting conditions. Curves 1 and 2 show that a higher surface texture (evaluated, for example, by the parameter Ra) is responsible for the decrease of the fatigue strength. It also shows that, depending on the orientation of grinding marks (surface texture effect), the fatigue resistance is different. Note that the surface roughness parameter Ra alone is not able to describe resistance to fatigue. Additional parameters such as Rku, RSk, *etc.* are more relevant.

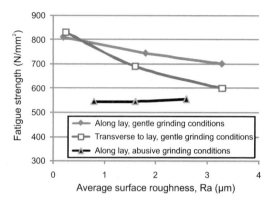

Figure 3.1. Influence of a grinding operation on the fatigue strength

However, the surface roughness criterion does not explain the results obtained for curve 3, which appears to be very poor. Moreover, with this manufacturing procedure, the surface roughness does not seem to influence the fatigue strength. It is clear that the sub-surface is as important as the surface texture. Especially, the microstructure and the residual stress state are strategic parameters. Indeed, this manufacturing procedure has probably induced dramatic modifications of the sub-surface, which is a very common problem in grinding, as described by a large number of publications [2–6].

Another investigation [7] has shown that the effect of the mechanical state of the sub-surface (the residual stress state) has much more influence on the fatigue resistance of a piece if its surface roughness is below a certain limit. This trend is also confirmed by [8]. This underlines that external parameters and internal parameters are both strategic, depending on the context of application. For that reason, the term *surface integrity* was introduced, which aims to describe the state of a surface (from external and internal points of view) with regard to its potential performance. The oiginal definition [9] of this term is "the inherent or enhanced condition of a surface produced in a machining or other surface generation operation". After some years of scientific works on the subject, a new definition has been proposed [10]: "the topographical, mechanical, chemical and metallurgical worth of a manufactured surface and its relationship to functional performance".

A surface can be defined as a border between a part and its environment. One particular piece belonging to a mechanical system has several surfaces. In common practice, engineers design pieces to satisfy a list of criteria in relation to the function of each surface. Some surfaces have mechanical contact with another part, whereas other surfaces just make contact with air or oil, *etc.* As a consequence, the specification of each surface is different depending on its function:

- Mechanical functions (capability of carrying mechanical loads)
- Thermal functions (heat resistance or temperature conductivity)

- Tribological functions (surface interaction with other surfaces: rolling, sealing, sliding, *etc.*)
- Optical functions (visible appearance, light reflection behaviour)
- Flow functions (influence on the flow of fluids)

From a global point of view, surfaces have to support: chemical attack, tribological solicitations, mechanical pressure, heat transfer, *etc.*

A surface is usually composed of several layers that differ from the bulk material in composition and structure. The surface composition is schematically illustrated in Figure 3.2. Indeed, as soon as a freshly machined surface is exposed, it will oxidize and adsorb. The adsorbate surface consists of water vapour and hydrocarbons from the environment (air, cutting fluids, *etc.*). Beneath this layer, there is an oxide. Its thickness can remain stable (for example, in the case of stainless steels or aluminium alloys) or may continue to grow with time (for example, in the case of low-carbon unalloyed steels). Beneath the oxide layer is the strained and metallurgically altered region (by the manufacturing processes), which is several orders of magnitude greater than the depth of the previous layers, several tenths of a millimetre.

The aircraft industry was among the first to consider the surface integrity of their pieces, since the consequences of a breakage are always dramatic from a human and economical point of view. Such an industry has a double objective: designing aircraft with a minimum weight (*i.e.*, with pieces having small sections) and producing pieces with a high degree of safety. Moreover, an additional objective is increasingly becoming important: economical competition, which induces pressure on production costs and obliges factories to produce more rapidly. The combination of these three objectives (thin, fast and safe) makes this job very difficult. In such a context, the surface integrity of their pieces is of primary importance.

Such objectives are also becoming increasingly critical in other industries, such as the automotive industry, because car manufacturers are engaged in a race with two objectives:

- weight reduction in order to reduce gas consumption,
- increase of engine power in order to satisfy pollution criteria, which leads to increased mechanical stresses that has to be supported by the pieces in the power transmission

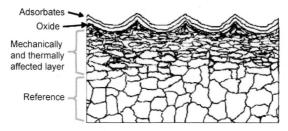

Figure 3.2. Schematic section of a machined surface

Failure cause	Surface physical properties					
	Yield stress	Hardness	Strength	Residual stress	Texture	Micro-cracks
Plastic deformation	++	++				
Scuffing/adhesion		++				
Fracture/cracking	+	+	+			+
Fatigue				++	++	++
Cavitation		+				+
Wear		++			+	
Diffusion					+	
Corrosion				+	++	++

++ : *large influence*
+ : *potential influence*

Figure 3.3. Interrelation between failure modes and surface properties

These two contradictory objectives give increasing importance to the impact of the surface integrity on the reliability of cars.

All companies manufacturing mechanical products have experience of a component breaking linked either to a poor design or to a manufacturing problem. Each company has developed some home-made specifications and the corresponding manufacturing procedure to ensure the reliability of their products. However, they have often carried out little investigation in order to correlate these specifications, their manufacturing procedures and the performance of the surfaces. Most of the time, they use the solutions developed in previous applications. When a problem occurs or when a serious change is introduced (for example, a new material), companies try to find a solution quickly without having all the information allowing solutions to be predicted reliably.

Some authors [10–11] have proposed a connection between the properties of the surface, the failure mode and the performance, as reported in Figure 3.3.

3.1.1   Link Between Surface Integrity and its Manufacturing Procedure

The evaluation criterion of a process depends on the functionality of the machined surface and on the economic efficiency of the process. Usually, the last machining operation is always suspected of being responsible for a breakage. In fact, it is very important to bear in mind that the state of the sub-surface is the consequence of the superposition of the individual stresses induced by all manufacturing sequences: from the purchase of the raw material to the superfinishing operation, including the machining in the low hardness state, the heat treatment, the semi-finishing operation in the hardened state, *etc.*

As an example, in the case of a synchrogear, one manufacturing procedure is illustrated in Figure 3.4. Three typical causes of failure are due to:

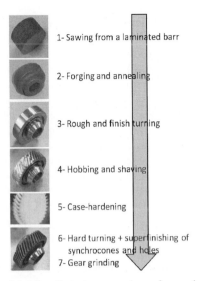

1- Sawing from a laminated barr

2- Forging and annealing

3- Rough and finish turning

4- Hobbing and shaving

5- Case-hardening

6- Hard turning + superfinishing of synchrocones and holes

7- Gear grinding

Figure 3.4. Manufacturing procedure of a synchrogear

- Poor surface texture generated by the super-finishing operation. This problem may lead to a breakage of the synchrocone due to the torsion supported by this surface when changing the speed of the gearbox.
- Poor grinding operation of the teeth (bad surface texture and/or a microstructural modification associated with tensile residual stresses). This problem may lead to the breakage of a tooth or to a rapid wear of the surface.
- Poor heat treatment. If the cooling rate is too high, phase transformations occur in the external layer, whereas the cooling of the bulk is not so rapid, which leads to strong internal stresses during a short period. If stresses exceed the strength, cracks may occur (Figure 3.5).

However, the hard turning operation of the synchrocone is potentially also a cause of breakage, since an abusive hard turning operation, involving an excessive cutting speed and/or worn tool, may induce microstructural modifications and tensile residual stresses, which may lead to fracture initiation in the sublayer. In deed, the final super-finishing operation is only responsible for the surface texture and for the initiation of a compressive state in the external layer (~10 μm), whereas hard turning can modify the microstructure deep beneath this layer [12, 13].

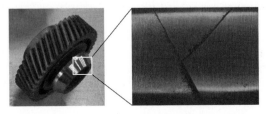

Figure 3.5. Cracks induced by the quenching of a gear

Another typical example concerns the fatigue resistance of crankshafts [5]. Some authors [14] have shown that the ultimate roller burnishing is able to delay the appearance of cracks in grooves and as a consequence improve the fatigue resistance. However, the final result depends strongly on the heat treatment made by induction, which is a very sensitive process [15].

Each machining process has his own signature on the surface integrity, since it removes layers from a workpiece with its specific mechanism. However, this signature has a spectrum of characteristics depending on the conditions of application (cutting conditions, lubrication, wear of cutting tools, *etc.*); for example, a gentle hard turning operation will generate a very smooth surface (Ra ~ 0.3 μm) and will induce compressive residual stresses, whereas an abusive hard turning operation will induce tensile residual stresses associated with microstructural modifications [12]. The main difference between these two configurations comes from the heat, strains and strain rates induced by the machining process.

As a consequence, it is impossible to give detailed informations about the surface integrity induced by each type of machining processes whatever the conditions of applications used by any end user. It is only possible to provide some general trends about the usual surface integrity observed in some current applications.

The most efficient method for designing/manufacturing companies is probably to characterize the surface integrity of each pair machining process/work material commonly applied in their shop floors. As shown previously, they should also consider this approach to validate new machining conditions (for example, a new cutting tool that is theoretically more productive or more wear resistant or a new cutting fluid that is more environmental friendly, *etc.*) and to optimise their production conditions with the best combination of productivity, wear and surface integrity parameters. Moreover these data would enable engineers to select the best machining process for a new application.

3.1.2 Impact of the Surface Integrity on the Dimensional Accuracy

Each machining sequence introduces a modification of the stress state of the piece. It produces relaxation inherent to the layer removal (modification of previous residual stresses) and induces additional stresses. Each production step can influence distortion by generating a distortion potential which is inherently stored in the workpiece and passed to the subsequent production steps. According to [16], distortions are influenced by:

- Steel production
- Metal forming
- Cutting
- Heat treatment
- Fine finishing

Between two manufacturing sequences, the surface residual stresses are balanced by bulk residual stresses of the opposite sign. Sometimes, if machining is only carried out on one side of a component with a large amount of material removal, the residual stresses can result in considerable distortion. Figure 3.6 shows the example of a semi-cylindrical piece produced with the following manufacturing procedure:

1. Rough machining from a monolithic bloc
2. Annealing heat treatment, which is supposed to cancel all previous residual stresses
3. Finish machining in order to obtain accurate dimensions. This is supposed to induce some residual stresses in a thin external layer: a few tenths of a millimetre. However these stresses are supposed to be negligible compared to the one induced in the next step
4. Cold forging (3 mm of plastic deformation), which induces strong residual stresses in the bulk material due to the plastification
5. Rough milling of a 20 × 20 mm slot

Figure 3.6 shows the residual stress state after the forging operation (following [17]). It appears that the residual stress level is very high (~ 800 MPa Von Mises stress). After the rough milling operation, a large distortion of the workpiece, due to the relaxation of the residual stresses, is observed: ΔA =1.35 mm.

Figure 3.7 shows another basic example typically observed after the machining of a bar in its drawn state (following [17]). The bar contains a large gradient of residual stresses. The milling operation modifies this field of residual stresses and leads to great distortion (more than 1 mm).

Another standard case study, *i.e.*, the manufacturing of a bearing cage, has been investigated in depth by [16], since it is a typical thin part, which is especially sensitive to distortions.

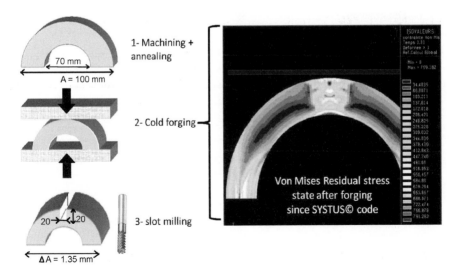

Figure 3.6. Distortion of a part due to relaxation of residual stresses

Bar with a rectangle section in its drawn state

Deformation after milling

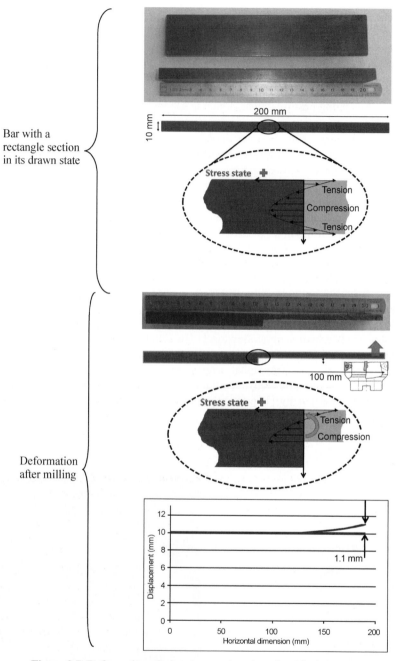

Figure 3.7. Deformation of a bar due to relaxation of residual stresses

It has been shown [18] that the residual stress state of the initial Al7440 T7651 block for aircraft structures can significantly affect the part distortion in the roughened stage (95% of chip removal), whereas the introduction of residual stresses in the finishing stage of thin sections also has an important effect on the case of thin structures.

Some authors have tried to make a link between the superposition of residual stresses produced by the various sequences of a manufacturing process and the relaxation induced by machining. For example, [19] has investigated the modification and evolution of the residual stress field, originating from welding, after chip-forming machining, such as milling and cutting.

3.1.3   Impact of the Surface Integrity on Fatigue Resistance

Residual stresses can have a wide variety of profiles depending on the manufacturing procedure. The magnitude and sign of the residual stress will have a significant effect on functional performance. A common idea is to prefer compressive residual stresses in the external layer because they tend to close surface cracks. However the fatigue resistance properties of a surface depend on the thermo-mechanical loading supported by the surface (bending, tension, torsion, rolling, etc.), for example, it is preferable that a rolling contact has a peak of compressive residual stresses in the sublayer, where the shear stresses are maximum. This would limit the pitting fatigue in some typical applications such as bearings, camshafts, etc. As an example, it has been shown by [20] that rolling fatigue of bearings is improved when hard turning is used instead of grinding. This improvement is explained by the large peak of compression in the sub-surface. In parallel, [21, 22] have shown that hard turning operations managed in gentle conditions with new tools can increase the rolling contact fatigue by up to six times compared to the same operation made with a worn tool. This result is explained by the modification of the residual stress profile and by microstructural modifications.

If a part is submitted to a bending loading (similar to a standard four-point bending test in a laboratory), the external residual stress state is of great importance. As an example, [7] investigated the influence of the hard turning process on the fatigue resistance of the case-hardened steel 16MnCr5 (AISI5115). It has been shown that a hard turning operation managed with a new tool leads to compressive stresses in the external layer and to high fatigue resistance. This fatigue resistance is significantly worse with flank wear of the c-BN insert.

A similar trend was observed for various applications by [23]: the fatigue resistance in four-point bending tests is directly correlated to the external residual stress state. A compressive residual stress is beneficial for fatigue resistance in the case of a 30NiCrMo16 bainito-martensitic steel manufactured by finish turning, or in the case of 7075 T7351 aluminium alloy manufactured by finish peripheral milling. On the contrary, a TiAl6V manufactured by finish turning, or a 7075 T7351 aluminium manufactured by face milling, do not seem to be sensitive to this parameter, but to the surface roughness only.

In a different context, [24] investigated the influence of the honing process on 12%Cr stainless steels and on Hastelloy X, showing that this process leads to compressive stresses in the surface and to a very smooth surface. Rotation bending fatigue tests and pulsating push–pull fatigue tests have shown a great improvement of the resistance compared to a shot peening operation. However this improvement was even better in an aggressive atmosphere (water + sodium chloride) than in air, which proves the improvement of the resistance against stress corrosion.

Finally, the effect of the application of a coolant during a machining operation should not be neglected [25]. The tribo-physical and tribo-chemical interactions between the cutting tool, workpiece, metalworking fluid and surrounding medium have an influence on the properties of the resulting surface. Defects can be generated by adsorption and reaction layers on the machined metal surface: dirt, oils, greases as well as residues from the machining process. In a study of the machining of 42CrMo4 steel the effect of various metalworking fluids on the results of the gas nitriding process used for surface hardening were reported. It was observed that using sulphur and phosphorus additives in the cutting fluid may lead to a decreased surface hardness after gas nitriding, which leads to a dramatic wear rate.

These complementary examples show that the correlation between the fatigue resistance of a part and its surface integrity depends strongly on the loadings supported by the surface (thermal, mechanical, chemical) and on the material and on the manufacturing process. It is very difficult to define universal ideas in this area.

In this chapter, surface integrity will be more detailed regarding its sub-surface state (metallurgical and mechanical states). Readers interested in information about the influence and characterization of surface texture are referred to [10, 26–28].

3.2   Material and Mechanical Aspects of Surface Integrity

Residual stresses are defined as mechanical stresses in a solid body, which is currently not exposed to forces or torques and which has no temperature gradient. The superposition of the residual stresses induced during the material manufacture and machining operations leads to the final residual stress distribution.

3.2.1   Mechanisms Leading to Material and Mechanical Modifications in Machining

In order to predict the performance of a surface, it is of great importance to characterize the signature of machining procedures in the field of material and mechanical state. A signature depends mainly on the combination of the mechanical, the thermal and the chemical loadings applied by the cutting process on the surface and on the sub-surface. Most of the processes are a combination of the three generating mechanisms. Just some processes can be clearly attributed to one single mechanism. As an example, electro-discharge machining (EDM) is clearly a thermal process, since there is no mechanical contact between the workpiece and the electrode. The temperature increases to the melting temperature because of electrical discharges, and then the material is solidified by the cooling effect of the dielectric. On the contrary, electro-chemical machining (ECM) is clearly a chemical process. It is not able to induce any residual stress in the material, nore any microstructural modifications. For other machining processes, the classifications are not so clear. It is evident that any processes inducing large strain and strain rates will also induce some heat due to the plastic deformation of the material and a large quantity of heat due to the friction between the cutting tool and the material. Modification of the mechanical state in the surface and sub-surface occurs systematically, inducing residual stresses. Depending on the amount of energy, the temperature reached and the time of exposure, micro-structural modifications may also occur in an external layer. A typical consequence in steel machining is the observation of so-called white layers, so named as they appear white in micrographs after etching. These are due to short-term metallurgical processes that occur under specific cutting conditions. Based on this statement, it is evident that each machining process involves more or less plastic deformation and friction, resulting in a large spectrum of consequences on the material and mechanical states of the machined surface. Moreover, even a defined machining process has a wide range of thermo-mechanical effects, since cutting tool geometries, cutting conditions, lubrication and tool wear are very sensitive parameters on this aspect of surface integrity.

Residual-stress-generating mechanisms can be simplistically represented by four models:

- Plastic deformation induced by mechanical load: the external residual stress is compressive because the surface layer is compacted by some form of mechanical action. There are no (or very limited) heating effects. This also applies to some other processes such as roller burnishing.
- Plastic deformation induced by thermal load (without phase transformation): the external residual stress is tensile because the surface expands greatly during heating, whereas the subsurface does not. The external surface is plastified by compression. When cooling, the external surface tends to recover its position, which is no longer possible due to the plastic deformation, leading to a tensile state.
- Plastic deformation induced by phase transformations: the residual stress may be caused by a volume change due to a phase transformation. If the phase change causes a decrease in volume (for example, the transformation of martensite into austenite), the surface layer wants to contract but the underlying bulk material will resist this. The result is that the surface layer is under tension, whereas the sub-layer is under compression. If the phase transformation causes an increase in volume (for example, the transformation of austenite into martensite), the residual stress will be compressive. This is the case with conventional heat treatment of steel. It also applies to nitriding or case hardening, in which the volume increase is caused by diffusion.
- Thermal/plastic deformation: in practice, a combination of the previous mechanisms occurs, leading to either more compressive or tensile stresses.

3.2.1.1 Mechanical Plastic Deformation

Plastic deformation occurs when the stress exceed the elastic limit, inducing strain hardening. A plastic flow occurs, which may lead to cracks. Depending on intensity and on the contact area, the mechanically affected zone can have various thicknesses. In some extreme cases, grains are so distorted that the structure is not observable in micrograph analysis and appears white after etching. Processes such as burnishing, reaming, honing and broaching produce large strain hardening with a limited amount of heat due to either low friction phenomena or low velocities. Processes such as turning, milling and drilling also include large plastic deformations at low cutting speeds, but thermal effects become predominant in high-speed cutting.

In mechanically dominated processes, the hardness constantly increases from bulk to surface, whereas in thermal processes this is not the case.

In order to discuss the mechanical impact of a cutting process, a turning operation will be described. As shown in Figure 3.8, mechanical effects are due to the pressure applied by the rake face and by the cutting edge radius. The workmaterial is plastically deformed by compression in front of the cutting tool, whereas it is submitted to tension behind the cutting tool. Figure 3.8 shows the compressive

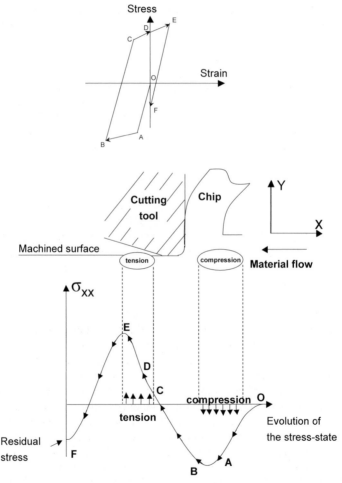

Figure 3.8. Principle of residual stress creation under a pure mechanical load in cutting

zone OAB and the tensile zone CDE. Finally, when the cutting tool has moved far from the surface, the unload zone EF appears. When the machined surface is not exposed to any forces, it appears that compressive residual stresses remain at the surface. Of course, to maintain the mechanical equilibrium of the system, tensile stresses exist in the sub-surface [29, 30].

If the plastic deformation is very high, the material may become so deformed that it is not possible to discern any microstructure. The external layer appears as white in a micrograph. If such so-called white layers remain during service operation, failure may occur by an excessive additional stress.

3.2.1.2 Thermal Effects Without Microstructural Changes

Thermal effects have a totally different action compared to mechanical effects. When crossing the primary shear zone in front of the cutting tool and when rubbing the clearance face of the cutting tool, the future machined surface is submitted to an intense heat flux (Figure 3.9). If the mechanical effects and the microstructural modifications are neglected, the machined layer is either in compression or in tension, depending on the expansion coefficient. Most of the time, for metals, the expansion coefficient has a positive value. As a consequence, in the zone OAB (Figure 3.9), the future machined surface is in compression. The zone BCD corresponds to the cooling of the surface by means of the bulk material or of the environment (air or coolant). During the exposure to intense heat fluxes, high temperature gradient exists, which may lead to a local plastification of the workmaterial. When the surface is returned to a steady state at room temperature, tensile stresses remain at the surface. Consequently compressive stresses exist in the sub-surface [31].

3.2.1.3 Thermal Effects with Microstructural Changes

As described previously, intense heat fluxes are applied on the machined surface in the primary shear zone and in the rubbing zone. A thermal load has a delayed consequence, since the heat transfer depends on the thermal properties of the material and of the cutting tool. This also depends on the heat exchange coefficient at the tool–work material interface. Anyway, the temperature can rise very quickly in the vicinity of the heat sources, whereas it takes time to raise some micrometres further in the sub-surface. For a defined amount of energy, microstructural modifications can occur. These changes are completely different to the one typically observed in the steady-state situation. Indeed, the heating rate and the cooling rate can be very high, which limits the possibilities of atoms diffusion and reconstruction of crystals. Example, in hard turning, a typical heating rate is around $10^6 \,^{\circ}C/s$ [32, 33], whereas a typical cooling rate in grinding is around $10^3 \,^{\circ}C/s$ [34, 35].

The case of the machining of treated steel will be more detailed in the rest of the section since it is the most common situation that has been largely discussed in the scientific literature [36]. In this context, the metallurgical modifications are often called white layers. This term refers to surfaces appearing white in a micrograph analysis, which means that its microstructure cannot be distinguished due to either a very thin structure or a lack of chemical reaction of the structure with the chemical reactor used (typically *nital* for steels). A white layer is often accompanied by

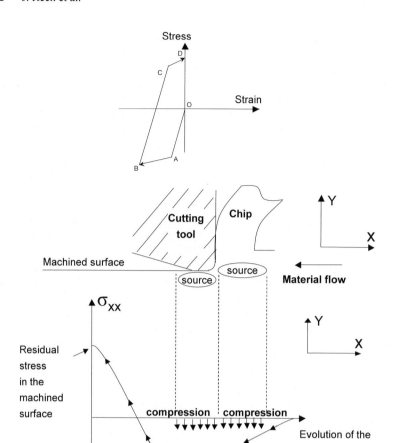

Figure 3.9. Principle of residual stress creation under a pure thermal load in cutting

untempered martensite and retained austenite with a very fine structure. Their hardnesses are usually higher than conventional martensite. The thickness depends on the quantity of energy (density and duration). White layers have been associated with grinding for many years because the energy density is very high in this process compared to other processes such as turning. So it is very easy to get white layers with inadequate grinding conditions, in contrast to with turning. Moreover, the surface appears in some extreme situation as burned (brown coloured) due to the oxidation of the surface besides the presence of coolant, which is almost impossible to obtain in any other processes even in critical conditions. White layers are assumed to be detrimental for the service performance of a piece. Abusive grinding operation can reduce the fatigue endurance limit of a piece by 40% [10]. The hard brittle layer can initiate cracks and is usually accompanied by tensile residual stresses, which facilitate the crack propagation.

In conventional heat treatment, TTT diagrams are used for the prediction of the phase transformation. However such diagrams suppose that the duration is long enough to obtain a homogeneous structure. In the case of machining processes, the heating rate is very high. As a consequence, in conventional heat treatment, the temperature necessary to reach similar transformations are much higher. Figure 3.10 shows how the heating rate modifies the limit Ac1 and Ac3 of an hypereutectoid steel. It appears that a tempered state enables the modification to occur at much lower temperatures and after a shorter duration. It is clear why hardened steel has a tendency to undesirable transformations even at a heat impact for a short time. After long heat treatment durations, transformations can take place via nucleation, crystal growth, diffusion, *etc*. At a definite critical heating rate no further shifts of the transformation lines Ac1 and Ac3 can be observed. In high-speed cutting, a diffusionless transformation occurs, which means that recrystallisation of different crystal modifications without atomic diffusion processes happens.

In short-time heat treatment processes, the quenching rate can be reached by rapid heat dissipation to the cooler core of the part or to the lubricant, so that a lasting heat impact on the newly generated martensite is suppressed. Hence, a very fine-grained structure is formed, which in micrographs can appear as white due to the limit of resolution of a light optical microscope. This induces the rehardening of the external layer. This specific performance is utilised in short-time heat treatment methods such as induction and laser hardening to carry out a local surface hardening while supplying relatively low quantities of energy.

When the amount of heat is higher due to a longer exposure or to a more intense heat flux, the sub-layer cannot evacuate rapidly the heat flux coming from the surface. Then the sub-layer may be tempered, causing softening. This layer is a so-called overtempered martensite structure (appearing black in the case of steel etched by nital), which has a lower hardness than bulk material made of conventional martensite (Figure 3.11). On micrographs, no clear transformation line can be seen between the heat-affected layer and the bulk.

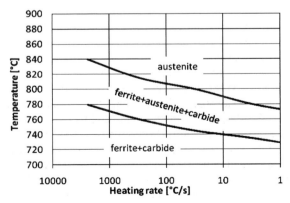

Figure 3.10. Influence of the heating rate on the TTA diagram for a hypereutectoid steel Cf53

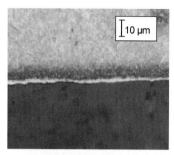

Figure 3.11. Microstructural modification induced by a hard turning operation on a 27MnCr5 case-hardened steel [12]

When this amount of heat becomes even higher, additional microstructural transformations (with diffusion) mavy occur, leading to untempered martensite and retained austenite, in parallel to an oxidation of the surface (burned surface). Such layers appear as white in micrographs, since austenite is not sensitive to nital etching.

Beside the aforementioned phase transformation, which occurs depending on the time–temperature profile and on the specific material properties, the thermal expansion caused by the generated heat as well as by the volume alterations caused by the phase transformations lead to an unstable stress field. This stress field is superposed onto the two previous ones induced by the pure mechanical and the pure thermal effects, which makes the prediction very complex in real cutting processes.

3.2.2   Modelling of Residual Stresses

Residual stresses in machined surfaces have been investigated since the early 1950s, leading to handbook data (experimental approach). More recently, finite element methods of machining have been used to predict residual stresses from computed stress and temperature distributions. However, such methods are highly time consuming and very costly. As a consequence, new approaches combining experimental, analytical and numerical models appeared recently in order to enable a rapid prediction of the residual stresses within a few minutes, making this approach usable for industrial applications.

The large majority of these researches are interested in predicting the residual stress state after orthogonal turning (or grinding), which is a 2D problem far from realistic cutting processes (3D turning, milling, drilling, *etc.*). The main limitation to moving towards complex cutting processes is central processing unit (CPU) time. Only few investigations dealt with 3D turning operation but with many more assumptions and uncertainties in order to limit the CPU time [37, 38]. Moreover such models do not consider microstructural modifications, which limit their applications. From this point of view, the cutting scientific community is behind the welding scientific community which has investigated the coupling of metallurgical–mechanical–thermal effects in 3D configurations for a long time [39, 40]. This section aims to provide some trends and references in residual stresses modelling in orthogonal cutting without considering metallurgical changes.

3.2.2.1 Numerical Modelling

Numerical models necessitate the application of standard finite element codes such as SYSWELD, ABAQUS, DEFORM, *etc*. In such approaches, two major types of parameters are strategic (Figure 3.12):

- The input data: mechanical properties of the workmaterial, thermal properties of the workmaterial and of the cutting tool, friction model at the tool–work material interface, *etc*.
- The numerical model:
  - Lagrangian, Eulerian or ALE techniques
  - Adaptative remeshing or none
  - Implicit or explicit formulation
  - Element type and size

The cutting tool geometry is provided by the tool manufacturer (rake and clearance angles, cutting edge radius, chip breaker geometry). Most of the time, authors consider a plane-strain configuration since they consider that the depth of cut is much larger than the feed.

The Lagrangian technique consists of tracking a discrete material point [41]. A predetermined line of separation at the tool tip is usually present, propagating a fictitious crack ahead of the tool in order to avoid severe mesh distortions (Figure 3.13). In this case, a failure criterion is required. The criterion is either based on a distance between the tool tip and the node, or based on a parameter depending on the stress state, on the strain rate and on the temperature at a certain distance ahead of the tool tip. In both cases, the separation occurs when a critical value is reached [42, 43]. However only sharp cutting tools can be modelled. Other kinds of Lagrangian techniques prefer the use of adaptive remeshing techniques to bypass the problem, which enables the modelling of blunt tools [44–46] (Figure 3.14). Of course, the CPU time becomes very high since a fine mesh is required around the cutting edge radius.

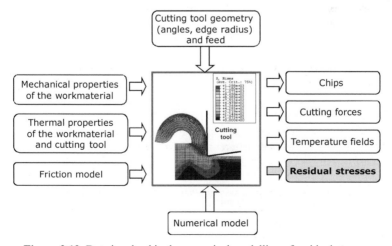

Figure 3.12. Data involved in the numerical modelling of residual stresses

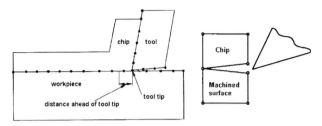

Figure 3.13. Principle of the Lagrangian technique applied to metal cutting modelling

Eulerian techniques consist of tracking volumes and do not induce problems of mesh distortion or require failure criterion [47, 48]. However the determination of free surfaces is critical, which necessitates some assumptions about the chip geometry (no segmented chips, *etc.*). Finally, the avoidance of elastic behaviour does not enable the estimation of residual stresses.

The arbitrary Lagrangian–Eulerian (ALE) technique is a relatively new modelling technique that represents a combination of the Lagrangian and Eulerian techniques without their drawbacks [49]. Figure 3.15 illustrates an application of this approach. Regions A, C and D were modelled as Lagrangian regions with adaptive meshing. So free surfaces can be modelled properly and boundary conditions can be applied in a simple way. Consequently, automatic chip formation takes place. Region B was modelled as an Eulerian region, where the mesh is fixed in space and the material flows through it. This solves the problems encountered around the tool tip.

In metal cutting simulations, authors usually apply explicit integration methods; although some works are available with implicit methods [44, 50]. In explicit integration, a system of decoupled differential equations is solved on an element-by-element basis, in which only the element stiffness matrix is formulated and saved without the need for the global stiffness matrix. On the other hand, the global stiffness matrix has to be formulated and saved in implicit integration, and the whole system of differential equations has to be solved simultaneously. Therefore, explicit methods are computationally more efficient, especially when non-linearity

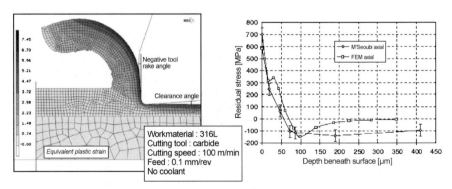

Figure 3.14. Example of Lagrangian model simulating the residual stresses of a 316L steel [46]

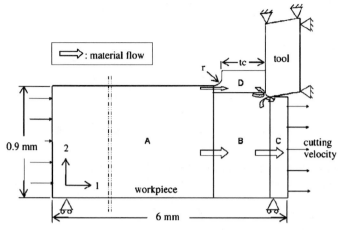

Figure 3.15. Principle of the ALE technique applied to metal cutting numerical modelling since [49]

is encountered. This becomes more evident in thermally coupled analysis, as in metal cutting, because structural and thermal variables are solved simultaneously. On the other hand, explicit integration is conditionally stable because the critical time step depends on the minimum element size and the speed of wave propagation, while implicit integration is unconditionally stable [51].

Concerning the input data of numerical models, the identification of the constitutive equations for the work material (flow stress model and damage model) remains an issue, since it requires the determination of material properties at high strain rates, large strains, high temperatures and high heating rates. The main problem originates from the strain rate achievable in standard mechanical tests (for example, Hopkinson's bar), which are about 100 times too slow compared to classical strain rates in metal cutting $\sim 10^5$–10^6 s^{-1}. A common practice consists of using the Johnsson–Cook model, including deformation hardening, thermal softening and rate sensitivity. Another major problem originates from the identification of the coefficients independently from each other, *i.e.*, the strain rate effect is identified at low temperature, the temperature effect is identified under low strain rates, *etc.* As a consequence, there is no way to validate that such models remain meaningful under the combination of high strain rates and high temperatures. Some authors have underlined the sensitivity of Johnsson–Cook parameters on the residual stresses predicted to prove the necessity of improvement of the identification methodology and of the constitutive models [44].

Friction is also one of the most critical parameter in metal cutting. Indeed, it is the most sensitive parameter for the determination of residual stresses, since a small modification of its value induces large modifications of the residual stress field [43, 52]. A lot of authors assumed that the Coulomb friction model is valid, with a constant coefficient irrespective of the pressure and of the temperature. Newly developed tribometres combined with numerical models enable the identification of more realistic friction models [53].

Due to the transient thermal and dynamic behaviour of the model, it is necessary to relax thermally and mechanically the workpiece after the cutting process in order to quantify residual stresses.

All these models provide relevant information in the range of application for which they have been developed for. Usually, scientific papers present results for a limited number of cutting conditions, without showing the capacity of the model to extrapolate results for other applications. An example is provided in Figure 3.14. However the CPU time required to obtain an acceptably accurate result is around 1 or 2 weeks even with a powerful computer, which does not currently enable its industrial application. However such models are useful for researchers, since the residual stress prediction can be made for several configurations, which facilitates the investigation of the sensitivity of these parameters. This also enables the development of cutting operations based on residual stress criterion by modifying:

- Cutting conditions: cutting speed, coolant, *etc.*
- Cutting tool geometry: edge radius, angles, *etc.*
- Cutting tool material: coatings, substrate, *etc.*

3.2.2.2 Analytical Modelling

The objective of analytical models is to predict residual stresses based on equations coming from mechanical and thermal properties of work materials. Such models are very efficient in terms of speed compared to experimental or numerical approaches. Moreover such models enable the effects of each parameter on the system to be understood in detail, which is necessary for any process optimization.

Pure analytical models are almost nonexistent since they are unable to model dynamic movement. Indeed, the cutting boundary changes continuously during machining. So modelling a turning process from the beginning to a steady state becomes complex. As a consequence, most authors apply the finite element technique to solve this problem. Compared to the previous approach, the numerical calculation is only used to facilitate and accelerate the resolution of the mechanical and thermal equations for each step. It especially enables to record the history of thermo-mechanical loading and unloading. From a practical point a view, authors have to perform three kinds of operations (Figure 3.16):

1. Quantifying the thermo-mechanical loadings supported by the machined surface
2. Moving the thermo-mechanical loadings with the velocity corresponding to the cutting speed
3. Removing the loadings and waiting for the relaxation of the thermal and mechanical field before quantifying the residual stress fields.

Based on this statement, each author has a different way to quantity the thermo-mechanical loadings supported by the machined surface. Some authors consider only thermal loadings and neglect mechanical effects [54]. Other authors prefer processing the thermo-mechanical loadings in two steps:

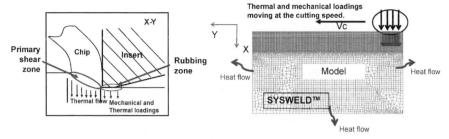

Figure 3.16. Principle of moving sources [52]

1. A first calculation aims to model the heat transfer between tool, chip and workpiece. Heat sources coming from the shear energy in the primary shear zone and from the friction energy produced at the tool rake face–chip contact zone. The heat generated from the friction at the flank face–workpiece interface is not considered by some authors [55], whereas other authors consider it to be of primary importance [52, 56]. The quantification of each parameter is always problematic, considering: the heat partition coefficient, heat exchange coefficient with the environment, amount of plastic deformation energy converted into heat, repartition of the heat flux density along the contact surface, availability of a relevant friction model (see previous section), influence of the coating etc. Among the articles using this approach, each author has their own choice depending on the data available in the literature. However, based on this thermal analysis, the temperature distribution within the workpiece is calculated. Temperature gradient leads to a non-homogeneous mechanical stress field.

2. A mechanical load is combined with the preliminary stress. At this step, two strategies exist. The first strategy consists of neglecting coupling effects [56], whereas the second strategy consists in taking into account this thermo-mechanical coupling [52]. Such an approach induces the absence of coupling between mechanical and thermal phenomena. However, the quantification of each parameter remains a problem: repartition of the pressure over the contact surface, area of contact surface, isotropic or kinematic hardening, *etc.* Depending on the author, various solutions exist. Some authors use pure theoretical models [55], whereas others prefer using experimental data obtained basic orthogonal cutting tests (cutting and feed force, flank face contact area, chip thickness, *etc.*) [52, 56].

Such analytical models enable one to obtain interesting results in terms of precision (example in Figure 3.17) and CPU time (some minutes according to [52]). Models combining analytical equations fitted by experimental results can be considered as a realistic way to model residual stresses before the development of efficient and rapid numerical models.

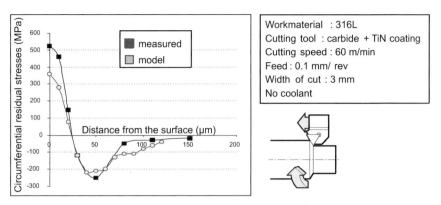

Figure 3.17. Example of results obtained by analytical models [52]

All these models (analytical and numerical) contribute to a better understanding of the phenomena leading to residual stresses. They also enable decision regarding the cutting tool design and the selection of proper cutting conditions to be taken. However, for the time being, all these models have been developed for one specific application: one work material, one kind of cutting tool material, one geometry, *etc.* and their validation is always restricted to few cases because of the cost for each residual stress measurement. As a consequence, there is still no software available on the market that is able to predict residual stress fields for a large range of applications (various work materials, various cutting tools, various cutting conditions, application of lubrication, *etc.*). The great improvement made during the last 10 years leads us to think that such software could appear in the next few years if some leading companies apply pressure and provide financial support to research laboratories in order to determine much more accurately the input data necessary for their models and if the CPU time continues to decrease in order to obtain accurate results within a reasonable time to enable industrial optimization. In the mean time, the experimental approach will probably remain the reference.

3.2.3   Experimental Approach

3.2.3.1 Introduction

The most widely used approach to characterize surface residual stresses induced by machining processes remains the experimental approach. It consists of characterizing the residual stress state and the microstructure before and after the machining operation. The X-ray diffraction technique and the micrograph analysis are the reference for such investigations. The experimental approach is very efficient to provide results, but it requires several days of characterization. Moreover, the cost of such equipment and the complexity of their application necessitates competent engineers, which leads to very high cost (several hundreds to thousands of US$ for each piece, depending on the number of directions investigated, the precision required and the number of points on the profiles).

However such results exhibit large deviations since a large number of cutting parameters have a strong influence on the residual stress profile. Parameters such as cutting speed, feed, *etc.* are very commonly known, but information about parameters such as cutting edge preparation, lubrication (nature, application) and wear of cutting tools is not always reliable and can vary significantly. Such parameters are well known to have a strong influence on the residual stress profile since they modify entirely the thermo-mechanical load supported by the surface [12, 13]. Most of the time, such detailed information is not clearly expressed, or controlled. As a consequence, it is necessary to carry out several characterizations for each case study in order to find the lower and upper bounds of the results, including standard industrial deviations. Based on this range of results, it is possible to take decisions regarding the capability of a process. This necessitates a lot of time and money in order to obtain reliable information. As a consequence, few companies are willing to perform such investigations, except the aircraft industry, the nuclear industry and the automotive industry for some strategic surfaces concerning safety parts. Usually, companies prefer to select their manufacturing conditions based on common criteria: cost, time, wear, accuracy, *etc.* and subcontract the characterization in order to validate (or not) the surface integrity. The main problem with this approach is that the probability of satisfying the specifications is rather poor since there are no basic reliable models available on the market that are able to predict the results (see the previous sections). As a consequence, companies are ill equipped to take a decision regarding the modification of one parameter among all the possible parameters. When the residual stresses have to be compressive in the external layer, the least hazardous solution consists of adding a super-finishing processes such as honing, belt finishing or roller burnishing. However such solutions are very expensive and delay the time to market of the product.

When companies decide to perform residual stresses measurements, some prefer measuring residual stresses on the external layer only. This practice is risky since the external residual stress level is very sensitive. A small variation of the process parameters can lead to large variations in its value. On the contrary, the shape of the residual stress profile is much more stable, so it is highly recommended to analyze residual stress profiles instead of single external values.

From the 1950s to the 1990s, the number of investigations was rather limited since the characterization methods were rather rare, expensive and difficult to manage. The development and diffusion of the X-ray diffraction technique have enabled a strong acceleration of these investigations in laboratories and industry. Since this time, the scientific literature has provided a large number of papers dealing with a wide range of machining techniques. It is almost impossible to provide a comprehensive description of all of these investigations. It is only possible to give an idea of the results provided for some key applications, which have driven large improvements in the basement knowledge.

The following sections will be devoted to the presentation of some typical cutting processes with common parameter (not extreme) values and the corresponding induced residual stress states.

3.2.3.2 Grinding

Among the machining technologies for which data dealing with residual stresses is available in the literature grinding is probably the most investigated process, for several reasons. Firstly it is a finishing technology and, as a consequence, is an important contributor to the fatigue and wear resistance of the surface. Second, this is a very sensitive technology involving a high concentration of energy dissipated through narrow surfaces, which may induce easily thermal damages (white layers, oxidation, burn, *etc.*). Moreover grinding is applied in almost all strategic parts of typical mechanisms: motor engines (crankshaft, camshaft, valves, *etc.*), gear boxes (gears, shaft, *etc.*), turbine engines (blades, rotors, *etc.*), and so on. As a consequence of the high potential risk of this technology, combined with the high stakes in such leading industries, a lot of investigations have been undertaken in order to qualify and model surface integrity in grinding.

Before presenting some typical residual stress profiles generated in grinding, it should be noted that there are a large variety of parameters:

- Work material: composition, microstructure
- Grinding wheel: material – c-BN, alumina, SiC- and structure
- Lubrication: composition, application (pressure, nozzles, *etc.*)
- Dressing conditions
- Grinding conditions
- Wear of the grinding wheel

For each configuration, the physical phenomena are very different, which leads to a large variety of residual stress profiles (tensile or compressive states in the external layer, with or without phase transformations). Surface integrity in grinding depends on the thermo-mechanical loadings supported by the work material and on its metallurgical properties. At the grain scale, different phenomena occur: microchip formation, ploughing, rubbing, *etc.* Grinding operations necessitate powerful machines. A large amount of the energy consumed is dissipated into heat because of the predominance of ploughing and rubbing phenomena (plastic deformation), which lead to a very poor energy efficiency ratio. So, grinding is a machining process that involves high concentrations of heat fluxes at the interface, combined with large contact areas, in comparison to other techniques (turning, *etc.*). The objective of the manufacturer is to dissipate this energy into the grinding fluid instead of into the work material (where it can cause surface integrity damage) or into the grinding wheel (where it can cause excessive wear and dimension variations). However, the complexity of the grinding wheel structure and the difficulty that the grinding fluid has in reaching the contact area due to the high tangential speed (from 30 m/s in conventional grinding to 200 m/s in high-speed grinding) lead to large deviations in the results, which classify this technology among unstable processes. A small difference in the parameters listed above can change the results dramatically.

When no phase transformation occurs, tensile stresses are always observed in the external layer because of the predominance of plastic deformation induced by the thermal expansion of the work material (Figure 3.18). However, an increase of

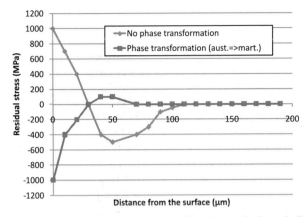

Figure 3.18. Typical residual stress profiles obtained after grinding

the grinding speed leads to high temperatures. When a defined temperature is reached, phase transformation may occur (for example, martensite ⇔ austenite). The rapid cooling of the surface leads to quenching and variation of volumes which may induce compressive residual stresses (austenite to martensite) or tensile stresses (martensite to austenite) (Figure 3.19).

A large number of authors have investigated the influence of grinding parameters on the surface integrity and it is almost impossible to mention them all [34–35, 42, 57–66, 30–31]. Among these investigations, authors have dedicated a large number of articles to the characterization of high-speed grinding involving c-BN wheels, since this enables the improvement of surface integrity (residual stresses, surface texture, wear rate, *etc.*) combined with a large gain in productivity.

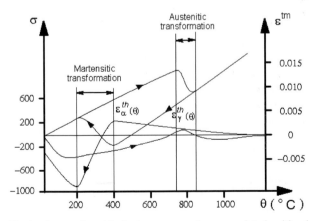

Figure 3.19. Mechanisms of residual stress generation associated with phase transformations [34, 35]

3.2.3.3 Turning

The second most widely investigated machining technology is probably turning. Since [67], various papers were published intermittently until the end of the 1990s. Companies manufacturing turbine engines associated with academic laboratories were the main contributors to this literature. Due to the difficulties in obtaining such data and the cost, only part of these investigations has been published. Moreover, most of the investigations were concerned with high-strength steels (*e.g.*, 39NiCrMo16), stainless steels (*e.g.*, 316L) and titanium alloys (*e.g.*, TA6V). Since the 1990s, the appearance of c-BN tools has completely changed the manufacturing processes in the automotive, aircraft and bearing industries. The possibility of replacing grinding operations by dry hard turning operations has brought about new opportunities for productivity improvements. In this period, the question was: is the surface integrity induced by hard turning appropriate for the fatigue/wear resistance of the machined surface. As a consequence, since this period, a large number of papers dealing with surface integrity in hard turning have appeared in scientific journals.

Turning of Austenitic Stainless Steels AISI316L with Carbide Tools [49, 68, 52]
Residual stress profiles induced by turning all have a similar shape (Figure 3.20). Tensile stresses are present on the external surface, followed by a large compressive peak in the sublayer. Among the strategic parameters, small nose radii and low cutting speeds are favourable to decrease the stress level and the affected layer thickness. Feed has a strong influence on the affected layer thickness.

Turning of Titanium Alloys (TA6V) in their $\alpha+\beta$ Structure with Coated Carbide Tools [23, 69]
External residual stresses are compressive. The affected layer can increase from 100 to 400 µm with increasing cutting speed. The fatigue resistance in four-point bending tests seems to be much more highly correlated with surface texture than to the residual stress state.

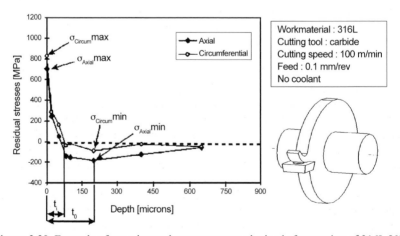

Figure 3.20. Example of experimental measurements obtained after turning of 316L [68]

Turning of Inconel718 in Their Hardened State with Mixed Ceramic Tools or c-BN Tools [70–71]
Ceramic tools induce larger tensile residual stresses compared to c-BN cutting tools when they are used in their own functioning domains (450 m/min for ceramic and 250 m/min for c-BN). With new tools, c-BN cutting tools generate compressive residual stresses, whereas wear changes it to tension. The residual stress is sensitive to cutting speeds. At low cutting speeds (~10–200 m/min) compressive residual stresses are generated, whereas tensile stresses are generated at high cutting speeds (~350–810 m/min). Round inserts and the application of coolant are much favourable to compressive stresses.

Turning of Inconel718 in Their Hardened State with Coated Carbide Tools [72]
As cutting speed increases the surface residual tensile stress drops while an increase in feed rate results in a slight increase in both the surface tensile stress and the depth of the compressive stress layer. The largest influence on the surface integrity generated is caused by tool wear.

Turning of Annealed Steels (C45, AISI1018) in Their Ferrito-Perlitic State with Carbide Tools [73–76]
Tensile stresses are obtained in the external layer, followed by a thick compressive layer. The parameters with the greatest influence seem to be the cutting speed and feed rate. An increase of the cutting speed seems to move residual stresses toward tension and to decrease the thickness of the affected layer. An increase of feed or wear makes the residual stresses more tensile, but increases the thickness of the affected layer.

Turning of Medium Carbon Steels (42CrMo4) in Their Bainito-Martensitic State with Coated Carbide Tools [77]
The most important parameters seem to be cutting speed, feed and wear. An increase of feed and wear seems to make the residual stresses more tensile, whereas a high cutting speed seems to make them more compressive. The application of lubrication seems to worsen residual stresses and increase the thickness of the affected layers. In contrast, the influence of the depth of cut does not seem to be significant.

Turning of High Resistant Steels (30NiCrMo16, 39NiCrMo16) in Their Martensitic State with Coated Carbide Tools [23, 78]
Tensile stresses are present in the circumferential and axial directions on the external surface, followed by a compressive layer below. The affected layer is lower than 400 µm thick. The effect of cutting parameters is controversial: [78] indicates that feed rate and nose radius are the most influential parameters, and that cutting speed has little influence; in contrast, [23] indicates that the external residual stress state decreases with increasing cutting speed and decreasing nose radius and feed rate. Additionally, external residual stresses seem to decrease with increasing depth of cut and seem to be minimum for cutting tools having a leading angle K_r of about 90°. The fatigue resistance in four-point bending tests seems to be corre-

lated with the residual stress state and not with the surface texture. In this context, the best fatigue performance for cutting applications is obtained using cutting tools having a small nose radius, a leading angle around 90° and a high cutting speed.

3.2.3.4 Hard Turning

The hard turning technique has been considered independently from the previous turning application, since this technique is in direct competition with grinding (work materials HRc > 55). Moreover, cutting tools for hard turning have a totally different substrate (c-BN instead of carbide) and their geometry is totally different (strong negative rake angles) from turning operations of steels with lower hardness. As discussed previously, this technique has been intensively investigated since the 1990s due to its high productivity.

The influence of some cutting parameters on residual stresses of a case carburized steel with hardness in the range HRc 58–62 has been investigated [20], showing that external residual stresses after hard turning are not significantly affected by depth of cut and feed rate. These observations and highlights have been confirmed [12], and it was reported that cutting speed and tool wear were the most influential parameters, leading to tensile stresses and microstructure modifications in the external layer. The influence of tool wear has also been confirmed by [6, 21–22, 62, 79–80]. There exists a tendency for the thrust force to increase gradually with the increase of the tool wear [80]. Moreover there is an increase of the contact area at the tool–workpiece interface. As a consequence, tool wear leads to higher thermal process energy related to the increase in friction, leading to a shift of the residual stresses towards tension in the external layer.

It has been reported that feed greatly influences residual stresses in the sublayer [81]. A high feed rate leads to higher compressive stresses into the material and moves the compressive peak deeper.

In parallel, some authors have investigated the influence of the geometry of cutting tools [20, 80, 82–84]. Thiele and Melkote [82] investigated the subsurface residual stresses in longitudinal turning of through hardened AISI52100 bearing steel with different tool edge preparations while keeping the cutting parameters fixed. By using X-ray measurements, they found that hone edge preparation produces more compressive residual stresses than the chamfer edge geometry after machining a workpiece hardened to HRc 57. Additionally, they showed that the penetration depth is more important with hone edges. This trend is expected for bearings in order to ensure satisfactory rolling contact fatigue, due to the fact that the maximal shear stress is not at the surface but some micrometres below.

Other investigations [20, 85] indicated that the size of the primary shear zone seemed to be of secondary importance, while the size of the plastic zone near the tool edge seemed to play a major role. They observed that the hone edge and the double chamfer geometries offered the greater subsurface penetration, and produced larger values of maximum compressive stress, compared to a sharp edge.

Liu *et al.* [80] explained the influence of the honed edge by the fact that a small depth of cut and low feed rates are chosen in hard turning. So, the undeformed chip thickness has the same order of magnitude as the radius of the cutting edge or the size of the edge chamfer. Consequently, the chips are formed along the nose of the tool in the region of the edge chamfer or the cutting-edge radius. Therefore, the cutting-edge geometry contributes to larger plastic deformation in the primary shear zone and around the edge, which is beneficial for compressive residual stresses.

Dahlman *et al.* [81] investigated the effect of the inclination angle of the chamfer during the machining of AISI52100 steels and revealed that a large negative angle provides greater compressive stresses as well as a deeper affected zone below the surface. With larger negative rake angles, the position of maximum stress is moved further into the material.

Other studies [86, 80] investigated the influence of the insert nose radius on the residual stresses induced by hard turning. They show that inserts with a small radii lead to much more compressive stresses in the external layer, compared to rounder insert. Liu *et al.* [80] explained this trend by a pure geometrical effect: a smaller contact area for an almost equivalent force increases the pressure at the tool–workpiece interface, leading to larger plastic deformation in the machined surface.

Rech and Claudin [84] revealed that a wiper nose geometry has a beneficial effect on residual stresses toward compression. Additionally cutting tools with a TiN coating and a low c-BN content are much more favourable for a compressive state.

Finally the service performance of hard turned surfaces compared to ground surfaces has been investigated [87, 20]. Hard turning can induce surface integrity, when machined in selected conditions, which can be without thermal damage and superior to grinding. The most exciting new opportunity for hard turning is its capability to produce compressive residual stress, in a wide range with sufficient magnitude and depth, which is not possible with conventional abrasive-based processes. It enables to reach desired pre-stresses. Experimental data has been reported to show that the fatigue life of hard turned surfaces is better than that of ground surfaces. However, the best turning parameters can result in as much as 47 times better fatigue life than that from the worst turning parameters with the same surface finish. Therefore, the method of selecting hard turning parameters is extremely important for obtaining the incredible benefit of hard turning in terms of fatigue life. However, for one set of cutting parameters and cutting tool, the fatigue resistance can be significantly worsened by excessive tool wear, as shown by [18].

3.2.3.5 Milling

In contrast to the two previous manufacturing techniques, milling has been poorly investigated. Indeed, surfaces machined with these processes are not critical parts for the fatigue resistance. The exception is parts involved in aircraft structures, mostly made of aluminium alloys. As shown by [18], milling operations of aircraft structures with small sections significantly affect distortions. The wide majority of investigations dealing with peripheral or face milling were concerned with aluminium from the 2000 or 7000 families.

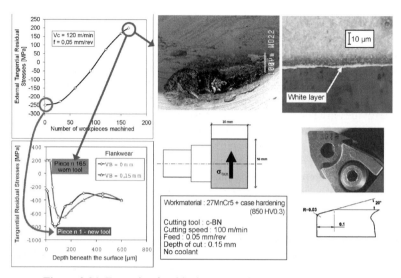

Figure 3.21. Example of residual stress profiles in hard turning [12]

Another field of research on surface integrity in milling is machining with end mills and ball nose mills. Such investigations were promoted by the development of the high-speed cutting (HSC) technology in two main areas: the production of dies and moulds, and the production of turbine blades made of super alloys (titanium, Inconel718, *etc.*), because such parts are very long and expensive to produce. Moreover they are submitted to intensive thermo-mechanical fatigue cycles. The expectation of industry was to finish parts directly from milling in order to avoid super-finishing processes (honing, belt finishing, *etc.*), which are well known as being favourable for surface integrity improvement.

Peripheral and Face Milling of Aluminium Alloys (7075 T7351, 2024 T351) [18, 23, 88, 89]
Residual stresses are affected by the cutting speed and the working mode. In particular, it is recommended to machine at low cutting speed under up-milling trajectories. Depth of cut is a secondary important parameter, for which a low value seems to be more favourable. The effect of coolant can be neglected since its effect is only important for low cutting speed, which is rather unusual in the aircraft industry. The effect of feed is not so clear. Indeed some authors [23, 89] recommend a low value, whereas others [18, 90] recommend high values in order to induce compressive stresses and thick affected layers. Finally a cutting tool with a large nose radius is recommended.

It should be noted that the residual stress state is very different in the direction parallel to the feed and in the perpendicular direction (an opposite sign may occur). The thickness of the affected layer is usually less than 0.2 mm. The fatigue resistance of aluminium parts in four-point bending tests seems to be more closely correlated to their residual stress state than to their surface texture.

Face Milling of Annealed Steels (C35, C45, C60, 42CrMo4) [91–94]
Compressive residual stresses are obtained much more easily if an up-milling strategy is selected. Cutting speed, feed and wear should be kept as low as possible in order to improve the surface integrity. However it appears that milling induces different stresses in the direction parallel to the feed and in the perpendicular direction (opposite sign may occur). Indeed, [93–94] expect tensile stresses in the direction parallel to the feed and compressive stresses in the opposite direction, whereas [91–92] expect compressive stresses. However the thickness of the affected layer is usually less than 0.15 mm.

Face Milling of Treated Steels (C35, C45, C60) [94]
Residual stresses are in traction at the surface and tend to become compressive inside the material.

Ball Nose Milling of Tool Steels (H13) or Inconel718 [95, 96]
High compressive stresses can be obtained with low cutting speeds, low feeds and low orientation angle of the tool (tool perpendicular to the surface). The trajectory is also a very sensitive parameter: a horizontal upwards cutter orientation (measured parallel to the feed direction) is critical to obtain compressive instead of tensile stresses when machining with a horizontal upwards strategy.

3.2.3.6 Super-finishing

From the previous sections, it is not surprising to announce that super-finishing processes have also been investigated since they are applied on critical surfaces submitted to fatigue and wear. The main difficulty with super-finishing techniques comes from their variety. A large number of super-finishing are available (honing, belt finishing, belt grinding, lapping, polishing, electro-chemical polishing, burnishing, roller burnishing, *etc.*). Moreover, the application of each super-finishing technique necessitates a lot of adaptation, which induces differences in surface integrity (surface texture and residual stresses). For example, the application of belt finishing to cylindrical parts such as bearings [97] is totally different from its application to crankshafts [98]. In the first case, the material removal is caused by the axial oscillation of a belt with small abrasive grains, whereas in the second case it is caused by the rotation of the workpiece but with larger grains. Additionally, the comparison between the surface integrity obtained after the belt finishing of crankshafts made of bainitic steel (*e.g.*, 35MnV7) or spheroid cast iron is very difficult. As a consequence, each scientific paper dealing with the surface integrity of super-finishing processes is usually difficult to exploit for other investigations, which does not facilitate the understanding of the fundamental mechanisms.

Belt Finishing/Belt Grinding on Hardened Steels [97, 99]
The belt finishing technique applied to hardened steels or on super alloys leads to compressive residual stresses in a thin layer ~10 μm in both the circumferential and axial directions. Such a technique never leads to thermally affected layers if it is used with plain oil lubrication.

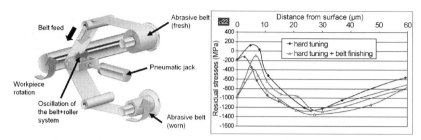

Figure 3.22. Residual stresses induced by belt finishing on a AISI52100 hardened steel

Lapping on a Hardened Steel [100]
Lapping induces strong compressive residual stresses in the surface and leads to great improvements in the fatigue strength of a martensitic stainless steel during plane bending fatigue tests, compared to conventional grinding operation.

Honing on Hardened Steel and Hastelloy X [20, 97, 24]
The honing technique shifts the external residual stress profile towards compression. Rotation bending fatigue tests and pulsating push–pull fatigue tests have shown a great improvement of the resistance compared to a shot peening operation. This improvement was even better in an aggressive atmosphere (water + sodium chloride) than in air, which proves the improvement of the resistance against stress corrosion.

Burnishing/Roller Burnishing of Hardened Steels [14, 27, 101, 102]
In roller burnishing operation, a hydrostatically borne ceramic ball rolls over the component surface under high pressure. The roughness peaks are flattened and the quality of the workpiece surface is improved. Roller burnishing delay the appearance of cracks in grooves. It has been shown that this process is able to generate deep compressive stresses in both the axial and circumferential directions. Hard roller burnishing does not generate any white layers. The affected layer is very thick (~5 mm).

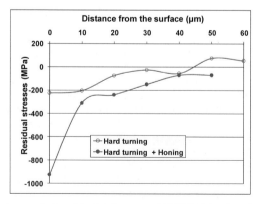

Figure 3.23. Influence of the residual stresses induced by honing on a AISI52100 hardened steel

Note that the residual stresses in reamed surfaces have some similarities with burnished surfaces. Reamed surface are compressive in nature. This is because of the rubbing and plastic deformation of the material caused by the sizing section of the reamer.

3.2.3.7 Hole Manufacturing

In contrast to the previous processes, machining processes such as drilling, reaming, *etc.* are poorly investigated. Small holes are usually not considered as critical surfaces in terms of fatigue resistance. Moreover, it is very difficult to investigate the surface texture of holes and almost impossible to investigate the residual stress fields, at least by X-ray diffraction. Problems with surface integrity of holes are rather recent. Indeed, critical surfaces have been intensively investigated and holes have been neglected. As a consequence, great improvements have been made on some surfaces whereas holes have not experienced any benefit. In parallel, the increase of the thermo-mechanical loadings supported by some parts such as crankshafts with the power increase of motor engines, or by some parts from turbine engines, have led to breakages. As a consequence, holes are starting to be the focus of attention of researchers. After the statement about the poor knowledge on surface integrity induced by drilling and reaming, some companies have decided to apply super-finishing processes such as roller burnishing or honing to induce compressive stresses and improve surface texture. This clearly shows how a lack of knowledge leads to abusive production costs.

References

[1]   Metcut research associates (1980) Machining data handbook. Institute of Advanced Manufacturing Sciences, Cincinnati, USA.

[2]   Barbacki A, Kawalec M, Hamrol A (2003) Turning and grinding as a source of microstructural changes in the surface layer of hardened steel. J Mater Process Technol 133:21–25.

[3]   Saravanapriyan SNA, Vijayaraghavan L, Krishnamurthy R (2003) Significance of grinding burn on high speed steel tool performance. J Mater Process Technol 134:166–173.

[4]   Guo YB, Sahni J (2004) A comparative study of hard turned and cylindrically ground white layers. Int J Machine Tools Manuf 44:135–145.

[5]   Silva FS (2003) Analysis of a vehicle crankshaft failure. Eng Failure Anal 10:605–616.

[6]   Wang JY, Liu CR (1999) The effect of tool wear on the heat transfer, thermal damage and cutting mechanics in finish hard turning. Ann CIRP 48/1:53–58.

[7]   Denkena B, Toenshoff HK, Muller C, Zenner H, Renner F, Koehler M (2002) Fatigue strength of hard turned components. Proc International Conference ICMEN, 3–4 October 2002, Sani, Greece.

[8]   Smith S, Melkote SN, Lara-Curzio E, Watkins TR, Allard L, Riester L (2007) Effect of surface integrity of hard turned AISI 52100 steel on fatigue performance. Mater Sci Eng A 459:337–346.

[9]   Field D, Kahles JF (1964) The surface integrity of machined and high strength steels. DMIC Rep 210: 54–77.

[10] Griffiths B (2001) Manufacturing surface technology. Penton, ISBN 1-8571-8029-1.

[11] Tonshoff HK, Brinksmeier E (1980) Determination of the mechanical and thermal influences on machined surfaces by microhardness and residual stress analysis. Ann CIRP 29/2:519–530.

[12] Rech J, Moisan A (2003) Surface integrity in finish hard turning of case hardened steel. Int J Machine Tool Manuf 43/5:543–550.

[13] Rech J, Kermouche G, Carcia-Rosales C, Khellouki A, Garcia-Navas V (2008) Characterization and modelling of the residual stresses induced by belt finishing and honing on a AISI52100 hardened steel. J Mater Process Technol, accepted for publication in 2008.

[14] Seiler W, Reilhan F, Gounet-Lespinasse F, Lebrun JL (1998) Détermination du profil de contraintes résiduelles dans les gorges galetées de vilebrequins. J Phys 8:139–146.

[15] Kristoffersen H, Vomacka P (2001) Influence of process parameters for induction hardening on residual stresses. Mater Des 22:637–644.

[16] Brinksmeier E, Solter J, Grote C (2007) Distortion engineering – Identification of causes for dimensional and form deviations of bearings rings. Ann CIRP 56/1:109–112.

[17] Rech J, Hamdi H (2005) Internal report about the modelling of part distortion due to residual stresses induced by thermo-mechanical loads, Ecole Nationale d'Ingénieurs de Saint-Etienne (France).

[18] Denkena B, Reichstein M, de Leon Garcia L (2007) Milling induced residual stresses in structural parts out of forged aluminium alloys, proceedings of the 6th international conference on HSM 2007, San Sebastian.

[19] Dattoma V, De Giorgi M, Nobile R (1986) On the evolution of welding residual stress after milling and cutting machining. Comput Struct 84:1965–1976.

[20] Matsumoto Y, Hashimoto F, Lahoti G (1999) Surface integrity generated by precision hard turning, Ann CIRP 48/1:59–62.

[21] Schwach DW, Guo YB (2005) Feasibility of producing optimal surface integrity by process design in hard turning. Mater Sci Eng A, 395:116–123.

[22] Schwach DW, Guo YB (2005) A fundamental study of surface integrity by hard turning on rolling contact fatigue. Trans NAMRI/SME 33:541–548

[23] Brunet S (1991) Influence des contraintes résiduelles induites par usinage sur la tenue en fatigue des matériaux métalliques aéronautiques. PhD Thesis, ENSAM Paris.

[24] Coulon A (1983) Effect of the honing drum upon the inducement of compressive residual stresses. J Mech Working Technol 8:161–169.

[25] Brinksmeier E, Lucca A, Walter A (2004) Chemical aspects of machining processes. Ann CIRP 53/2:685.

[26] Mainsah E, Greenwood JA, Chetwynd DG (2001) Metrology and properties of engineering surfaces. Kluwer Academic, Norwell, ISBN 0-412-80640-1.

[27] Stephenson DA, Agapiou JS (2006) Metal cutting theory and practice. Taylor and Francis, London, ISBN 978-0-8247-5888-2.

[28] Enache S (1972) La qualité des surfaces usinées, Dunod, Paris.

[29] Mahdi M, Zhang LC (1999) Residual stresses in ground components caused by coupled thermal and mechanical plastic deformation. J Mater Process Technol 95:238–245.

[30] Mahdi M, Zhang L (1999) Applied mechanics in grinding. Part 7: residual stresses induced by the full coupling of mechanical deformation, thermal deformation and phase transformation. Int J Mach Tools Manuf 39:1285–1298.

[31] Mahdi M, Zhang L (1998) Applied mechanics in grinding. Part 6: residual stresses and surface hardening by coupled thermo-plasticity and phase transformation. Int J Mach Tools Manuf 38:1289–1304.

[32] Poulachon G, Moisan A, Jawahir IS (2001) On modelling the influence of thermo-mechanical behaviour in chip formation during hard turning of 100Cr6 bearing steel. Ann CIRP 50/1:31–36.

[33] Poulachon G, Moisan A, Dessoly M (2002) A contribution to the study of the cutting mechanisms in hard turning. Mécanique Ind 3/4:291–299.

[34] Hamdi H, Zahouani H, Bergheau JM (2004) Residual stresses computation in a grinding process. J Mater Process Technol 147:277–285.

[35] Hamdi H, Dursapt M, Zahouani H (2004) Characterization of abrasive grain's behaviour and wear mechanisms. Wear 254/12:1294–1298.

[36] Brinksmeier E, Brockhoff T (1999) White layers in machining steels, Proceedings of the 2nd International Conference on HSM, Darmstadt.

[37] El-Wardany TI, Kishawy HA, Elbestawi MA (2000) Surface integrity of die material in high speed hard machining – part 2: microhardness variations and residual stresses. Trans ASME 122:632–641.

[38] Attanasio A, Ceretti E, Giardini C (2007) 3D simulation of residual stresses in turning operations, Proceedings of the 1st international conference on sustainable manufacturing, 17–18 October 2007, Montreal (Canada).

[39] Leblond JB, Pont D, Devaux J, Bru D, Bergheau JM (1997) Metallurgical and mechanical consequences of phase transformations in numerical simulations of welding processes. Modelling in Welding, Hot Powder Forming and Casting, Chapter 4, edited by Pr Lennart Karlsson. ASM International, pp. 61–89.

[40] Devaux J, Mottet G, Bergheau JM, Bhandari S, Faidy C (2000) Evaluation of the integrity of PWR bi-metallic welds. ASME J Pressure Vessel Technol 122/-3:368–373.

[41] Obikawa T, Sasahara H, Shirakashi T, Usui E (1997) Application of computational machining method to discontinuous chip formation. J Manuf Sci Eng 119:667–674.

[42] Chen L, El-Wardany TI, Harris WC (2004) Modelling the effects of flank wear and chip formation non residual stresses. Ann CIRP 53/1:95–98.

[43] Shet C, Deng X (2003) Residual stresses and strains in orthogonal metal cutting. Int J Machine Tools Manuf 43:573–587.

[44] Umbrello, D. M'Saoubi, R. Outeiro, J.C. (2007) The influence of Johnsson-Cook material constant on finite element simulation of machining of AISI 316L steel. Int J Machine Tools Manuf 47:462–470.

[45] Marusich TD, Ortiz M (1995) Modelling and simulation of high speed machining. Int J Numer Methods Eng 38:3675–3694.

[46] Salio M, Berruti T, De Poli G (2006) Prediction of residual stress distribution after turning in turbine disks. Int J Mech Sci 48:976–984.

[47] Strenkowski JS, Athevale SM (1997) A partially constrained eulerian orthogonal cutting model for chip control tools. J Manuf Sci 119:681–688.

[48] El-Wardany TI, Kishawy HA, Elbestawi MA (2000) Surface integrity of die material in high speed hard machining – part 2: microhardness variations and residual stresses. Trans ASME 122:632–641.

[49] Nasr MNA, Ng EG, Elbestawi MA (2007) Modelling the effects of tool-edge radius on residual stresses when orthogonal cutting AISI316L. Int J Machine Tools Manuf 47:401–411.

[50] Mamalis AG, Horvath M, Branis AS, Manolakos DE (2001) Finite element simulation of chip formation in orthogonal metal cutting. J Mater Process Technol 110:19–27.

[51] Sun JS, Lee KH, Lee HP (2000) Comparison of implicit and explicit finite element methods for dynamic problems. J Mater Process Technol 105:110–118.

[52] Valiorgue F, Rech J, Hamdi H, Gilles P, Bergheau JM (2007) A new approach for the modelling of residual stresses induced by turning of 316L. J Mater Process Technol 191:270–273.

[53] Zemzemi F, Rech J, Ben Salem W, Kapsa P, Dogui A (2007) Development of a friction model for the tool-chip-workpiece interface during dry machining of AISI4142 steel with TiN coated carbide cutting tools. Int J Mach Machinability Mater 2/3-4:361–367.

[54] Kevin Chou Y, Evans CJ (1999) White layers and thermal modelling of hard turned surfaces. Int J Mach Tools Manuf 39:1863–1881.

[55] Ulutan D, Erdem Alaca B, Lazoglu I (2007) Analytical modelling of residual stresses in machining. J Mater Process Technol 183:77–87.

[56] Yu XX, Lau WS, Lee TC (1997) A finite element analysis of residual stresses in stretch turning. Int J Machine Tools Manuf 37/10:1525–1537.

[57] Lortz W (1979) A model of the cutting mechanism in grinding. Wear 53:115–128.

[58] Wang H, Subhash G (2002) An approximate upper bound approach for single-grit rotating scratch with conical toll on pure metal. Wear 252:911–933.

[59] Matsuo T, Toyoura S, Oshima E, Ohbuuchi Y (1989) Effect of grain shape on cutting force in superabrasive single-grit tests. Ann CIRP 38/1:323–326.

[60] Brinksmeier E, Giwerzew A (2003) Chip formation mechanisms in grinding at low speeds. Ann CIRP 52/1:253–258.

[61] Sosa AD, Echeverria MD, Moncada OJ, Sikora JA (2007) Residual stresses, distortion and surface roughness produced by grinding thin wall ductile iron plates. Int J Machine Tools Manuf 47:229–235.

[62] Abrao AM, Aspinwall DK (1996) The surface integrity of turned and ground hardened bearing steel. Wear, 196:279–284.

[63] Grum J (2001) A review of the influence of grinding conditions on resulting residual stresses after induction surface hardening and grinding. J Mater Process Technol 114:212–226.

[64] Klocke F, Brinksmeier E, Evans C, Howes T, Inasaki I, Minke E, Toenshoff HK, Webster JA, Stuff D (1997) High-speed grinding: fundamentals and state of the art in Europe, Japan, and the USA. Ann CIRP 46/2:715–724.

[65] Moulik PN, Yang HTY, Chandrasekar S (2001) Simulation of thermal stresses due to grinding. Int J Mech Sci 43:831–851.

[66] Xu XP, Yu YQ, Xu HJ (2002) Effect of grinding temperatures on the surface integrity of a nickel-based superalloy. J Mater Process Technol 129:359–363

[67] Henrisken EK (1951) Residual stresses in machined surface. Trans ASME 73:69–76.

[68] M'Saoubi R, Outeiro JC, Changeux B, Lebrun JL, Dias AM (1999) Residual stress analysis in orthogonal machining of standard and resulfurized AISI 316L steels. J Mater Process Technol 96:225–233.

[69] Franz HE (1979) Bestimmung des Eignespannungen an deflieniert bearbeiteten oberflâchen des Werkstoffe TiAl6v and TiAl6VSn2. Techn Mitt 34:1–37.

[70] Arunachalam RM, Mannan MA, Spowage AC (2004) Residual stress and surface roughness when facing age hardened Inconel 718 with CBN and ceramic cutting tools. Int J Mach Tools Manuf 44:879–887.

[71] Schlauer C, Peng RL, Oden M (2002) Residual stresses in a nickel based superalloy introduced in turning. Mater Sci Forum 404–407:173–178.

[72] Sharman ARC, Hughes JI, Ridgway K (2006) An analysis of the residual stresses generated in Inconel 718 when turning. J Mater Process Technol 173:359–367.

[73] Tsuchida K, Kawada Y, Kodama S (1975) A study on residual stresses distribution by turning. Bull JSME 18/116.

[74] Jouan M (1986) Influence des paramètres d'usinage de tournage sur les contraintes résiduelles mesurées par diffraction de rayons X. Mémoire CNAM.

[75] Liu CR, Barash MM (1982) Variable governing patterns of mechanical stresses in a machined surface. J Eng Ind 104:257–264.

[76] Outeiro JC, Dias AM (2006) Influence of workmaterial properties on residual stresses and work hardening induced by machining. Mater Sci Forum 524–525:575–580

[77] Schreiber E, Schlicht H (1986) Residual stresses after turning of hardened components. Proc ICRS1 – Garmisch 2:853–860.

[78] Capello E (2005) Residual stresses in turning part 1: influence of process parameters. J Mater Process Technol 160:221–228.

[79] Agha SR, Liu CR (2000) Experimental study on the performance of superfinishing hard turned surfaces in rolling contact. Wear 244:52–59.

[80] Liu M, Takagi J, Tsukuda A (2004) Effect of nose radius and tool wear on residual stress distribution in hard turning of bearing steel. J Mater Process Technol 150:234–241.

[81] Dahlman P, Gunnberg F, Jacobson M (2004) The influence of rake angle, cutting feed and cutting depth on residual stresses in hard turning. J Mater Process Technol 147:181–184.

[82] Thiele JD, Melkote SN (1999) The effect of tool edge geometry on workpiece subsurface deformation and through-thickness residual stresses for hard turning of AISI52100. Trans NAMRI/SME 27:135–140.

[83] Zhou JM, Walter H, Andersson M, Stahl JE (2003) Effect of Chamfer angle on wear of PCBN cutting tool. Int J Mach Tools Manuf 43:301–305.

[84] Rech J, Claudin C (2008) Influence of cutting tool constitutive parameters on residual stresses induced by hard turning. Int J Mach Machinability Mater.

[85] Hua J, Shivpuri R, Cheng X, Bedekar V, Matsumoto Y, Hashimoto F, Watkins TR (2005) Effect of feed rate, workpiece hardness and cutting edge on subsurface residual stress in the hard turning of bearing steel using chamfer + hone cutting edge geometry. Mater Sci Eng 394:238–248.

[86] Choi, Y. Liu, C.R. (2006) Rolling contact fatigue life of finish hard materials surfaces. Part 2. Experimental verification. Wear 261:492–499.

[87] Liu CR, Mittal S (1998) Optimal pre-stressing the surface of a component by superfinish hard turning for maximum fatigue life in rolling contact. Wear 219:128–140.

[88] Scholtes B (1987) Residual stresses induced by machining. Adv Surf Treat 4:59–71.

[89] Fuh KH, Wu CF (1995) A residual-stress model for the milling of aluminium alloy (2014-T6). J Mater Process Technol 51:87–105.

[90] Rao B, Shin YC (2001) Analysis on high-speed face-milling of 7075-T6 aluminium using carbide and diamond cutters. Int J Mach Tools Manuf 41:1763–1781.

[91] Hoffmann J, Starker P, Macherauch E (1982) Bearbeitunsgeigenspannungen, Eigenspannungen und Lastspannungen. Carl Hanser Verlag, Munich, pp. 84–91.

[92] Weber H, Lutze HG, Zimmermann P, Höfer U (1980) Einfluss des Bearbeitungsparameter auf die Bearbeitungsspannungen, Wissenschaftliche Zeitschift der Technischen Hochschule Karl Marx Stadt, pp. 15–24.

[93] Bouzid Saï W, Ben Salah N, Lebrun JL (2001) Influence of machining by finishing milling on surface characteristics. Int J Mach Tools Manuf 41:443–450.

[94] Chevrier P, Tidu A, Bolle B, Cezard P, Tinnes JP (2003) Investigation of surface integrity in high speed end milling of a low alloyed steel. Int J Mach Tools Manuf 43:1135–1142.

[95] Axinte DA, Dewes RC (2002) Surface integrity of hot work tool steel after high speed milling – experimental data and empirical models. J Mater Process Technol 127:325–335.

[96] Aspinwall DK, Dewesa RC, Ng EG, Sage C, Soo SL (2007) The influence of cutter orientation and workpiece angle on machinability when high-speed milling Inconel 718 under finishing conditions. Int J Mach Tools Manuf 47:1839–1846.

[97] Rech J, Kermouche G, Carcia-Rosales C, Khellouki A, Garcia-Navas V (2008) Characterization and modelling of the residual stresses induced by belt finishing on a AISI52100 hardened steel. J Mater Process Technol
doi: 10.1016/j.jmatprotec.2007.12.133

[98] El Mansori M, Sura E, Ghidossi P, Deblaise S, Dal Negro T, Khanfir H (2007) Toward physical description of form and finish performance in dry belt finishing process by a tribo-energetic approach. J Mater Process Technol 182:498–511.

[99] Axinte DA, Kritmanorot M, Gindy NNZ (2005) Investigations on belt polishing of heat-resistant titanium alloys. J Mater Process Technol 166:398–404.

[100] Yahata N (1987) Effect of lapping on the fatigue strength of a hardened 13Cr-0.34C stainless steel. Wear 115:337–348.

[101] Mathews PG, Shunmugam MS (1999) Neural-network approach for predicting hole quality in reaming. Int J Mach Tools Manuf 39:723–730.

[102] Klocke F, Liermann J (1998) Roller burnishing of hard turned surface. Int J Mach Tools Manuf 38/5–6:419–423.

4

Machining of Hard Materials

Wit Grzesik

Department of Manufacturing Engineering and Production Automation, Opole University of Technology, P.O. Box 321, 45-271 Opole, Poland. E-mail: w.grzesik@po.opole.pl

This chapter presents basic knowledge on the special kind of the machining process in which a workpiece material hardened to 45–70 HRC hardness or more is machined with mixed ceramic or CBN tools. An extended comparison with finish grinding, as well with other abrasive finishing processes, is carried out. Specific cutting characteristics, including cutting forces, chip formation mechanisms and tool wear modes with relevant interface temperatures are discussed in terms of process conditions. Currently developing finite element (FE) and analytical modelling is overviewed. A complete characterization of surface integrity including geometrical features of hard-machined surfaces, along with specific microstructural alterations and process-induced residual stresses, is provided. Finally, the state of the art of hard cutting technology is addressed for many cutting operations to show how manufacturing chains can be effectively utilized and optimized in practice.

4.1 Basic Features of HM

4.1.1 Definition of Hard Machining

Basically, hard turning, which is the dominant machining operation performed on hardened materials, is defined as the process of single-point cutting of part pieces that have hardness values over 45 HRC but which are more typically in the 58–68 HRC range. The world-leading manufacturer of cutting tools, Sandvik Coromant, defines hard materials as those with hardness of above 42 HRC up to 65 HRC.

Commonly, hard-machined materials include white/chilled cast irons, high-speed steels, tool steels, bearing steels, heat-treatable steels and case-hardened steels. Sometimes, Inconel, Hastelloy, Stellite and other exotic materials are classified as hard-turned materials.

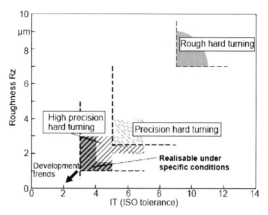

Figure 4.1. Achievable surface roughness and ISO tolerance in hard turning [1]

As shown in Figure 4.1, values of 1 μm Rz (equivalently 0.1 μm Ra) in CBN high-precision machining and correspondingly IT3 dimensional tolerance are possible. However, for extremely tightly toleranced parts, hard turning can also serve as an effective pre-finishing operation, followed by finishing grinding. Their applications have spread over such leading industrial branches as the automotive, roller bearing, hydraulic, and die and moulds sectors. Gear wheels, geared shafts, bearing rings and other transmission parts are typically machined by turning, while high-speed milling dominates the die and mould industry.

In general, hard turning can provide a relatively high accuracy for many hard parts but sometimes important problems arise with surface integrity, especially with undesirable patterns of residual stresses and the changes of subsurface microstructure, so-called *white layer*, which reduces the fatigue life of turned surfaces. This problem will be discussed in the following sections.

4.1.2 Comparison with Grinding Operations

Traditionally, the finishing operations on machine parts in a highly tempered or hardened state with hardness value in excess of 60 HRC are grinding processes, but recently hard cutting operations using tools with geometrically defined cutting edges have become increasingly capable of replacing them and guaranteeing comparable surface finish. Grinding and turning are machining operations so opposite that their full substitution is not always easy or possible.

Some of the inherent differences between these machining processes are as follows [2]:

1. Hard turning is a much faster operation because it can be done in one setup and pass under dry conditions.
2. Lathes offer more production flexibility.
3. Rough and finish operations can be performed with one clamping using a CNC lathe.

	hard cutting	grinding	legend:
	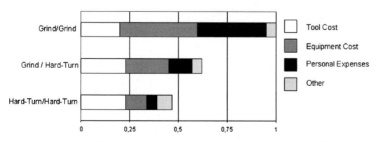		(+) positive valuation (n) neutral (-) negative valuation (?) valuation not possible
economical aspects	+ -	+ -	processing time
	+	-	material removal rate
	n +	n -	cost of acquisition
	-	+	tool costs
flexibility	+	-	multi-face machining
	+	- n	profile machining
ecological aspects	+	-	power requirement
	+	-	coolant
	+	-	chip recycling
quality	?	+	workpiece quality
	?	+	process reliability
	?	+	surface integrity

Figure 4.2. Criteria used in comparison between grinding and hard cutting operations [1, 3]

4. Multiple turning operations are easier to automate through tool changes on turning centre or turning cell.
5. Since hard turning is done dry, there are no costs for coolant, its maintenance or disposal.

In particular, the hard cutting process performed with ceramic or CBN tools can often cut manufacturing costs, decrease production time (lead time), improve overall product quality, offer greater flexibility and allow dry machining by eliminating coolants (Figure 4.2).

There are many opportunities for substituting grinding by turning operations when finish-machining of hardened ferrous materials. In general, hard turning reduces both equipment cost and personal expenses because it can be performed in one pass using one setup. On the other hand, as shown in Figure 4.3, the tool cost for finish-turning a gear blank of approximately 62 HRC hardness with CBN cutting material is almost 50% of the overall cost.

Figure 4.3. Cost comparison of turning versus grinding [4]

4.1.3   Technological Processes Including Hard Machining

The advantages of hard machining specified in Section 4.1.2 lead to substantial shortening of the traditional technological chain with heat treatment and finish grinding after rough operation, as illustrated in Figure 4.4.

With the development of super-hard cutting materials, the technology of HSM of hardened steels has created considerable interest for die and mould manufacturing. It is expected that about 50% of traditional machining operations can be replaced by HSM operations, mainly milling ones. In particular, high-volume-fraction (90%) CBN tools are recommended for milling hardened steel with cutting speeds of about 1000 m/min [5]. Figure 4.5 illustrates the substantial reduction of the production time due to decreasing hand polishing and eliminating the EDM process.

The technological process in which the ring is immediately quenched in a salt bath just after forming of the rough part is illustrated in Figure 4.6. Such optimized technology leads to about 45% energy saving and 35% reduction of costs. As reported by DMG, Germany, the integration of roughing HSC-milling at rotary speeds of up to 42,000 rpm and finishing laser machining can be very profitable for hard part machining. The elimination of EDM operations and the use of laser shaping result in shortening of the production cycle time by about sixfold. This technology is especially suited for complete machining of small and precise parts made of both metallic and non-metallic materials.

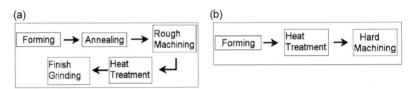

Figure 4.4. Technological chains for conventional production process (a) and production process with hard turning operations (b)

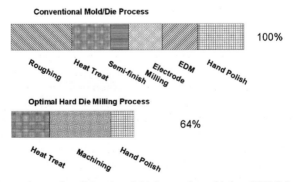

Figure 4.5. Comparison of traditional and high-speed machining (HSM)-based processes used in die and mould manufacture

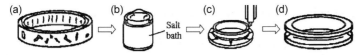

Figure 4.6. New method of production of bearing rings: (a) hot forging, (b) quenching in salt bath, (c) hard turning and (d) finish product

4.2   Equipment and Tooling

4.2.1   Machine Tools

It was proven by modern machine shops that the greatest success in hard machining is achieved from machine tools that address several key issues in design and construction. In general, the degree of machine ridigity and damping characteristics dictate the degree of hard machining accuracy and the quality of surface finish. It is well known from practice that machine systems that operate with lower vibration levels can exploit the capability of the CBN cutting materials better. Typically, high dynamic stiffness, which determines low levels of vibration over a wide frequency range, is increased by addding damping.

The next critical machine attribute is the motion capability and accuracy of the machine tool. These required a number of construction features, including composite-filled bases (polymer composite reinforcment), direct-seating collected spindles that locate the spindle bearing close to the workpiece and hydrostatic guideways, to be integrated in machining centres for hard turning or milling. Moreover, a hard turning process needs rigid spindle tooling and rigid tool holders. Maximizing system ridigity means minimizing all overhangs, tool extension and part extension, as well as eliminating shims and spacers. For turning centres the goal is to keep everything as close to the turret as possible.

Figure 4.7 shows a CNC mould and die miller with a patented self-adjusting preload spindle capable of high-speed hard milling of material hardened to HRC40, HRC50, or even HRC62 at a maximum spindle speed of 20,000 rpm, a maximum rapid feed of 200 m/min (800 ipm), and a maximum feed rate of 125 m/min (500 ipm). Also, it is equipped with a thermal distortion stabilizing system due to the danger that temperature fluctuations of the machine shop and self-generated heat from the machining process may impact performance. This system circulates a temperature-controlled fluid through the main components of the machine, minimizing the thermal distortion of the machine structure. Controlling distortion is essential for optimum machining accuracy of die and mould parts, especially for finishing operations that require long-duration cutting (several hours) with the same cutter, and for high-precision machining applications. As a result, positioning accuracy of +/–0.002 mm and repeatability of 0.001 mm can be achieved.

The spindle self-adjusts and maintains optimum pre-load (spindle rigidity) throughout the entire spindle range. This guarantees a large preload at low speed and reduces the preload according to the heat generated by higher speeds. In addition, the direct drive system in which the spindle and drive motor are connected

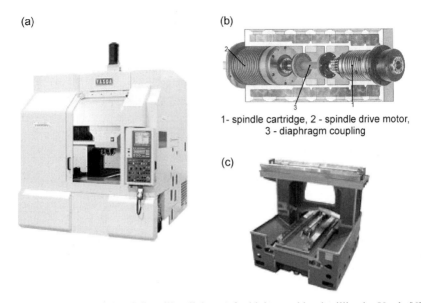

(a)

(b)

1- spindle cartridge, 2 - spindle drive motor,
3 - diaphragm coupling

(c)

Figure 4.7. CNC mould and die miller (jigborer) for high-speed hard milling by Yasda [6]: (a) general view, (b) spindle and drive motor and (c) monoblock bridge-type concrete construction

coaxially by the diaphragm coupling without any backlash (Figure 4.7(b)) is designed to isolate vibration and beat from the spindle drive motor, and enhances machining accuracy, cutter life performance and surface finish. The diaphragm coupling allows the load inertia from the spindle drive motor to provide the spindle cartridge with a smooth, vibration-free and rotationally accurate ride.

Part distortion is a serious problem for thin-walled parts for which spring-back to the original out-of-round condition occurs when using traditional clamping methods. This negative effect can be eliminated by using multiple contacts on the chuck (for example, by using the shape-compliant chuck by Hardinge [6]) and gripping the part without forcing its diameter to become round. In the case of hard milling, magnetic work holding allows for complete 3D (five-axis) machining in a single setup with improved accuracy and better surface finish due to the provision of sufficient clamping force and consistent part location.

4.2.2   Cuting Tools and Materials

Hard machining can be realized in a number of machining operations (turning, milling, drilling, broaching, reaming and threading) performed with coated carbide, cermet, ceramic, PCBN and PCD tools. In general, solid carbide tools, such as drills, taps and milling cutters (end-mills and ball-nosed cutters), coated with TiNAl (recently also with supernitrides) and TiCN layers can be used to machine hardened materials up to 65 HRC, also for high-speed cutting. Cermet (solid titanium carbide) works well for continuous cutting of case-hardened materials.

The ceramic types suitable for machining hard materials are the aluminium-oxide based, mixed and reinforced (whiskered) grades, and the silicon-nitride-based grades. They have excellent characteristics including high wear resistance, high hot hardness and good chemical stability. The mixed-type grade ceramic with TiC content and micrograined structure is used most widely in continuous or slightly interrupted hard machining of steels and cast iron. Normally, ceramics is not recommended when tolerances are tighter than ±0.025 mm (±0.001 inches). Machine tool condition and performance, methods and the insert types, as well as edge preparation, are also important for the final machining effect. Edge rein-forcement with –20° chamfer is typically applied when machining hard steels.

Polycrystalline CBN is an ideal cutting tool material for machining iron-based workpiece materials, but in a production environment cost per piece is one of the ultimate considerations. Excelent surface finish can be obtained in good, stable machining conditions, and the harder the workpiece material is, the more advanta-geous the use of CBN will be. As a rule, CBN tools are recommended for hard-nesses above 50 HRC up to about 70 HRC to generate finishes down to $Ra = 0.3$ μm. Low-content CBN (45–65%), in combination with a ceramic binder, has better shock and wear resistance and chemical stability, and is better suited to hard steel components. Oppositely, higher-content CBN, which is tougher, is more suitable for hard cast-iron and high-temperature alloys. A sufficiently large tool radius and suitable edge reiforcement are also important. Honing of the cut-ting edge reduces risk of microchipping. A typical S-edge treatment combines a 0.1 mm × 20° chamfer with a radius on the cutting edge. Recently, both mixed ceramic and CBN inserts are offered in so-called wiper configuration with special smothing micro-edges or $Xcel$ geometry with the smaller approach angle resulting in a reduced chip thickness relative to the feed rate [6].

Some newly designed CBN inserts are shown in Figure 4.8. Figure 4.8(a) shows the petite inserts (NEW PETIT CUT by Mitsubishi Carbide) in which the CBN tip is directly brazed to the host carbide insert. This results in a stronger CBN blank and allows more of the generated heat to be absorbed. All negative CBN inserts produced by Sandvik Coromant are equipped with mechanically interlocked CBN brazed corners (called Safe-Lok), as shown in Figure 4.8(c). This design gives suprior strength and security of the cutting edges, especially beneficial when machining up to shoulders, undercuts and in other profiling operations. In order to simplify detection of used edges, the insert is coated with a thin, golden TiN film.

(a)   (b)   (c)   (d)   (e)

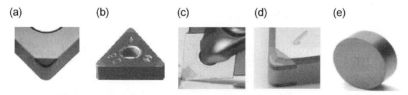

Figure 4.8. Examples of CBN inserts: (a) CBN tip brazed directly to the host carbide insert, (b) double-sided, multicorner insert, (c) insert with mechanically interlocked CBN solid corners brazed far from the hot tool–chip contact, (d) CBN insert equipped with chip breaker and (e) solid PCBN insert coated with cooper coloured (Ti,C) Al layer [6]

4.2.3   Complete Machining Using Hybrid Processes

Most applications processed across a turning centre and grinder do not require grinding on all surfaces. Motor shafts, for example, need to be ground on bearing or wear surfaces. For the rest of the features, hard turning is more than sufficient. For some applications a multifunctional machine has the potential to reduce part cycle times by as much as 25%, mainly by eliminating rough grinding steps.

In this manufacturing sector combined/simultaneous machining operations, involving hard turning and CBN grinding, are performed on gear wheels and bearing components using one machine tool equipped with two machining stations [7, 8]. This specific type of complete machining is shown in Figure 4.9. As shown in Figure 4.9 complete machining of a gear in the hardened state is successively performed on four machining stations: two for hard turning and two for grinding or super-finishing. After hard turning operations with CBN tools at a cutting speed of 300 m/min (workstations 1 and 2 in Figure 4.9(b)) only a small machin-

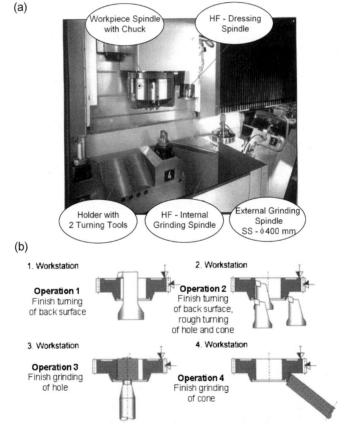

Figure 4.9. Turning and grinding machining centre: (a) working area with two separate stations and (b) complete machining of a gear in the hardened state

ing allowance of 20–30 μm remains for finishing grinding at the extremely high speed of 100 m/s, using CBN grinding wheels.

Another example is a turn-grinding centre by Index [9] equipped with a counter spindle, an outside diameter (OD) grinding spindle mounted at an angle of 15° and an inside diameter (ID) grinding spindle for producing a wide range of toolholding fixtures with the HSK63 interface, assuring high precision and high process safety. The technological process, previously performed on four single machines, combines centring, external grinding, hard turning ($v_c = 150$ m/min, $f = 0.1$ mm/rev) and bore (internal) grinding. After external grinding, the taper surface reaches 1 μm Rz, and the roundness and form tolerance less than 1 μm. More advanced machine tools for complete machining by Index and Junker [10] accommodate various machining modules, such as turning and milling modules along with OD and ID grinding. A laser unit can also be mounted for in-process work hardening.

4.3   Characterization of Hard Machining Processes

4.3.1   Cutting Forces

Hard machining is performed under unique technological and thermo-mechanical conditions and, as expected, the cutting process mechanisms (chip formation, heat generation, tool wear) differ substantially from those observed in machining soft materials. As noted in [1–4, 11], HM is also performed as a dry and HSM process. In particular, while small depth of cuts (0.05–0.3 mm) and feed rates (0.05–0.2 mm/rev) are used, small values of both the undeformed chip thickness and the ratio of the undeformed chip thickness to the radius of the cutting edge are obtained in such processes. These geometrical relationships lead to an effective rake angle of –60° to –80° and as a result extremely high pressure is generated to remove material in the vicinity of the cutting edge. Moreover, a large corner radius causes the components of the resultant cutting force to be high in conjunction with extremely high thermal stresses, as shown in Figure 4.10. It can be observed in Figure 4.10(c) that cutting forces increase drastically when machining materials with hardness higher than about 45 HRC (this value is often refered to as the lower limit of HPM).

In particular, larger negative rake angle and tool corner radius, which influene the passive force F_p, increase remarkably, meaning that an absolute stable and rigid process has to be provided. This requirement has to be especially kept when using super-hard tools with smoothing, multi-radii geometry, so-called wiper tools.

4.3.2   Chip Formation

The formation of saw-tooth chips (Figure 4.11) is one of the primary characteristics in the machining of hardened steels with geometrically defined cutting tools. Catastrophic failure within the primary shear zone during saw-tooth chip formation is usually attributed to either cyclic crack initiation and propagation or to the occurrence of a thermo-plastic instability [12, 13]. For example, for the orthogonal

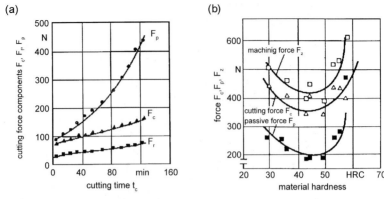

Figure 4.10. (a) Time dependence of cutting forces in HT of 16MnCr5 (AISI 5115) steel of 60–62 HRC hardness with CBN tools at cutting speed of 145 m/min and $a_p = 0.2$ mm [4] and (b) the influence of steel hardness on cutting forces ($v_c = 90$ m/min, $f = 0.15$ mm/rev, $a_p = 0.9$ mm) [11]

machining of a through-hardened AISI 52100 bearing steel of 50–65 HRC with PCBN tools the onset of chip segmentation due to adiabatic shear was observed at relatively low cutting speeds below 1 m/s [14]. In addition, these shear bands are formed at frequencies in the range of 50–120 kHz when the cutting speed was varying from 0.35 to 4.3 m/s and the segment spacing becomes more periodic as cutting speed is increased. The production of saw-tooth chips in orthogonal cutting of the 100Cr6 steel of HV730 hardness at cutting speeds of 25–285 m/min and feed rates of 0.0125–0.2 mm/rev was confirmed by Poulachon *et al.* [15]. Moreover, Shaw *et al.* [16] reported, that in face milling of case carburized AISI 8620 steel (61 HRC) with PCBN tools at $v_c = 150$ m/min, $f = 0.13$ and 0.25 mm/rev and $a_p = 0.13$ and 0.25 mm, the chip formation is of a cyclic saw-tooth type.

Figure 4.11 illustrates the cyclic mechanism of the formation of chip segments due to crack initiation (numbered successively 1 and 2) when the undeformed chip thickness is higher than 0.02 mm (for very small undeformed chip thickness less than $h < 0.02$ mm continuous chips are formed).

When machining hardened 100Cr6 bearing steel, the direct stresses σ_{VB} (Figure 4.12(a)), which reach approximately 4000 MPa independent of flank wear, result in extended high mechanical and thermal stresses on the machined surface of the workpiece. Thermal stresses result mainly from the friction between the flank wear land and the workpiece, which for a friction coefficient of 0.2–0.3 causes high tangential stress [18]. The temperature field due to friction when assuming a semi-infinite moving body with an adiabatic surface and a heat partition to the workpiece of 80% is shown in Figure 4.12(b).

When the temperature near the machined surface exceeds the γ–α transition temperature, martensite produced by friction development can form a so-called white layer observed in chip micrographs.

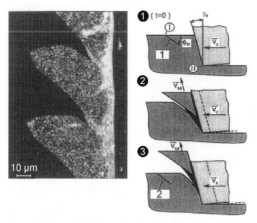

Figure 4.11. Chip formation mechanisms for hardened 100Cr6 (60–62 HRC) steel and undeformed chip thickness of 0.05 mm (when $h > 0.02$ mm) when using a PCBN tool [17]

The characteristic phenomenon of material side flow generated during hard turning operations is shown in Figure 4.13. According to many investigations, this is attributed to the squeezing effect of the workpiece material between the tool flank and the machined surface when the chip thickness is less than a minimum value h_{min}. On the other hand, it may also originate from the flowing of plastified material through the worn trailing edge to the side of the tool [19]. This causes substantial deterioration of the surface finish because the squeezed, flake-like, hard and very abrasive material (right upper image) is loosely attached to the generated surface along the feed marks. This negative effect is more significant for higher cutting speed and larger tool nose radii, and becomes more intensive with progressive tool wear.

The material behaviour when the uncut chip thickness h is less than the minimum chip thickness h_{min} at point I defined by the stagnation angle $\theta \approx 25°$ is illustrated in Figure 4.13(b). Because chip formation does not occur, elastic-plastic

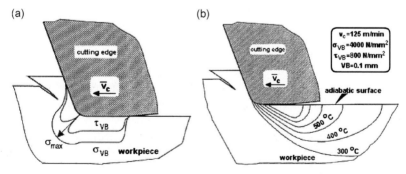

Figure 4.12. Stress distribution before crack propagation (a) and temperature field (b) for the machining case from Figure 4.11 (cutting speed of 125 m/min and tool wear VB = 0.1 mm) [18]. Workpiece material: 100Cr6 (63 HRC)

(a)

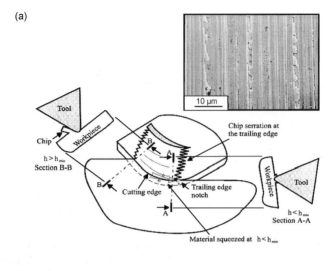

(b)

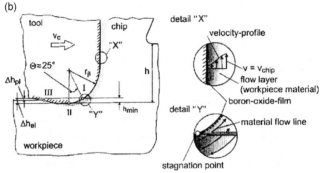

Figure 4.13. Mechanism of material side flow during hard turning [4, 19]

deformation of the surface layer is observed as depicted in detail Y. At point II the elastic component Δh_{el} springs back after the moving tool and behind point III the plastic deformation component Δh_{pl} leads to the final deformation of the surface layer.

4.3.3 Cutting Temperature

Thermal consideration of hard machining processes is very important for tool wear mechanisms and heat penetration into the subsurface layer, which leads to the formation of the white layer and determines the distribution of residual stresses. It is maintained by various researchers that the cutting temperature in hard machining depends not only on the cutting conditions (predominantly influenced by cutting speed and the tool nose radius) but also on the hardness of the workpiece material.

Figure 4.14(a) shows how work material with increasing hardness affects the tool edge temperature in the vicinity of the flank face measured by means of

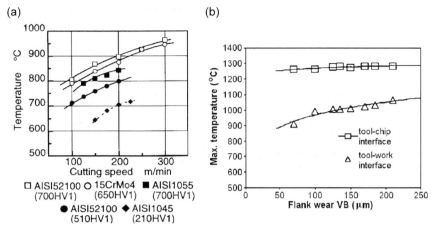

Figure 4.14. The influence of work material on the contact temperature for: (a) materials with defined hardness [20] and (b) flank wear for AISI 52100 steel [21]. Cutting conditions: (a) depth of cut 0.1 mm, feed rate 0.1 mm/rev; (b) v_c 100 m/min, depth of cut 0.089 mm

a two-colour pyrometer with a fused fibre coupler. There is a close relation between the tool temperature and the hardness of the work material used, and the cutting speed causes a substantial increase of the temperature [11, 22]. In a higher range of hardness, an increase of the material hardness leads to an increase of the cutting force. With increasing cutting force, the cutting energy becomes higher and results in elevated temperatures. As shown in Figure 4.14(a) the influence of cutting speed on the temperature of CBN tool is considerable. For machining high-carbon chromium AISI 52100 bearing steel with the highest 700V1 hardness with low-content CBN (60% vol.) the temperature is about 800°C at $v_c = 100$ m/min and it increases to about 950°C at a cutting speed of 300 m/min. On the other hand, the influence of feed rate and depth of cut is substantially less intensive than the effect of the cutting speed. In general, hard turning with a tool with larger nose radius results in higher friction and thus higher cutting temperature, while a smaller nose radius leaves deeper white layers (the oppositely for worn tools) [23].

It was revealed by Wang and Liu [21] that tool flank wear is a major cause of thermal damage in the subsurface layer in finish hard machining. In particular, it alters the heat partition coefficient and the tool–chip and tool–work interfaces. As a result, the maximum tool–work interface temperature increases as flank wear progress (reaching about 1150°C for VB = 200 μm) while the tool–chip interface temperature remains relatively constant at about 1300°C.

Thermographical measurements of workpiece subsurface temperature during turning of 60–62 HRC AISI 4130 steel ($v_c = 180$ m/min, $f = 0.08$ mm/rev, $a_p = 0.1$ mm) indicate that the maximum temperature on the component surface develops at the beginning of the contact area near the material seperation and if the minimum undeformed chip thickness (UCT) is exceeded [11]. The temperature of the workpiece surface measured by an infrared (IR) camera was about 350°C.

4.3.4   Wear of Ceramic and PCBN Tools

Figure 4.15 illustrates typical tool wear features observed in finish turning on CBN tools with the predominant wear effect: VB_{max} and VB_C localized on the clearance face and tool corner, respectively. In many hard-finishing turning operations with mixed ceramic tools the notch wear is also concentrated on the active secondary cutting (trailing) edge, as shown in Figure 4.16(b). This is responsible for the profile sharpening effect [24], causing the blunt irregular initial peaks that are transformed into final sharp individual peaks from the side of the trailing edge by copying the notch grooves.

The contact area between the worn PCBN tool, the workpiece and the chip is divided into five characteristic sections, as shown in Figure 4.16(a). In the outer zones 1 and 5 no tribological effects occur. Similarly, in zone 3, where a stable flow layer prevents tool wear they were not observed. Probably, a protective layer consisting of boron oxide films (details Y and X in Figure 4.13(b)) is formed at high contact temperatures. The flank wear concentrated in zone 4 with characteristic scouring areas is affected by severe abrasive wear. In zone 2 the predominant wear mechanisms are material deformation and continuous sliding friction caused by the chip. The bottom side of the saw-tooth chips shows a developed white layer whose scale is almost independent of tool wear progress.

4.3.5   Modelling of Hard Cutting Processes

Recently, more numerical investigations dealing with FE modelling of fundamental characteristics (deformations, forces, temperatures) under orthogonal hard cutting conditions have appeared [25–27], as well as much narrower considerations focussed on the prediction of surface quality, residual stresses and rolling contact for hard-machined components [28, 29]. In addition, it is proposed to

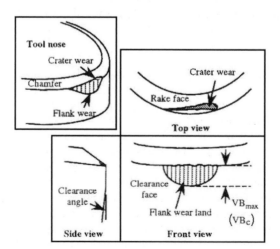

Figure 4.15. Typical wear types in CBN finish hard turning [25]

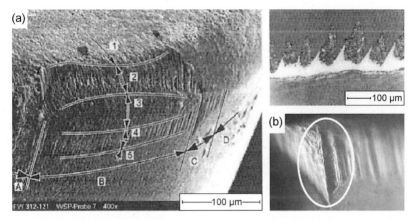

Figure 4.16. (a) Typical wear pattern of a PCBN tool corresponding to saw-tooth chip formation [11] and (b) the developed notch wear on the secondary flank face of mixed ceramic tool [24]. Cutting conditions in case (a) were: 16MnCr5 of 60–62 HRC, $\alpha_0 = 7°$, $\gamma_0 = -7°$, $\kappa_r = 95°$, $\varepsilon_r = 80°$, $r_\beta = 130$ μm

model surface finish and subsurface residual stresses using multiple regression models [30] and an artificial neural network (ANN) approach [31], respectively.

Figures 4.17(a,b) show the simulated results of chip formation and temperature distribution in the cutting zone when orthogonally machining AISI H13 tool/die steel treated to about 49 HRC with PCBN tools using the ABAQUS/Explicit™ FEM package. Figure 4.17(a) shows the deformed finite element mesh for a shear-localized segmental chip which agrees well with the photomicrograh of the chip section obtained from actual machining experiment (inserted alongside the mesh plot). In particular, the shear angle for the segmental chip was 50°. The predicted temperature map generated during segmental chip formation, in which the shear zone temperature ranges between 600 and 700°C, is shown in Figure 4.17(b). In contrast, for continuous chips this temperature was only 194–243°C [25].

It is noteworthy that FEM with the J–C equation predicts higher tool–chip interface temperatures when using a tool with lower CBN content [27]. Numerical simulations with the commercial FE code MARC have been performed [26] to investigate the variables of tool edge radius and cutting speed and to consider various material constitutive models for the prediction of stress and temperature fields for precision hard cutting of the ball bearing steel 100Cr6. It was found that a material model with isotropic work hardening, temperature and strain-rate effects predicts the maximum tool–chip temperatures close to those experimentally measured (about 550°C and 700°C for cutting speeds of 60 and 120 m/min, respectively). The 3D FE cutting models satisfactorily predict the cutting forces but significantly underestimate the thrust force.

The simulation of bearing rolling contact with the input of process-induced compresive residual stress profile was performed for AISI 52100 steel with 62 HRC hardness using the ABAQUS FEM package [29]. Based on the simulation of evolutions of equivalent plastic strain for the hard-turned and ground surfaces the

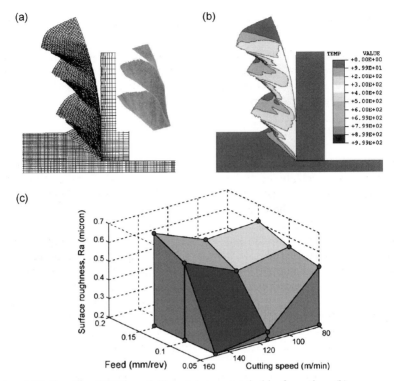

Figure 4.17. Results of FEM modelling: (a) segmental chip formation, (b) coresponding temperature distribution and (c) surface finish prediction by multiple linear regression models [30, 32]

effects of the residual stress profile on the contact stress and strain were found to be of secondary importance. On the other hand, it was shown that a negative slope of a compressive residual stress profile reduced fatigue damage and increased fatigue life. Moreover, rolling contact tends to reduce initial compressive residual stress due to friction. By using special adaptive-meshing techniques in the FEM method, which improves the stability of saw-tooth chip formation, it is possible to predict the stresses, strains and temperatures in the subsurface generated by hard machining more accurately [28].

Residual stresses as well as surface roughness and tool flank wear can be predicted using neural network or/and multiple linear regression models [30, 31]. Figure 4.17(c) shows the hypersurface describing the effect of feed rate and cutting speed on the roughness average Ra when hard turning AISI D2 cold work steel (60±1 HRC) using wiper ceramic inserts. It can be observed in Figure 4.17(c) that for this case (and also for cutting force and flank wear) the optimal cutting speed is 120 m/min. The ANN predictions of both tangential and axial components of residual stresses induced into AISI 52100 specimens after PCBN hard turning reveal good correlations with experimental data and FEM simulations (with prediction errors that do not exceed 10%) [31].

4.4   Surface Integrity in Hard Machining Processes

The performance of ceramic and CBN cutting tools and the quality of the surfaces machined are highly dependent on cutting conditions, including cutting speed, feed rate, depth of cut and tool nose radius, which all significantly influence surface roughness, white layer formation, residual stress profile and tool wear.

4.4.1   Surface Roughness

Surface roughness is the most frequently investigated characteristic of hard machining processes, mainly due to the continuous competition between hard turning and grinding, and appropriate data can be found in numerous references, for example, [3, 11, 33–36]. These data deal with different grades of hardened steels (alloy, bearing, tool and cold-work steels) mostly used in the automotive and die/mould industries. However, the 2D and 3D characterization range of surface roughness seems to be incomplete, when considering the service properties that are demanded [37]. Figure 4.18 presents some characteristic surface profiles and corresponding 3D topographies for hard turning with CBN (*left*) and mixed ceramic (*right*) tools at the same cutting parameters ($v_c = 100$ m/min, $f = 0.1$ mm/rev and $a_p = 0.2$ mm). When comparing the relevant surface profiles they contain sharper (CBN HT) or blunter (MC HT) peaks with characteristic lateral smaller flashes (more pronounced in MC HT) resulting from the side flow effect shown in Figure 4.13(b).

Characteristic surface profiles generated by conventional and wiper ceramic tools are magnified in Figure 4.19. It was observed that, due to the smoothing effect, wiper tools produce blunt irregularities with the root-mean-square (RMS) profile slope ranging from $R\Delta q = 1.5°$ to $5.5°$. On the other hand, sharper profiles

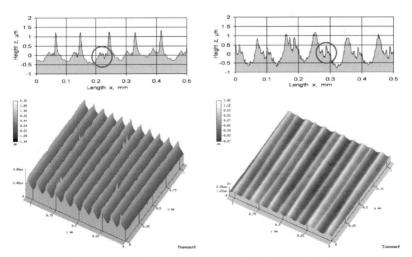

Figure 4.18. Typical surface profiles and corresponding 3D visualizations produced in (a) CBN and (b) MC turning operations. 3D images by permission of J. Rech

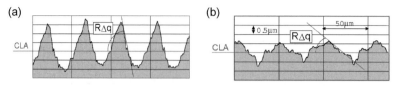

Figure 4.19. Characteristic shapes of the profiles generated in turning with conventional (a) and wiper (b) ceramic tools for a constant feed rate of 0.1 mm/rev; Vertical magnification 7000×, Horizontal magnification 200×

with RΔq values of 5–10° were recorded [33]. As can be expected distinct differences in surface profile shapes will result in corresponding bearing area curves (Figure 4.24(a)) and further in their contact capabilities. Figure 4.20 presents the deterioration of surface roughness Ra during natural wear testing performed at cutting speed of 100 m/min during 30 min. It is evident that for both fresh tools Ra values of 0.5–0.6 μm were obtained.

Only minor changes of the Ra parameter, around 10%, were recorded during the first 15 min of wear testing, with corresponding VB$_C$ nose wear of about 0.1 mm. For the lowest feed rate of 0.04 mm/rev and conventional ceramic tools the values of the height parameters Ra, Rz and Rt were 0.25, 1.60 and 1.80 μm respectively. In general, the values of the average peak spacing RSm were approximately equal to the appropriate feed rates used.

4.4.2 Residual Stresses

In order to substitute grinding process, the profile of residual stress induced in hard turning process becomes an important factor for assuring the quality and fatigue life of the machined components, especially roller bearings. Numerous studies [11, 18, 38, 39] including measurements and predictions, about the process-induced residual stresses have been conducted to determine the relationship between the cutting variables (cutting speed, feed rate, depth of cut, tool edge preparation, tool nose radius and tool wear) and the magnitude and the entire profile of stresses. In general these variables depend on the thermoplastic deformation of the workpiece.

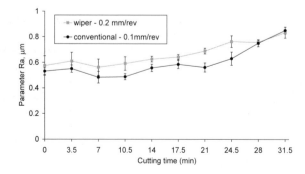

Figure 4.20. Changes of the Ra parameter with time [33]

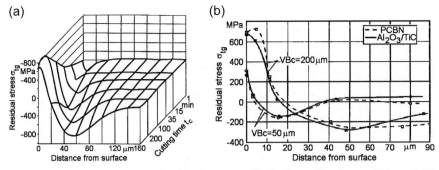

Figure 4.21. (a) Modification of residual stress profile in the surface layer within tool life period [18]; work material 16MnCr6 case-hardened steel of 62 HRC, tool material CBN; (b) Comparison of residual stress distribution when turning with PCBN and Al$_2$O$_3$-TiC ceramic tools [11]

The changes of the distribution of residual tangential stress in the sublayer with cutting time are shown in Figure 4.21(a). It should be noted that the tensile stresses occur near the machined surface and their values increase progressively with tool wear. In addition, a corresponding displacement of the point of maximum compressive stresses (about –800 MPa) deeper beneath the surface is observed. The effect of the influence of tool flank wear was confirmed for Al$_2$O$_3$-TiC ceramic tools and the differences in the shapes of stress distribution curves for these two cutting tool materials are marginal.

The combination of martensite and extremely fine-grained austenite structure constituents can cause different residual stresses in white layer. In particular, the residual stresses of the dominant austenite material components are clearly shifted towards compressive stresses. In contrast, the martensite tensile residual stresses result from lower specific volume and consequently higher density of austenite [11]. Compressive residual stresses induced by hard turning were found to improve rolling contact fatigue (RCF). Hard turned surfaces may have more than 100% longer fatigue life than ground ones, with an equivalent surface finish of 0.07 μm [40] but this positive effect can be significantly reduced by the presence of a white layer.

4.4.3 Micro/Nanohardness Distribution and White-Layer Effect

The corresponding microhardness distribution (using a Knoop indenter and 100 g load) in the surface layer is shown in Figure 4.22(a). It should be noted that the hardness is similar in the white layer of a few microns in thickness, whereas the dark layer of overtempered martensite is substantially softened. The total machining thermally affected zone is about 20 μm thickness, which is about one tenth of the cut depth ($a_p = 200$ μm). Typically, the depth of the white layer increases rapidly with tool wear and cutting speed but for the bearing steel tested saturation is observed at $v_c = 3$ m/s [41].

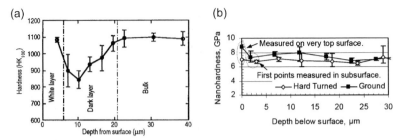

Figure 4.22. Microhardness (a) and nanohardness (b) distribution in hard turned and ground surfaces [41, 42]

Distribution of nanohardness measured by means of a diamond Berkovich indenter with a maximum load of 8 mN along the subsurface layers for hard turned and ground surfaces on a 61–62 HRC heat-treated AISI 52100 steel is shown in Figure 4.22(b). It was found that the surface nanohardness for the ground specimen is about 25% higher than of the turned sample (8.8 versus 7.0 GPa). It remains higher to a depth of about 30 μm. The trend in nanohardness data agrees with the measured microhardness data.

White layers consist of over 60% austenite which is a non-etching white component, in contrast to dark martensite scoring [11]. Due to this fact, Klocke *et al.* [7] suggested that the term *white layer* is misleading and proposed to distinguish between two groups of this structure's appearance in micrographs. When applying special etching chemicals or increased etching time, white layers of the first group remain white. On the other hand, white layers of the second group are affected by chemicals and a fine-grained martensitic structure becomes visible. Both types of white layer are distinguished by high brittleness and susceptibility to cracking and are therefore classified as damage of the machined components.

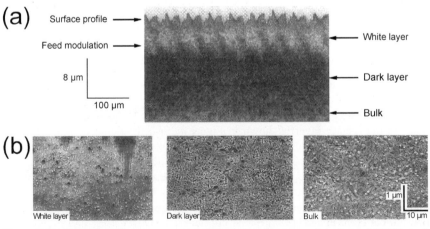

Figure 4.23. Examples of microstructural changes in a hard turned surface of 52100 steel: **(a)** optical micrograph, and **(b)** SEM micrographs of three characteristic zones [41]

In the contact zone, the austenite temperature of the workpiece material is reached after an extremely short time of approximately 0.1 ms and structural changes must be expected [11]. A metallurgically unetchable structure, called the *white layer*, followed by a dark etching layer, results from microstructural changes observed in hard machining, for example, on AISI 52100 steel surfaces of 60 HRC hardness (Figure 4.23). The three lower scanning electron microscopy (SEM) images show the white layer, the underlying dark layer and the bulk material, consisting of carbides distributed in a martensitic matrix. In general, carbide size and its distribution in both the white and dark layers indicate no difference from the bulk.

4.4.4 Modification of Surface Finish in Hybrid Processes

In general, as shown in Figure 4.24(a), the shapes of the bearing (Abbot's) curves obtained with conventional CBN and MC cutting tools are not satisfactory and the surfaces machined by hard turning require further improving by finishing belt grinding or super-finishing operations. Additionally, it is possible to generate, using a special wiper tools with varying chamfer geometry, the flatness of the bearing area curve (BAC) comparable to honing effect [7].

The residual stress profiles for the four finished surfaces on a 58–62 HRC carburized medium carbon steel are presented in Figure 4.24(b). It can be observed in Figure 4.24(b) that, although the surface compressive stress of the HT bearing is less than the HN bearing, its area of compressive stress is much wider, which leads to higher relative bearing life. In contrast, the GD surface has the lowest surface stress and unsuitable surface finish (Table 4.1), which decrease the fatigue life of such bearing surfaces.

In order to compare the four differently machined bearing surfaces some key 2D and 3D roughness parameters are given in Table 4.1. In contrast, IF generates the lowest average roughness Ra and GD the highest one. This trend is also observed for the maximum roughness height Rt. The values of the plasticity index Ψ exceed unity (indicating that the asperities are deformed plastically) in all cases

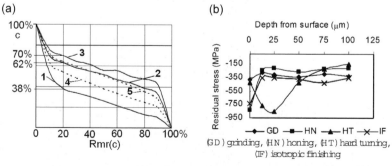

Figure 4.24. Comparison of hard turned and abrasive finished surfaces. (a) Differences in BAC curves obtained in CBN (1) and MC (4) hard turning and finish abrasive processes (preliminary (2) and finish (3) belt grinding and super-finishing (5)) after [33]. (b) Distributions of residual stresses [37]

Table 4.1. Comparison of 2D and 3D surface roughness parameters for grinding (GD), honing (HN), hard turning (HT) and isotropic finishing (IF) [21]

	GD	HN	HT	IF
Ra (μm)	0.51	0.24	0.29	0.05
Rt (μm)	3.81	1.72	2.37	0.32
Rsk	−0.66	−1.6	0.22	−1.5
Δq (°)	15.7	10.7	8.7	2.0
Ψ	4.7	3.2	2.7	0.6
Sq (μm)	0.52	0.26	0.32	0.06
Sal (μm)	18.2	8.8	44.8	43.8
Sds (μm^{-2})	636	2130	92	73

instead of the IF process. The average slope Δq differs substantially and profiles with blunter peaks (plateau) are generated in both HN and HT processes (Figure 4.19(b)). The values of the 3D RMS roughness height are close to the Ra values. The fastest decay autocorrelation length Sal, which identifies the direction of low areal autocorrelation function (AACF), has a large value for turned surfaces and a several times smaller for honed surfaces. The density of summits Sds, which provides an estimate of the average number of asperities per a unit sampling area, is highest for HN and GD surfaces (which implies a larger contact area and corresponds well with the negative value of kurtosis Rsk) and are significantly lower when using HT and IF operations.

4.4.5 Cutting Errors and Dimensional Accuracy

The achievable dimensional accuracy in precision hard turning using CNC machine tools with high static and dynamic stiffness, thermal stability and high precision motion control is standardised in ISO IT5 [5, 43]. As shown in Figure 4.25 many factors affect the part accuracy and process stability and reliability.

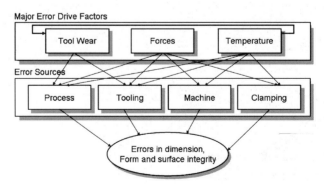

Figure 4.25. Major error drive factor and error sources in precision hard turning [40]

In order to minimize possible errors in precision hard turning magnetic chucks, a solid carbide tool holder and active error compensation and special monitoring systems are proposed. Three main factors, namely tool wear, cutting forces and cutting temperature, are the driving factors providing the error sources in process, tooling, machine structure and clamping devices. For instance, when flank wear of a CBN tool reaches VB = 0.2 mm, the dimensional errors on 100Cr6 (60–62 HRC) steel parts measured in the radial/axial directions increase to 25 μm.

Moreover, because the cutting temperature is at least 800°C, substantial thermal expansion of both the tool shaft and the workpiece is observed. It is assumed that thermal effects can contribute to more than 50% of the overall error of the machined parts. In particular, for the same machining conditions (v_c = 160 m/min, f = 0.05 mm/rev, a_p = 0.05 mm) thermal expansion on the CBN tool tip and workpiece can reach up to 10 and 15 μm, respectively [40].

4.5   Applications of Hard Machining Processes

Hard machining has been found to be very efficient in many branches of industry, including the automotive, aerospace, bearing, hydraulic and die/moulds sectors.

4.5.1   Hard Turning

A range of machining operations performed on lathes on hardened transmission parts including longitudinal continuous and interrupted turning, facing and boring, and grooving and threading are shown in Figure 4.26. Further improvement of the hard turning process can be achieved by using special single-edged tools which performed orthogonal cuts, known as hard plunge turning [34, 44].

The cutting speed for HT of 60 HRC steels with CBN inserts varies from 100 to 200 m/min. Conventional values of feed rates vary between 0.05 and

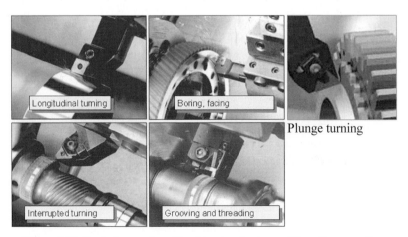

Figure 4.26. Typical examples of hard turning, boring and threading operations

0.20 mm/rev. Ceramics and cermets are suitable for lower hardnesses of 40–50 HRC at cutting speed of 300–350 m/min and surface finish down to Ra = 0.6 μm.

4.5.2   Hard and High-Speed Milling of Dies and Moulds

In recent years, hard milling (HM) technology has helped many manufacturers eliminate EDM operations and benchmarking time for more profitable die and mould making. In general, hard milling can be viewed as a process complementary to EDM in the manufacture of mouldes and dies. Hard milling is a one-step process that makes complex parts such as moulds/dies to net shape (or near-net shape) with minimal scrap. Obviously, hard milling cannot completely replace EDM, which is still necessary to make features such as internal corners and deep grooves.

However, the key to succeful and economical transition of HM technology is putting together the right combination of machine, tooling, programming and suppliers. Today, the machining of hard steels is often combined with high-speed machining, especially when it comes to milling operations. Definitions for hard milling and HSM vary from company to company and from application to application. A more conventional way of defining hard milling/HSM is machining material with 45–64 HRC at spindle speeds of 10,000 rpm and higher [45]. The depth of cut is typically 0.2 mm or less in both the radial and axial directions. Figure 4.27 illustrates two exemplary hard roughing milling operations on convex (Figure 4.27(a)) and a milling of complex structure on a five-axis CNC milling machine (Figure 4.27(b)) using solid and indexable insert ball-nose end mills. CNC machine tools for hard machining have provided a very competitive alternative to grinding and EDM.

Figure 4.28 shows some examples of parts machined on high-speed machining centres. Total cutting time for these parts was 1 hour 14 min (a), 5 hours 10 min (b) and 11 hours 45 min (c). In particular, the workpiece from Figure 4.28(c) was produced on two Hi-Net vertical machining centres, specifically designed for hard milling and high-speed, net-shape machining of complex structure, when roughing (9 hours) and finishing (2 hours 45 min) respectively. It should be noted that the new generation of Hi-Net high-speed machining centres (the AI NANO HPCC-high-precision contour control) equipped with the fastest control available (which delivers 150,000 blocks/min) uses nanotechnology for the ultimate in precision and speed.

(a)   (b)

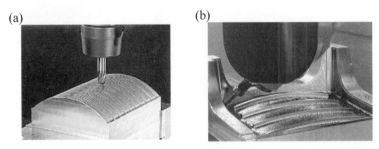

Figure 4.27. Examples of hard milling (a) a convex surface and (b) five-axis hard machining

Figure 4.28. Examples of dies and moulds [45]: (a) punch from SKD11 steel with HRC 60, (b) buckle mould from STAVAX with HRC 52 and (c) D2 tool steel workpiece with HRC 60

4.5.3   Hard Reaming

Recently, fine hole making (ISO IT4 dimensional accuracy) by hard reaming has been applied as an alternative to grinding and honing. Figure 4.29 shows one- (a) and four-edge (d) reamers equipped with CBN inserts. The new design of small-diameter reamer (Figure 4.29(c)) with a special guide pad can stabilize itself in the hole and differs from a classical boring tool (Figure 4.29(b)), which sustains a deflection due to the action of the high passive force F_p. For instance, it is possible to obtain a surface finish of $Rz = 0.8$–1.0 μm and roundness deviation of less than 4 μm of the 6.4 mm hole at rotational speed of 5,000 rpm and a feed velocity of 200 m/min [46]. In the design shown in Figure 4.29(d) four CBN blades are brazed in the head body and the clamping system used educed the radial run-out down to 3 μm. As a result, a surface roughness of $Rt = 1.0$–1.3 μm was produced.

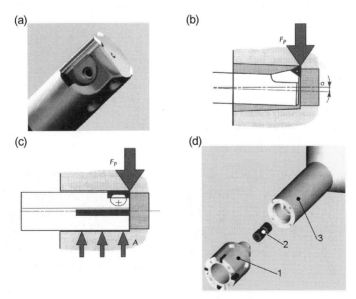

Figure 4.29. Hard reaming with one-edge (a–c) and four-edge (d) reamers [46]. 1: four-edge reaming head with short taper, 2: adjusting pin, 3: tool holder

4.5.4   Hard Broaching

Figure 4.30 shows a hard broaching machine (a) along with a diamond broaching tool (b) for making internal splines in a hardened gear wheel (c). The broach is coated with metallically bonded diamond grains. It consists of a steel body that also contains the mounting flange. The centring piece (labelled 2 in Figure 4.30(a)) positions the workpiece according to the profile to be shaped and checks the permissible broaching stock. It is possible to obtain a surface finish of Ra about 0.3 μm.

Typical applications of HB include finishing of a variety of internal profiles such as prismatic internal profiles, internal gear flanks, multiple key ways, polygons and spline profiles. It has also been reported [7] that a PCBN grade with a high PCBN content can be successfully used in dry broaching operations, but from the performance point of view (cutting distance is the limiting factor) cutting speeds are substantially lower than in hard turning (60 to a maximum of 90–100 m/min, versus 300 m/min for hard turning).

4.5.5   Hard Skive Hobbing

A new tendency in gear manufacturing is to use skive hobbing as a hard finishing process, possibly on a standard hobbing machine, instead of profile or generating grinding [7]. This is especially interesting if only a small number of gears are to be hard finished in the workshop. Unfortunately, this involves long machining times, which results in frequent regrinding of the tool. This is because the maximum feed rate is limited by the tolerable profile deviations due to the process generating feed scallops.

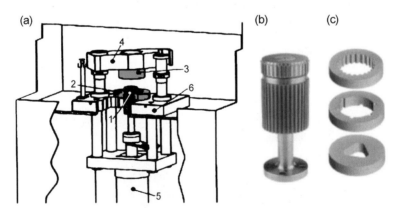

Figure 4.30. Hard broaching of a gear wheel [43]: (a) scheme of the broaching machine, 1-broach, 2-workpiece holder, 3-hold-down, 4-hold-down assembly, 5-hydraulic piston, 6-table; (b) PCD coated broach and (c) typical internal profiles

4.5.6 Optimization of Hard Machining Processes

This section does not reproduce obviously known recommendations of cutting tool manufacturers and end-users of hard machining technology but provides some new ideas by which the process chains can be enhanced in many production aspects by hard cutting operations. Generally, the functional behaviour of machined parts is strongly influenced by the finishing processes that are performed as the final step in many process chains. High flexibility and the ability to manufacture complex workpieces in a single setup are decisive advantages of hard turning over grinding. Nowadays, hard cutting operations can produce a workpiece quality comparable to grinding processes but their lower process reliability is still an important issue. The residual stress values on the subsurface of the workpiece are mainly influenced by the friction between the workpiece material and the tool tip, and consequently by the thermoplastic deformation of the workpiece. Because the profile of residual stress induced in hard turning process becomes an important factor for assuring the quality and fatigue life of the machined components, tool wear should be precisely controlled (monitored) because progressively worn tools cause tensile residual stresses and the appearance of a white layer with substantial microstructural alterations [3, 47].

Hence, the key to producing surfaces with minimal microstructural alterations and compressive residual stresses is to limit the rate of cutting tool wear. Similarly, the surface roughness is sensitive to the increase of tool wear, especially if it exceeds some specific range, for example in CBN precision, when boring of small holes (≤5 mm) in a 60-HRC case-hardened 16MnCr5 steel, the average roughness Ra stabilizes at 0.2 μm when the flank wear land is smaller than 0.1 mm [3].

Basic rules for optimization of single hard cutting operations (turning, milling reaming, boring) are given successively in Section 4.5. Apart from using the most suited process for each functional face of a workpiece, a two-step approach for one functional face, which optimizes the take-over condition between hard turning as a roughing operation and grinding as a finishing process, can also improve the quality and economy of the machining. Figure 4.31 illustrates the essential operating time for both individual processes and for the combined process over the hard turning feed; a sum of these single processes shows a minimum value at a hard turning feed of $f = 0.4$ mm/rev.

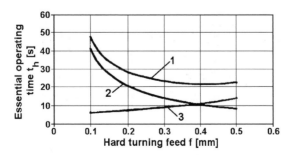

Figure 4.31. Optimization of combination of rough hard turning and finish grinding [3]: 1-process combination; 2-CBN hard turning ($v_c = 125$ m/min); 3-CBN grinding

As an alternative to the sequential process chains described in Section 4.1.3, it is also possible to perform two or even more processes, such as example hard turning and high-speed grinding, simultaneously on chuck-clamped shaft components [3]. The problem that arises in this case is the overlap of the workpiece rotational speeds between hard turning and grinding. On the one hand, the normal forces generated in grinding, if the workpiece deflects, can overload the cutting edge and increase tool wear. On the other hand, chatter generated in rough hard turning can be a risk for grinding stability. Moreover, as indicated in Section 4.2.1, there are special requirements regarding the stiffness of the machine components, the construction of the spindle and appropriate damping characteristics. The challenge for simultaneous machining is to optimize the process chain in such a way that it minimizes the impact of one process on the other and maximizes time savings resulting from parallel performance of machining operations.

References

[1]   Byrne G, Dornfeld D, Denkena B (2003) Advancing cutting technology. Ann CIRP 52/2, 483–507.
[2]   Erdel BP (2003) High-Speed Machining. SME, Dearborn.
[3]   Klocke F, Brinksmeier E, Weinert K (2005) Capability profile of hard cutting and grinding processes. Ann CIRP 54/2, 557–580.
[4]   Eredel BP (1998) New dimensions in manufacturing, Hanser Gardner, Cincinnati
[5]   Dewes CR, Aspinwall DK (1996) The use of high speed machining for the manufacture of hardened steel dies. Trans NAMRI/SME 24, 21–26.
[6]   www.coromant.sandvik.com;   www.secotools.com;   www.mitsubishicarbide.com, www.yasda.com, www.hardinge.com (accessed 2007).
[7]   Klocke F, Brinksmeier E, Weinert K (2005) Capability profile of hard cutting and grinding processes. Ann CIRP 54/2, 557–580.
[8]   Mushardt H, Manger P (2001) Komplettbearbeitung mit Hartdrehen und Schleifen. Werkstatt und Betrieb. 134/1–2, 37–40.
[9]   Beyer R (2002) Hard turning and grinding of HSK tool holders, MAV, No. 10, www.hskworld.com.
[10]   Köpfer Ch, When does hard turn/grind make sense?, www.mmsonline.com/articles/040203/html (accessed 2007).
[11]   Tönshoff HK, Arendt C, Amor R Ben (2000) Cutting of hardened steel. Ann CIRP 49/2, 547–566.
[12]   Barry J, Byrne G (2002) The mechanisms of chip formation in machining hardened steels. Trans ASME J Manuf Sci Eng 124, 528–535.
[13]   Davies MA, Burns TJ, Evans CJ (1997) On the dynamics of chip formation in machining hard metals. Ann CIRP 46/1, 25–30.
[14]   Davies MA, Chou CJ, Evans CJ (1996) On chip morphology, tool wear and cutting mechanics in finish hard turning. Ann CIRP 45/1, 77–82.
[15]   Poulachon G, Moisan A, Jawahir IS (2001) On modelling the influence of thermomechanical behavior in chip formation during hard turning of 100Cr6 bearing steel. Annals CIRP 50/1, 31–36.
[16]   Shaw MC, A Vyas (1993) Chip formation in the machining of hardened steel, Ann CIRP 42/1, 29–33.
[17]   König W, Klocke F (1997) Fertigungsverfahren 1. Drehen, Fräsen, Bohren, Springer, Berlin.

[18] König W, Berktold A, Koch KF (1993) Turning versus grinding-a comparison of surface integrity aspects and attainable accuracies. Ann CIRP 42/1, 39–43.

[19] Kishawy HA, Elbestawi MA (1999) Effects of process parameters on material side flow during hard turning. Int J Mach Tools Manuf 39, 1017–1030.

[20] Ueda T Huda M Al, Yamada K, Nakayama K (1999) Temperature measurement of CBN tool in turning of high hardness steel. Ann CIRP, 48/1, 63–66.

[21] Wang JY, Liu CR (1999) The effect of tool flank wear on the heat transfer, thermal damage and cutting mechanics in finish hard turning. Ann CIRP, 48/1, 53–58.

[22] Dewes RC, Ng E, Chua KS, Newton PG, Aspinwall DK (1999) Temperature measurement when high speed machining hardened moul/die steel. J Mater Process Technol 92–93, 293–301.

[23] Chou YK, Song H (2004) Tool nose radius effects on finish hard turning. J Mater Process Technol 148, 259–268.

[24] Grzesik W, Krol S, Wanat T, Zalisz Z (2007) Wear behaviour of mixed ceramic tools and deterioration of surface finish in the machining of a hardened alloy steel, Proceedings of the 4th International Conference on Advances in Production Engineering, Warsaw, Poland, 267–274.

[25] Chou YK, Evans ChJ (1997) Tool wear mechanism in continuous cutting of hardened tool steels. Wear 212, 59–65.

[26] Mamalis AG, Branis AS, Manolakos DE (2002) Modelling of precision hard cutting using implicit finite element method. J Mater Proces Technol 123, 464–475.

[27] Huang Y, Liang SY (2003) Cutting forces modelling considering the effect of tool thermal property-application to CBN hard turning. Int J Mach Tools Manuf 43, 307–475.

[28] Wen Q, Guo YB, Todd BA (2006) An adaptive FEA method to predict surface quality in hard machining. J Mater Process Technol 173, 21–28.

[29] Guo YB, Yen DW (2004) Hard turning versus grinding-the effect of process-induced residual stress on rolling contact. Wear 256, 393–399.

[30] Özel T, Karpat Y, Figueira L, Davim JP (2007) Modelling of surface finish and tool flank wear in turning of AISI D2 steel with ceramic wiper inserts. J Mater Process Technol 189, 192–198.

[31] Umbrello D, Ambrogio G, Filice L, Shivpuri R (2007) An ANN approach for predicting subsurface residual stresses and the desired cutting conditions during hard turning. J Mater Process Technol 189, 143–152.

[32] Ng E-G, Aspinwall DK (2002) Modelling of hard part machining. J Mater Process Technol 127, 222–229.

[33] Grzesik W, Wanat T (2006) Surface finish generated in hard turning of quenched alloy steel parts using conventional and wiper ceramic inserts. Int J Mach Tools Manuf 46, 1988–1995.

[34] Grzesik W, Wanat T (2005) Hard turning of quenched alloy steel parts using conventional and wiper ceramic inserts. Trans NAMRI/SME 33, 9–16.

[35] Rech J, Moisan A (2003) Surface integrity in finish hard turning of case-hardened steels. Int J Mach Tools Manuf 43, 543–550.

[36] Lima JG, Avila RF, Abrao AM, Faustino M, Davim JP (2005) Hard turning: AISI 4340 high strength alloy steel and AISI D2 cold work tool steel. J Mater Process Technol 169, 388–395.

[37] Hashimoto F, Melkote SN, Singh R, Kalil R (2007) Effect of finishing methods in surface characteristics and performance of precision components in rolling/sliding contact, Proceedings of the 10th CIRP International Workshop on Modeling of Machining Operations, Reggio Calabria, Italy, 21–26.

[38] Dahlman P, Gunnenberg F, Jacobson M (2004) The influence of rake angle, cutting feed and cutting depth on residual stresses in hard turning. J Mater Process Technol 147, 181–184.

[39] Liu M, Takagi J-I, Tsukuda A (2004) Effect of tool nose radius and tool wear on residual stress distribution in hard turning of bearing steel. J Mater Process Technol 150, 234–241.

[40] Zhou JM, Anderson M, Ståhl JE (2004) Identification of cutting errors in precision machining hard turning process. J Mater Process Technol 153–154, 746–750.

[41] Chou YK, Evans ChJ (1998) Process effects on white layer formation in hard turning. Trans NAMRI/SME 26, 117–122.

[42] Hashimoto F, Guo YB, Warren AW (2006) Surface integrity difference between hard turned and ground surfaces and its impact on fatigue life. Ann CIRP 55/1, 81–84.

[43] Hard broaching, www.faessler-ag.ch (accessed 2006).

[44] Huddle D (2002) Plunge turning can be a cost-effective grinding alternative. Manuf Eng 128/4, 76–81. CIRP Ann 55/1 (2006) pp. 81–84.

[45] Mickelson D (2007) Guide to Hard Milling & High Speed Machining. Industrial, New York, NY.

[46] Kress D (2001) Erfolge furs Hartreiben und Hartfräsen. Werkstatt und Betrieb 133/1–2, 64–65.

[47] Klocke F, Kratz H (2005) Advanced tool edge geometry for high precision hard turning. Annals CIRP 54/1, 47–50.

5

Machining of Particulate-Reinforced Metal Matrix Composites

A. Pramanik, J.A. Arsecularatne and L.C. Zhang

School of Aerospace, Mechanical and Mechatronic Engineering,
The University of Sydney, Sydney, NSW 2006, Australia.
E-mail: a.pramanik@usyd.edu.au
E-mail: j.arsecularatne@usyd.edu.au
E-mail: l.zhang@usyd.edu.au

The presence of hard reinforce particles in two phases materials, such as metal matrix composites (MMCs), introduces additional effects, such as tool–particle interactions, localised plastic deformation of matrix material, possible crack generation in the shear *plane etc.*, over the monolithic material during machining. These change the force, residual stress, machined surface profile generation, chip formation and tool wear mechanisms. Additional plastic deformation in the matrix material causes compressive residual stress in the machined surface, brittle chips and improved chip disposability. Possible crack formation in the shear plane is responsible for low machining force and strength and higher chip disposability. Tool–particle interactions are responsible for higher tool wear and voids/cavities in the machined surface.

This chapter presents the effects of reinforcement particles on surface integrity and chip formation in MMCs. The modelling of cutting is also discussed. Finally, tool wear mechanisms are described.

5.1 Introduction

Appropriate selection of high-performance materials for any engineering application is crucial for its success. From the beginning of industrialization and with the continuing advancement of technology, engineers and scientists have made enormous efforts in developing materials to satisfy technical requirements.

The extensive use of aluminium alloys in manufacturing aerospace and automobile structures is well recognised today. This is due to some superior properties of these materials such as high strength-to-weight ratio, excellent low-temperature performance, exceptional corrosion resistance, high machinability index and comparatively low cost [1]. Nevertheless, aluminium alloys cannot meet all engineering requirements in the advanced fields of science and technology. Their main weaknesses are poor high-temperature performance and low wear resistance. To overcome these problems, new engineering materials have been developed by reinforcing aluminium alloys with ceramic particles/whiskers; these are known as metal matrix composites (MMCs). Various properties of MMCs, such as strength-to-weight ratio, thermal stability, wear and corrosion resistance, are superior to those of their constituents [2]. Owing to their characteristics, MMCs have become key materials in many technical fields, including nuclear power stations, aerospace, aviation and defence. They are also used in the automotive industry for manufacturing engine-connecting rods, propeller shafts, brake disc and, in the leisure industry, items such as tennis racquets. Whatever their applications, efficiency in use is improved by the composites' higher strength-to-weight ratio [3, 4].

Commonly used matrix metals are aluminium and magnesium, which exhibit properties such as light weight, high ductilty, corrosion resistance, *etc*. On the other hand, frequently used ceramic reinforcements are silicon carbide (SiC) and alumina (Al_2O_3) of various shapes and sizes. Those based on a aluminium alloy matrix and reinforced with either SiC or Al_2O_3 are currently attracting most attention because of their availability and enhanced mechanical properties [3, 5, 6].

The applications of MMCs can be traced back to the 1970s, when they were investigated and applied successfully in the aeronautic and aerospace industries [7, 8]. In the middle of the 1980s, these materials reached the automobile industry and nowadays their use, though not very wide, is gaining importance. Initially, structural aluminium matrix sheets reinforced with larger fibres were developed. Since then research has especially focussed on aluminium matrix composites with discontinuous reinforcements such as SiC and Al_2O_3 [7] because of their comparatively low manufacturing cost, ease of production and macroscopically isotropic mechanical properties compared to continuously reinforced MMCs [9–11]. According to the variation of the reinforcements shape, discontinuously reinforced MMCs can be divided into two main categories: particulate-reinforced and whisker-reinforced materials. The latter possess higher elastic modulus and strength. The former has characteristics such as light weight, high specific strength and stiffness, lower thermal expansion coefficient, high thermal conductivity and excellent resistance to abrasion and corrosion [12]. As these composites contain very high-hardness strengthening reinforcements, cutting tools are apt to wear severely during their processing, resulting in low accuracy, poor surface quality and high machining cost [3].

It seems that research on the machining of MMCs was first reported in 1985 by Burn *et al.*, [13] who investigated the performance of various tool materials during machining of an aluminium alloy reinforced with 40 vol% SiC particles (Al/40%SiC MMC) and concluded that edge cracking due to mechanical chipping was the main cause of tool wear and that PCD was superior to any other tool material for machining MMCs. Also in 1985, the research committee of Japan Society for Precision Engineering (JSPE) started a cooperative research program on cutting and grinding of MMCs and published a summarized report in 1989 [14]. Comprehensive research on machining of aluminium-alloy-based MMCs started from the 1990s.

In recent years, the problems associated with precision and efficient cutting of composite materials have become an important issue in the domains of materials and manufacturing. Consequently investigations on machining, in particular, tool wear mechanisms, performance of different tools with different coatings and the effects of cutting parameters and MMC compositions on tool wear and surface finish have been reported [8].

There is no doubt that the presence of reinforcement makes MMCs different from monolithic materials and leads to the superior physical properties of MMC. On the other hand, these reinforcement particles are responsible for complex deformation behaviour, high tool wear and inferior surface finish when machining MMCs [15–19]. Thus the application of these materials has been severely limited in many fields. MMCs are yet to make major inroads into high-volume automotive and aerospace applications. This is mainly due to difficulties in fabrication and machining compared to monolithic materials. Even with near-net-shape manufacturing methods such as squeeze casting, the need for machining cannot be completely eliminated [6].

This chapter will discuss and investigate issues related to the machining of particulate-reinforced MMCs such as the mechanisms of deformation and material removal, tool wear, surface generation and chip formation.

5.2   Effect of Reinforcement Particles on Surface Integrity and Chip Formation

Investigations that explore machining mechanisms are very few because of the complex deformation behaviour of MMCs. To exploit MMCs completely, it is essential to understand the machining mechanism. In order to carry out machining efficiently and to minimize machining cost, comprehensive investigations on chip formation, surface integrity, *etc.* are urgently required.

During machining of MMC, short and long chips are formed, depending on the cutting conditions and constituents of the MMC. Sharp tools, in particular those made of diamond, produce long chips but blunt or worn tools produce short chips [20]. Low feeds (0.05–0.1 mm/rev), speeds in the range 100–800 m/min and depths of cut of 0.25–1 mm result in long chips during machining of an MMC (reinforced by 20 vol% SiC particles) by a sharp diamond tool [21]. Chips undergo very severe plastic deformation in the shear zones. The formation of small voids/cracks due to particle debonding that coalesce to form large cracks in front of the tool has been noted [8, 22–24]. However, the deformation of grain boundaries along the shear plane (similar to in ductile monolithic metals) and alignment of reinforced particles along the shear plane have also been observed [8, 25].

At low cutting speed (*e.g.*, 20 m/min) a built-up edge is generally formed during machining of an MMC, but with increasing speed and volume percentage of reinforcement, the formation of the built-up edge reduces and eventually does not form (around speed 100 m/min and 10 vol% of reinforcement) [6, 26, 27].

Particle size also plays an important role in the cutting mechanism of an MMC [28]. For those materials reinforced with course particles, higher stress intensity near the particle occurs due to higher obstruction to plastic deformation; this causes easier particle fracture. In addition, the coarse particles themselves may have more defects, which will result in higher tool wear and worse surface finish during machining of MMCs. A large quantity of harmful dust after machining MMCs with larger particle sizes has been reported, although this disappears when machining MMCs with a smaller particle size.

The effect of particles on processes parameters during machining of MMCs can be explained explicitly by comparing machining outcomes and those of the corresponding matrix material. These are described in the following sections.

5.2.1   Strength of MMC During Machining

The chip formation forces during turning depend on the strength of the material, cutting conditions and tool geometry. Speed and feed influence the strength of the workpiece material in the deformation zones through temperature, strain and strain rate [29, 30]. The strength of the non-reinforced aluminium alloy is nearly insensitive to strain rate at low strain and strain rate [31–33]. However, at higher strains (more than 1) and strain rates (10^3 s^{-1} or higher), i.e., as experienced during turning, the strength is considerably dependent on strain rate, increasing with increasing strain rate [34–39]. In the following work, due to a lack of data, the effects of strain, strain rate and temperature on shear strength are not considered explicitly. However, it was found that the measured chip formation forces in the cutting and thrust directions (F_{cc}^1 and F_{tc}, respectively), and shear angle (ϕ) depend on the cutting conditions. Hence, the experimental shear strength values, τ_s, for both the MMC and corresponding aluminium alloy at different machining conditions were determined using Equation (5.1) following the procedure described in [29, 34].

$$\tau_s = \frac{[(F_{cc}\cos\phi)-(F_{tc}\sin\phi)]\sin\phi}{A_c} \tag{5.1}$$

The shear strength values of MMC and non-reinforced alloy for different machining conditions are presented in Figure 5.1. It can be seen that the strength of MMC is significantly lower than that of non-reinforced alloy for all the machining conditions considered. At low feeds, the strength of MMC and non-reinforced alloy decreases with increasing feed (Figure 5.1(a)). However, at higher feeds, τ_s does not vary with feed significantly. Similarly speed does not influence the strength of the MMC (Figure 5.1(b)). At lower speed ranges, the strength of non-reinforced alloy increases with increasing speed, but after a certain speed it decreases with further speed increases.

At low feed (cut thickness), the cut area is small and the entire cut area may have been work-hardened by previous tool passage. This will result in a higher τ_s value at lower feeds than at higher feeds. Increased percentage of particle fracture/debonding forces (as will be seen in Figure 5.25) indicates higher tool–particle interactions at low feed for MMCs, which may be another reason for in-

---

[1] Chip formation forces were calculated by subtracting the ploughing and particle fracture/debonding forces from total machining forces.

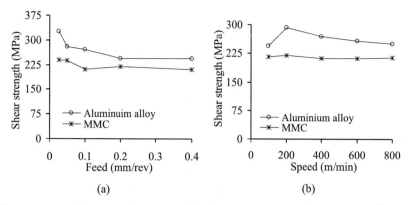

Figure 5.1. Variation of the shear strength with: (a) feed (at a speed of 400 m/min and depth of cut of 1 mm); (b) speed (at a feed of 0.1 mm/rev and a depth of cut of 1 mm)

creased strength at low feed [19]. Consequently, higher strength of workpiece materials is noted at low feeds. However, with increasing feed, work hardening decreases and temperature increases, cancelling out the net variation of the strength of the MMC and non-reinforced alloy.

Note that the strength of the two workpiece materials decreases with increasing feed (at feeds below 0.2 mm/rev). However, the cutting forces increase due to increasing area of cut.

For the non-reinforced alloy, at low cutting speeds, temperature generation is low [40]; hence the increase of strength with cutting speed is likely to be due to the influence of increasing strain rate. With further increases of cutting speed, the machining temperature increases, and consequently thermal softening of workpiece material occurs. However, the increase in strain rate will increase the strength of the material [31]. It seems that, above a certain speed, thermal softening becomes dominant over strain hardening, resulting in a decrease in strength [1, 31].

Similar to the non-reinforced alloy, work hardening of MMC increases with increasing strain rate and decreases with increasing temperature [37]. Researchers have found that composite materials may display considerably greater strain rate sensitivity (*i.e.*, increase in forces with cutting speed) than that of non-reinforced material [41, 42]. However, the lower strength (Figures 5.1(a,b)) of MMCs during machining may be a result of cracks generated due to presence of particles in the shear plane and tool–chip interface [16, 22, 26].

5.2.2 Chip Shape

The types of chips formed are related to the material properties and cutting parameters such as speed, feed, *etc*. Compared to the non-reinforced alloy, chips of different shapes are noted during machining of the MMC. The effect of reinforcement particles on chip shape under different machining parameters is discussed in the following sections.

For MMCs, chip shape varies over the considered range of feeds as shown in Figure 5.2. At a feed of 0.025 mm/rev, chips are very short and irregular in shape. With increasing feed long chips were formed. At feeds of 0.05 and 0.1 mm/rev, long spiral and straight chips, respectively, were observed. With further increases of feed (0.2 and 0.4 mm/rev), all chips became short and of C-shape. Though at medium feeds chips were very long, they did not entangle with the tool or work-piece and were easily breakable. For non-reinforced alloys, in general, the chip

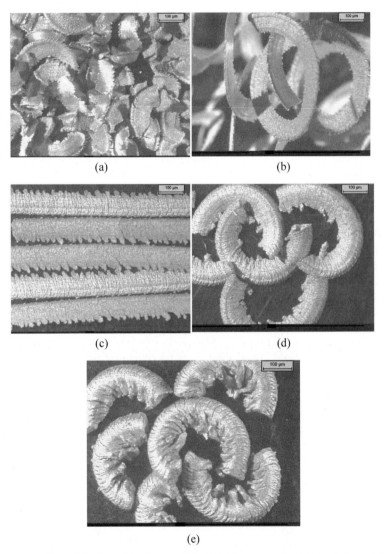

Figure 5.2. Effect of feed on chip shapes of the MMC at a speed of 400 m/min and depth of cut of 1 mm: (a) 0.025 mm/rev; (b) 0.05 mm/rev; (c) 0.1 mm/rev; (d) 0.2 mm/rev; (e) 0.4 mm/rev

shape did not change significantly with increasing feed (Figure 5.3). At all feeds, chips were long, slightly twisted and had a tendency to entangle with the tool and workpiece, which damaged the newly generated surface.

With variation of cutting speed, very long and brittle chips were formed for the MMC (Figure 5.4). At lower speed (100 and 200 m/min) all the chips were of spiral shape but at higher speeds (400 and 600 m/min) chips became straight. With further increases of speed (at 800 m/min), some tightly curled chips were formed together with long straight chips. For the non-reinforced alloy, at all cutting speeds, chips were long and/or large spirals, which entangled with the workpiece and/or the tool (Figure 5.3).

Continuous chips are forced to curl during formation due to unequal strain occurring across the plastic zone [43]. This curl depends on the ductility/brittleness of the chips. Chips of brittle materials have little or no tendency to curl whereas those of ductile materials may form long spiral chips. Shapes of chips are influenced by the uniformity of deformation and shear localization [44]. During deformation of the MMC, stress concentrations and local deformations are experienced due to the presence of the reinforcement particles [16, 19].

As the MMC experiences high strain while passing through the primary and secondary shear zones, some particles are debonded, initiating cracks and work hardening the matrix material [16, 22, 26, 45]. This makes chips brittle and easy to fracture, resulting in the formation of short chips. At lower feeds, the deformation

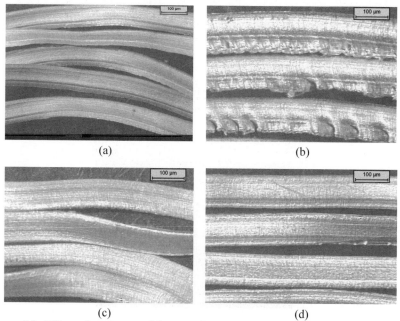

(a)	(b)
(c)	(d)

Figure 5.3. Effect of cutting conditions on shapes of the aluminium alloy at a depth of cut of 1 mm: (a) feed 0.025 mm/rev and speed 400 m/min; (b) feed 0.4 mm/rev and speed 400 m/min; (c) speed 200 m/min and feed 0.1 mm/rev; (d) speed 800 m/min and feed 0.1 mm/rev

of the chips is more homogeneous across its thickness which may lead to the formation of longer chips. However, it seems that, if the feed is very low, chips become very thin and may break due to failure of the highly strained particle–matrix interfaces. On the other hand, at higher feeds, considerable non-homogeneous deformation occurs due to higher cut/chip thickness, which contributes to the generation of shorter chips. Similarly, at low cutting speed, the strain rate effect is prominent, which may cause inhomogeneous deformation, resulting in the formation of spiral chips. However, with increasing speed, thermal effects may reduce these inhomogeneous deformation and increase the ductility of the matrix [46], which produces straight chips.

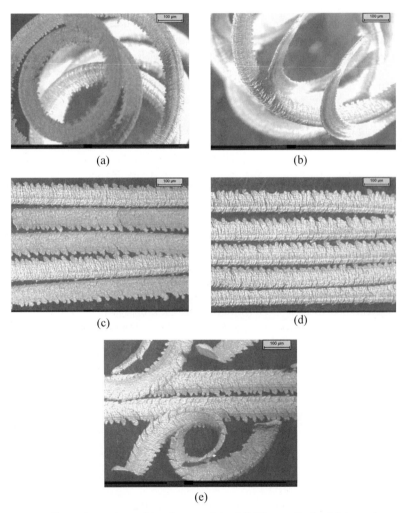

(a)   (b)

(c)   (d)

(e)

Figure 5.4. Effect of speed on chip shapes of the MMC at a feed of 0.1 mm/rev and a depth of cut of 1 mm: (a) 100 m/min; (b) 200 m/min; (c) 400 m/min; (d) 600 m/min; (e) 800 m/min

All the chips formed during machining of the non-reinforced alloy were long and less breakable because of its high ductility and deformation without formation of cracks due to absence of particles.

A harder material generally exhibits better chip disposability, *i.e.*, shorter chips with brittle fracture which may, however, affect the machined surface. On the other hand, ductile material produces very long chips with poor disposability. Long chips can damage the newly generated surface. Ductile cutting with short chips are normally desired to obtain an undamaged surface [47]. It seems that hard reinforcement particles in the MMC introduce disposability to highly ductile matrix material.

5.2.3   Surface Integrity

Surface roughness, residual stress and hardness are three important measures to describe integrity of a machined surface. It is well recognised that the surface quality depends on cutting parameters and workpiece material properties.

5.2.3.1 Surface Roughness

During machining of an MMC, defects such as voids and cavities are formed on the surface due to tool–particle interactions and resulting pull-out/fracture and debonding of particles [16, 48–50]. The tool–particle interactions are influenced by feed. The surface roughness of machined MMC surface is influenced by both feed and tool–particle interaction. When the particle size is approximately equal to the cut thickness (feed) then particle pull-out/fracture and debonding dominate the surface finish. Otherwise it is controlled by feed as well as particle pull-out/fracture and debonding. Hence, for an MMC, surface roughness decreases with increasing feed up to a certain limit, but with further increases of feed, it increases [21, 49]. The increase of the volume percentage and diameter of the reinforcement particles in an MMC increases tool–particle interactions and creates greater damage on the machined surface, which causes inferior surface finish [50, 51]. At cutting speeds of 10–100 m/min and low/medium feeds (0.1–0.48 mm/rev), surface roughness decreased with increasing speed but the decreasing trend was less marked at higher speeds [52].

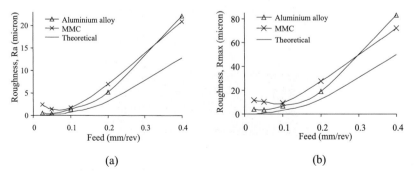

Figure 5.5. Effect of feed on surface roughness at a speed of 400 m/min and a depth of cut of 1 mm: (a) R_a and (b) R_{max}

The profile of surface generated can be considered as successive movements of the tool cutting edge profile at intervals of feed. Figures 5.5(a,b) show the variation of the measured surface finish (R_a and R_{max}) with feed. As expected, surface roughness is low at low feeds and increases with increasing feed for both reinforced and non-reinforced materials. At low feeds, roughness for the MMC is higher than that for non-reinforced alloy, but above a feed of 0.3 mm/rev the reverse trend is observed. The theoretical roughness values are lower than the experimental values for both materials, though the deviation of the experimental from the theoretical roughness value is much smaller at low feeds.

The machined MMC surfaces in Figure 5.6 show that the feed marks are not noticeable at lower feeds and that the surface texture is very irregular likely due to

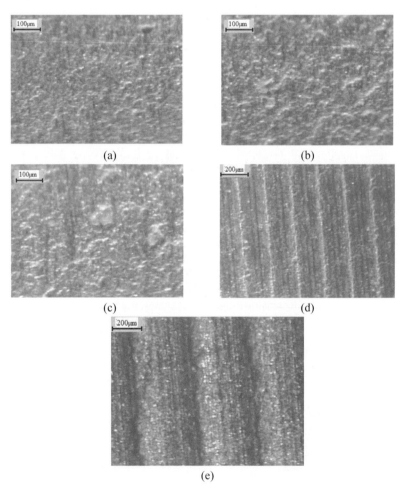

Figure 5.6. Effect of feed on machined surface of the MMC at a speed of 400 m/min and a depth of cut of 1 mm: (a) 0.025 mm/rev; (b) 0.05 mm/rev; (c) 0.1 mm/rev; (d) 0.2 mm/rev; (e) 0.4 mm/rev

the presence of particles. On the other hand feed marks are very clear on the non-reinforced alloy surface at all feeds and some material build-up is clearly visible at the higher feed (0.4 mm/rev, Figure 5.7).

No noticeable influence of speed on the machined MMC and non-reinforced alloy surface roughness values is noted for the range of speeds considered (Figure 5.8).

For the MMC, the absence of feed marks at low feed may be due to pull out and fracture of particles from the machined surface and indentation by particles. These are considered to be dominant factors that influence the texture of the newly generated surface [16, 23, 48, 46, 53]. For a given length of cut, at low feed, the distance between two successive tool paths is less and hence a greater number of tool–particle interactions will occur than at higher feed. The relatively high parti-

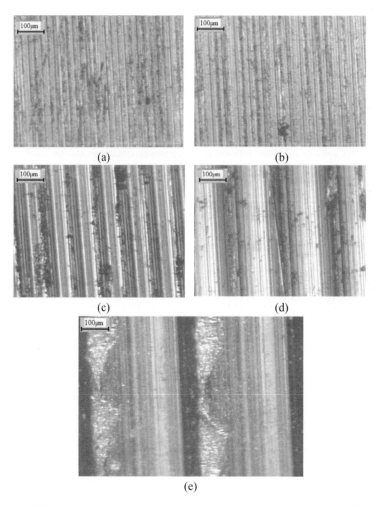

Figure 5.7. Effect of feed on machined surface of aluminium alloy at a speed of 400 m/min and a depth of cut of 1 mm: (a) 0.025 mm/rev; (b) 0.05 mm/rev; (c) 0.1 mm/rev; (d) 0.2 mm/rev; (e) 0.4 mm/rev

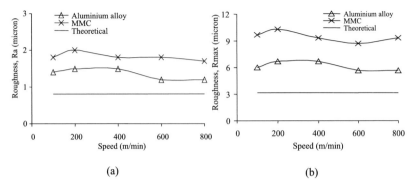

Figure 5.8. Effect of speed on surface roughness (at feed 0.1 mm/rev, depth of cut 1 mm):
(a) R_a; (b) R_{max}

cle fracture/debonding force at lower feed (as will be seen in Section 5.3.1) also indicates higher tool–particle interactions. These will cause greater surface damage at low feed. In the case of non-reinforced alloy no such damage is expected at low feeds, which accounts for its superior surface finish. However, at higher feeds, the crest (due to side flow of material) on feed mark ridges of the surface is likely to exist due to its high ductility (Figure 5.7, at the feed of 0.4 mm/rev). In the case of the MMC, those may not exist due to its lower ductility and tendency to fracture (Figure 5.6). These may cause higher roughness of the non-reinforced alloy surface compared to that of the MMC.

This can be further investigated using the profiles of machined MMC and non-reinforced alloy surfaces which, at various feeds, are given in Figures 5.9 and 5.10,

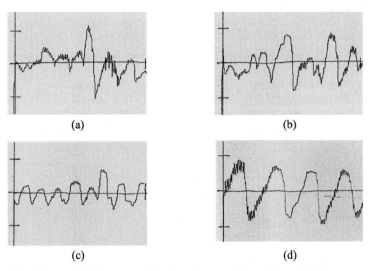

Figure 5.9. Effect of feed on machined surface profile of the MMC at a speed of 400 m/min and a depth of cut of 1 mm: (a) 0.025 mm/rev; (b) 0.05 mm/rev; (c) 0.1 mm/rev; (d) 0.2 mm/rev

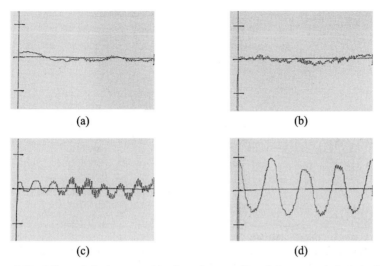

Figure 5.10. Effect of feed on machined surface profile of the non-reinforced alloy at a speed of 400 m/min and a depth of cut of 1 mm: (a) 0.025 mm/rev; (b) 0.05 mm/rev; (c) 0.1 mm/rev; (d) 0.2 mm/rev

respectively. It can be seen that the MMC surface profile is very irregular at low feeds but, with increasing feed, the feed marks are clearly recognised in the surface profile. On the other hand, for the non-reinforced alloy, very smooth surface profiles are noted at low feeds. However, surface profile is dominated by feed marks at higher feeds. For the MMC, it is noted that at low feeds (0.025–0.1 mm/rev) the magnitude of $R_{\max}$ varies from 7 to 12 μm (Figure 5.5(b)), which is in the range of the particle sizes (6–18 μm). It appears that the surface roughness of the MMC is influenced by particle size at low feeds.

5.2.3.2 Residual Stress

An important parameter of a machined component's surface integrity is the residual stress distribution, which determines the fatigue life, *etc*. Residual stresses are related to the incompatibility between a surface layer and the bulk material which is generated by any mechanism that generates a variation in the geometry of the surface layer [54]. These stresses depend on the workpiece material and machining parameters such as the cutting speed and feed. Only a few studies on turning-induced residual stress of monolithic (non-reinforced) materials have been reported to date [54–56]. These suggest that both mechanical and thermal effects are responsible for the generation of residual stresses on the machined surface. Considering that the machining and deformation mechanisms of an MMC are more complicated and different to those of a monolithic material, the mechanisms of residual stress generation are also likely to be more complex. As a result, the effects of machining parameters on surface residual stress may not be similar when the reinforcement particles are present. The effects of reinforcement particles and varying machining parameters on the residual stress are compared and discussed in the following sections.

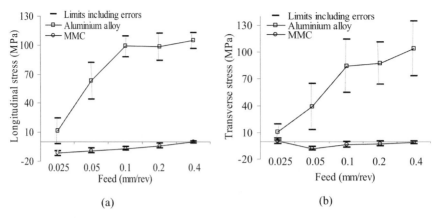

Figure 5.11. Effect of feed on residual stress at a speed of 400 m/min and a depth of cut of 1 mm: (a) longitudinal and (b) transverse

Figure 5.11(a) shows that the longitudinal (parallel to the axis of the machined bar) residual stress on the machined non-reinforced alloy surface is tensile for the whole range of feeds considered but that it is compressive for the MMC. The magnitude of the tensile residual stress (10–140 MPa) is much larger than the compressive one (0–16 MPa). The residual stress of the non-reinforced alloy is low at low feed, but with increasing feed, it increases at a very high rate. After a certain feed, this rate decreases and very little further increase in residual stress is noted. For the MMC, the compressive residual stress decreases and moves towards neutral at a low rate with increasing feed. The transverse (perpendicular to the axis of the machined bar) residual stress (Figure 5.11(b)) for the non-reinforced alloy shows a similar trend to the longitudinal one. The transverse residual stress for the MMC is nearly neutral and does not vary significantly with feed (Figure 5.11(b)).

Figure 5.12(a) shows that the residual stress (longitudinal) is tensile (10–100 MPa) in machined surface of the non-reinforced alloy and compressive (3–12 MPa) in that of the MMC for the considered range of speeds. The longitudinal residual stress for non-reinforced alloy is low at lower cutting speed and increases at a high rate with increasing speed before reaching a constant value. The influence of speed on longitudinal residual stress on the MMC surface is negligible. The transverse residual stress in the non-reinforced alloy is also tensile and increases at an almost constant rate increasing speed (Figure 5.12(b)). Similar to the longitudinal residual stress, the transverse residual stress for the MMC does not vary significantly with the variation of speed for the range considered.

From the above discussion it is clear that the residual stress in the machined non-reinforced alloy surface is tensile but that it is compressive in the MMC for all the conditions considered. Capello [54] divided the mechanisms of residual stress generation into three categories: mechanical (plastic deformation), thermal (thermal plastic flow) and physical (specific-volume variation). Tensile residual stresses are caused by thermal effects and compressive stresses by mechanical effects related to the machining operation. The relatively small compressive stress

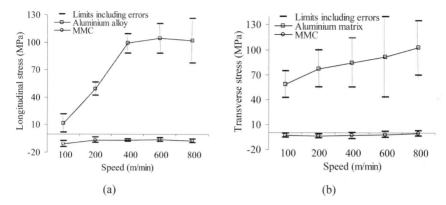

Figure 5.12. Effect of speed on residual stress at a feed of 0.1 mm/rev and depth of cut of 1 mm: (a) longitudinal and (b) transverse

measured on the MMC surface indicates marginally higher influence of the mechanical factor compared to the thermal factor. On the other hand, thermal effects play a prominent role over mechanical effects in residual stress generation when the reinforced particles are absent. The influence of the thermal factor for the non-reinforced alloy increases with increasing feed/speed.

For the MMC, based on the machining and indentation investigations [16, 19], it appears that three factors are mainly responsible for excessive mechanical deformation on the machined surface that take over the thermal effects: (a) restriction of matrix flow due to the presence of particles, (b) indentation of particles on the machined surface and (c) high compression of matrix in between particles and the tool. At low feed, these factors become very prominent. Increased percentage of particle fracture/debonding forces (as will be discussed in Section 5.2.6) indicates higher tool–particle interactions at low feed. However, with increasing feed, the indentation effects of particles as well as the tool–particle interaction decrease for the same length of machined workpiece. Additionally, the effects of temperature increase with increasing feed. Thus high compressive residual stress values at low feed and lower stress values at higher feed can be expected.

The influence of temperature is comparatively small at low cutting speed but increases its influence with increasing speed [46]. The influence of mechanical factors also increases due to increasing strain rate. With varying speed, it appears that mechanical and thermal effects balance out, resulting in a negligible compressive residual stress on the machined MMC surface.

5.2.3.3 Hardness

Similar to monolithic metal, microhardness of the surface/subsurface of machined MMC increases beyond the bulk hardness due to the plastic deformation, causing an increase in dislocation density that is controlled by particle size and volume fraction, and machining parameters [49]. The microhardness is higher near the machined surface layer and decreases with the depth of machined sub-surface due to a decrease in the work hardening of the matrix material beneath the surface layer. The microhardness beneath the machined surface increases with increasing particle size and volume fraction. The initial matrix hardness greatly influences the extent of sub-surface damage. The lower the matrix microhardness, the deeper the plastic deformation zone beneath the machined layer will be [57]. The overall increase of hardness of machined MMC surface is more marked for composites reinforced by fine particles [28].

5.2.4 Shear and Friction Angles

Shear and friction angles are associated with machining forces, efficiency of metal removal process, surface roughness and tool wear. Generally a ductile material will have a low shear angle and a brittle material will have a large shear angle [23]. This parameter can be determined using the chip thickness ratio, which is a measure of plastic deformation in metal cutting. The relation to calculate the shear angle ϕ is:

$$\phi = tan^{-1}\left(\frac{r_c \cos\gamma}{1 - r_c \sin\gamma} \right) \tag{5.2}$$

where r_c is chip thickness ratio (defined as the cut thickness divided by the chip thickness) and γ is the tool rake angle.

The friction angle controls the temperature generation at the tool–chip interface and hence crater wear. This parameter can be derived from the associated cutting and thrust forces using the equation:

$$\beta = tan^{-1}\left(\frac{F_{cc} + F_{tc}\,tan\gamma}{F_{cc} - F_{tc}\,tan\gamma} \right) \tag{5.3}$$

where β is the mean friction angle and F_{cc} and F_{tc} are the chip formation forces in the cutting and thrust directions, respectively. A higher friction angle will result in the generation of a higher temperature at the tool chip–interface and hence high tool wear. The applicability of Equations (5.2) and (5.3) for MMC machining was considered in [57, 58].

Figure 5.13(a) shows the effect of reinforced particles on the shear angle with the variation of feed. The shear angle increases with the increase of feed for both workpiece materials. Initially its rate of increase is very high and the shear angle for the MMC is higher than that for the non-reinforced alloy. After a certain feed it becomes higher for the non-reinforced alloy. Then the variation of shear angle with feed reduces for both materials. According to Equation (5.2) a higher shear angle means a lower chip thickness or higher chip thickness ratio (r_c). During machining, the chip undergoes complete deformation across its thickness in the primary shear zone, but in the secondary shear zone deformation is restricted to the tool–chip interface region. Therefore, with increasing cut thickness the thickness of the secondary deformation of chips reduces compared to the total chip thickness. This causes inhomogeneous deformation of chips and the generation of well-broken C-shaped chips for the MMC.

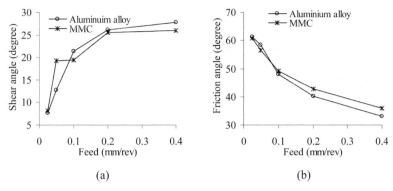

(a)   (b)

Figure 5.13. Effect of feed on shear and friction angles at a speed of 400 m/min and a depth of cut of 1 mm: (a) shear angle and (b) friction angle

The effect of reinforced particles on friction angle with the variation of feed is presented in Figure 5.13(b). The friction angle curves have a hyperbolic shape for the MMC and non-reinforced alloy. Initially the friction angle for the non-reinforced alloy is little higher than that for the MMC and they start to decrease at a high rate with increasing feed. Above a certain feed the friction angle for the non-reinforced alloy becomes lower than that for the MMC. With a further increase of feed this angle continues to decrease at a reduced rate for both materials.

The variation of shear angle with speed due to presence and absence of particles is depicted in Figure 5.14(a). Shear angles for the non-reinforced alloy are higher than those for the MMC over the range of speeds considered in this investigation. For the non-reinforced alloy, the shear angle initially decreases with increasing cutting speed, then it starts to increase at a small rate with further increases of speed. This reflects the initial increase of force with speed and then decrease after certain speed. The shear angle for the MMC continuously increases with the increase of speed at a low rate.

Figure 5.14(b) shows the effect of particles on the friction angle with variations of the cutting speed. Unlike the shear angles, friction angles for the MMC are

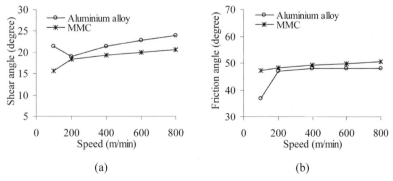

(a)   (b)

Figure 5.14. Effect of speed on shear and friction angles at a feed of 0.1 mm/rev and a depth of cut of 1 mm: (a) shear angle and (b) friction angle

higher than those for the non-reinforced alloy over the whole range of speeds considered. At low speed, a comparatively low friction angle is noted for the non-reinforced alloy, but it increases rapidly with increasing speed and reaches a constant value with further increases of speed. A small increase of friction angle for the MMC is noted with increasing speed.

5.2.5 Relation Between Shear and Friction Angles

As noted earlier the shear angle ϕ is associated with the geometry of chip formation and hence cutting forces, *etc*. The theoretical relations obtained for ϕ by Merchant [60] and Lee and Shaffer [61] are:

$$\phi = \begin{cases} \dfrac{\pi}{4} - \dfrac{1}{2}(\beta - \gamma) & \textit{Merchant} \\[2ex] \dfrac{\pi}{4} - (\beta - \gamma) & \textit{Lee and Shaffer} \end{cases} \qquad (5.4)$$

where β *and* γ are the friction and rake angles, respectively.

However, the experimental results obtained by investigators such as Kobayashi and Thomson [62] and Pugh [63] for a wide range of (monolithic) work materials and cutting conditions show that the following relation is more appropriate for ϕ.

$$\phi = B - C(\beta - \gamma) \qquad (5.5)$$

where B and C are constants that depend on the work material.

Figures 5.15(a and b) show the experimental values of ϕ plotted against $(\beta - \gamma)$ for the MMC and aluminium alloy. The linear regression lines for the data are also shown in the figures. It can be seen that the experimental results fall

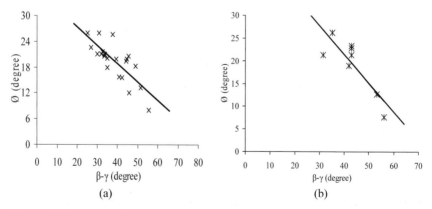

Figure 5.15. ϕ versus $(\beta - \gamma)$ relationship: (a) for the MMC; (b) for the aluminium alloy

close to the lines represented by Equations (5.6) and (5.7) for the MMC and aluminium alloy, respectively.

$$\phi = \frac{\pi}{5} - \frac{1}{2}(\beta - \gamma) \quad for\ MMC \tag{5.6}$$

$$\phi = \frac{\pi}{4} - \frac{1}{2}(\beta - \gamma) \quad for\ aluminium\ alloy \tag{5.7}$$

It can be seen that, similar to the case of cutting monolithic materials discussed above, there also exists a linear relationship between ϕ and $(\beta - \gamma)$ even for the MMC. In the case of the non-reinforced alloy, the relationship is similar to Merchant's equation (Equation (5.4)). The notable difference is that the value of B (Equation (5.5)) for the MMC is not identical to that for the matrix material (Equations (5.6) and (5.7)).

5.2.6 Forces

The measured cutting and thrust forces at different feeds are presented in Figure 5.16. It can be seen that cutting force for the non-reinforced aluminium alloy is slightly larger than that for the MMC (Figure 5.16(a)). For the two materials, the experimental cutting forces increase more or less linearly with the increase in feed and the rate of increase is almost similar. Thrust forces increase at a lower rate than the cutting forces (Figure 5.16 (b)). At lower feeds, the thrust force for the non-reinforced alloy is higher than that for MMC, but above a certain feed the opposite trend is noticed. At this stage, a similar rate of increase of forces is noted for the two materials. Thrust forces are higher than cutting forces at lower feeds (below 0.1 mm/rev) but the opposite is observed at higher feeds.

Figures 5.17(a, b) present the variation of cutting and thrust forces with cutting speed for the MMC and aluminium alloy. In the case of the MMC, the speed does not influence the two forces significantly. For the non-reinforced alloy, both forces are lower than those of the MMC at low cutting speed but with increasing speed the forces increase and at a certain point are higher than those of the MMC. With

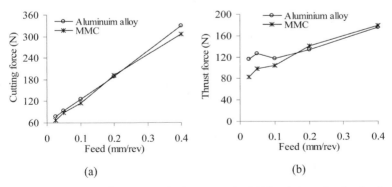

(a)   (b)

Figure 5.16. Variation of forces with feed at a speed of 400 m/min and a depth of cut of 1 mm: (a) cutting force and (b) thrust force

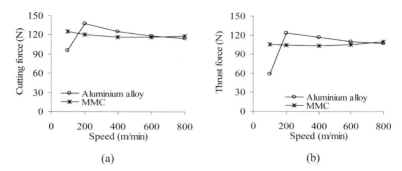

Figure 5.17. Variation of forces with speed at a feed of 0.1 mm/rev and a depth of cut of 1 mm: (a) cutting forces and (b) thrust forces

further increases of the speed, the forces start to decrease (due to thermal softening) and at a certain point again become lower than those of the MMC. The cutting forces are higher than the thrust forces for both materials at all cutting speeds considered in this investigation.

To study the influence of tool–particle interactions on force generation, force signals from dynamometer were investigated. Force signals at different cutting conditions for the MMC and the non-reinforced alloy during cutting are presented in Figure 5.18. No significant influence of these interactions on the force signals is noted, as the signals are similar for the MMC and non-reinforced alloy. This may be due to smaller interparticle distance in the MMC. It is estimated that for an MMC with 20 vol% (uniformly distributed spherical particles) of reinforcement (size 12 μm) the interparticle distance is approximately 4.5 μm. At the lowest cutting speed used (100 m/min) the cutting tool will travel this distance in only 0.0027 s and it seems that the data acquisition system is not fast enough to detect individual tool–particle interactions. In addition, at a depth of cut of 1 mm, since the length of the active cutting edge is over 1 mm and particles are more or less uniformly distributed in the MMC, continuous tool–particle interactions will occur along the cutting edge. Hence, the effect of individual tool–particle interaction on the force signal is not likely to be distinguishable.

From this discussion, it is clear that the strength of MMC did not vary significantly with feed and speed compared to that of corresponding matrix material. However, the strength of the MMC was lower than that of the matrix material because of the possible cracks generated during deformation. Chip breakability was found to improve due to the presence of the reinforcement particles in the MMC. Short chips were formed under almost all conditions. Both the longitudinal and transverse residual stresses on the matrix surface were tensile and increased with increasing speed and feed. On the other hand, the presence of the reinforcement particles induced compressive residual stresses on the machined MMC surface due to their interaction with the cutting tool. The surface roughness of the MMC was controlled by particle fracture and/or pull out at low feeds, but at higher feeds it was controlled by the feed. On the other hand, the surface roughness of the matrix material was mainly controlled by the feed. The variations of the shear and friction angles with feed for the MMC were similar to those for the matrix mate-

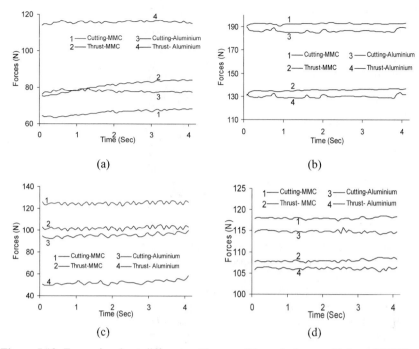

Figure 5.18. Force signals at different cutting conditions during machining of MMC and aluminium alloy at a depth of cut of 1 mm: (a) feed 0.025 mm/rev and speed 400 m/min; (b) feed 0.4 mm/rev and speed 400 m/min; (c) speed 100 m/min and feed 0.1 mm/rev; (d) speed 800 m/min and feed 0.1 mm/rev

rial, *i.e.*, with increasing feed, the shear angle increased and the friction angle decreased though the rate of variation depended on the feed. However, the influence of speed on the shear and friction angles was less marked for both materials. For the MMC, the shear and friction angles increased very little with increasing speed. On the other hand, for the aluminium alloy, initially the shear angle decreased and the friction angle increased at low speed, but after a certain speed, the shear angle increased and the friction angle remained constant with further increases of speed. The relationship between ϕ and $(\beta - \gamma)$ for the matrix material, *i.e.*, $\phi = B - C (\beta - \gamma)$, still holds for the MMC. The value of C is 1/2 for both materials and the values of B are $\pi/5$ and $\pi/4$ for the MMC and matrix material, respectively.

5.3   Modelling

5.3.1   Forces

A few models are available for cutting force prediction for MMCs. These models are now considered.

Kishawy *et al.* [45] developed an energy-based analytical force model to predict force in the cutting direction for orthogonal cutting of an MMC using a ceramic tool at low cutting speed. The total specific energy for deformation was estimated from the energy consumed in the primary and secondary shear zones (which depended on the matrix matrial) and the energy due to debonding/fracture of the particles (which depended on the MMC properties). Only the force in the cutting direction was calculated form the total energy consumed in machining. This model has several weaknesses. Firstly, the energy in the secondary deformation zone was taken to be one-third of that in the primary deformation zone. This assumption was based on the results obtained for (monolithic) steel work materials which may not be applicable to MMCs. Secondly, the width of crack and, initial and final crack lengths of ceramic particles were 1 μm, 1 μm and the circumference of a particle respectively. However, no information was given to justify these assumptions. Thirdly, the model was only verified at low cutting speed (60 m/min) because of the use of ceramic tools in their tests. In addition, energy due to ploughing was not considered.

Pramanik *et al.* [29] developed a mechanics model for predicting the forces of cutting particle reinforced MMCs based on material removal mechanism. The force generation mechanism was considered to be due to three factors: (a) chip formation, (b) ploughing and (c) particle fracture/debonding. The chip formation force was obtained by using Merchant's analysis but those due to matrix ploughing deformation and particle fracture were formulated, respectively, with the aid of the slip-line field theory of plasticity and the Griffith theory of fracture. This model is now discussed.

As noted earlier, there are similarities in the chip formation mechanism of MMCs to that of monolithic materials such as aluminium or steel. It should be noted that, during machining, shearing occurs in a zone rather than on a plane. However, at higher cutting speeds, the thickness of the shear zone reduces and hence it can be approximated by a shear plane [22]. Due to the simplicity of the shear plane models and relatively high cutting speeds normally used when machining MMCs with PCD tools, Pramanik *et al.* [29] selected Merchant's analysis[2] [60] to determine the chip formation forces. In Merchant's analysis, the chip is considered as a separate body in equilibrium under the action of two equal, opposing forces: the force which the tool exerts on the back surface of the chip and that which the workpiece exerts on the base of the chip at the shear plane (AB in Figure 5.19). Thus the force components acting on the tool in the direction of cutting, F_{cc}, and that in the direction of feed (thrust), F_{tc}, were determined using the equations given below [60]:

$$
\left.
\begin{aligned}
F_{cc} &= \tau_s \, A_c \, \frac{\cos(\beta - \gamma)}{\sin\phi \, \cos(\phi + \beta - \gamma)} \\
F_{tc} &= \tau_s \, A_c \, \frac{\sin(\beta - \gamma)}{\sin\phi \, \cos(\phi + \beta - \gamma)}
\end{aligned}
\right\}
\qquad (5.8)
$$

---

[2] The theory is limited to orthogonal cutting with plane face tools having a single straight cutting edge.

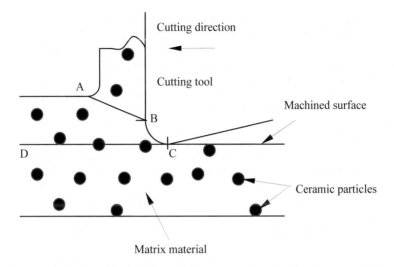

Figure 5.19. Machining process of MMC (from Pramanik *et al.* [29])

where A_c is the cross-sectional area of cut, τ_s is the shear strength of MMC, β is the mean friction angle, γ is the tool rake angle and ϕ is the shear angle, which is given by Equation (5.2). The shear angle, ϕ, and the mean friction angle, β, are shown to depend on machining conditions, workpiece material, *etc.*

Since the particle fracture was considered separately in this model, only the matrix metal was assumed to be subjected to ploughing. In order to determine the force components due to metal matrix ploughing, the equations given in [64], which are based on a slip-line field model for a rigid wedge sliding on a half-space and for orthogonal cutting, were used. With the above model, the ploughing force components acting on the tool in the direction of cutting, F_{cp}, and that in the direction of feed, F_{tp}, were determined as

$$
\left.
\begin{aligned}
F_{cp} &= \tau_{sm} l r_n \tan\left(\frac{\pi}{4}+\frac{\gamma}{2}\right) \\
F_{tp} &= \tau_{sm} l r_n \left[1+\frac{\pi}{2}\right] \tan\left(\frac{\pi}{4}+\frac{\gamma}{2}\right)
\end{aligned}
\right\}
\tag{5.9}
$$

where l is the active cutting edge length, r_n is the cutting edge radius and τ_{sm} is the shear strength of the matrix material.

Noting that the width of cut is the active cutting edge length which encompasses straight and round parts of the cutting edge involved in cutting, l is given by

$$
l = r_\varepsilon \left[\kappa_r + \sin^{-1}\left(\frac{f}{2r_\varepsilon}\right)\right] + \frac{a - r_\varepsilon \left[1 - \cos\left(\kappa_r\right)\right]}{\sin\left(\kappa_r\right)}
\tag{5.10}
$$

where r_ε is the tool nose radius, κ_r is the approach angle, f is the feed and a is the depth of cut.

While the tool moves in the direction of cutting, the hard particles in the ploughing zone are fractured or displaced. Thus the cutting edge of the tool can be considered to be responsible for particle fracture/displacement which, therefore, occurs along the cutting edge of the tool. Pramanik *et al.* [29] calculated the total energy for particle fracture during orthogonal cutting of MMC as

$$E_{ft} = \mu_g l \qquad (5.11)$$

where μ_g is the average fracture energy per unit cutting edge length in orthogonal cutting. If the force in the cutting direction due to particle fracture and displacement is F_{cf}, then $F_{cf} L = E_{ft} = \mu_g l$, where L is the cutting distance. F_{cf} is then determined by

$$F_{cf} = \frac{\mu_g l}{L} \qquad (5.12)$$

In order to determine the force due to particle fracture/displacement in the thrust direction, the direction of the resultant force must be determined. It was assumed that the reinforcement particles are uniformly distributed and spherical with uniform average diameter, and that the interaction of particles with the cutting edge could be represented by a particle placed at the middle of the ploughing zone, as shown in Figure 5.20. Neglecting the frictional effects during the interaction, the direction of the resultant force was determined from the relation

$$\sin \delta = \frac{H}{d_p + 2r_n} \qquad (5.13)$$

where δ is the angle of resultant force measured from the cutting direction, d_p is the average diameter of particles and H is the width of the ploughing zone, which is given by

$$H = r_n (1 + \sin \gamma) \qquad (5.14)$$

Then the component of the force due to particle fracture and displacement acting in the thrust direction is

$$F_{tf} = F_{cf} \tan \delta \qquad (5.15)$$

The forces F_{cf} and F_{tf} can be determined from the above equations when μ_g is known. The total forces acting on the tool in the cutting and thrust directions F_c and F_t are therefore given by

$$\left. \begin{aligned} F_c &= F_{cc} + F_{cp} + F_{cf} \\ F_t &= F_{tc} + F_{tp} + F_{tf} \end{aligned} \right\} \qquad (5.16)$$

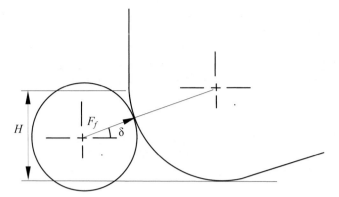

Figure 5.20. Interaction of a particle with cutting edge (from Pramanik *et al.* [29])

In order to determine the forces due to particle fracture and displacement using Equations (5.11)–(5.15), the value of μ_g is required. In order to determine μ_g, the following approach was adopted.

It was noted that Yan and Zhang [65] calculated the specific energy for particle fracture during scratching of aluminium-alloy-based MMCs. In [66] Griffith's theory[3] was applied to estimate the energy for particle fracture under a pyramid indenter. The specific energy for particle fracture was determined by subtracting the scratching energy of the corresponding aluminium matrix material from the total specific energy of the composites reinforced with SiC (10 and 20 wt%) and Al_2O_3 (10 and 20 wt%) particles. Pramanik *et al.* [29] used the results in [65] to determine μ_g because in deriving Equation (5.11), the tool edge was considered responsible for particle fracture.

For a given groove depth h, the specific particle fracture energy μ_f can be obtained from the results given in [65]. The total energy of particle fracture E_f for scratching is

$$E_f = \mu_f AL = \mu_f h^2 \tan\theta\, L \qquad (5.17)$$

where A is the cross-sectional area of groove, θ is the indenter apex angle between opposite faces and L is the length of scratch. Note that the volume of material removed is $AL = h^2 \tan\theta\, L$. The effective indenter edge length L_{ei} causing particle fracture is given by

$$L_{ei} = \frac{2h}{\cos^2\theta} \qquad (5.18)$$

Hence, the energy of particle fracture per unit cutting edge length L_{ei} is

$$\mu_g = \frac{E_f}{L_{ei}} = \frac{\mu_f\, h\, \sin(2\theta)\, L}{4} \qquad (5.19)$$

---

[3] According to this theory, the energy required to propagate a crack is equal to the surface energy of the crack surface created.

The value of μ_g was interpolated from a range of data available in [65], the obtained value of μ_g was 0.01 J/mm for SiC (20 vol%). Using the given data and Equations (5.11)–(5.15), the cutting and thrust force components due to particle fracture and displacement can be determined for different cutting conditions.

The predicted forces for chip formation, ploughing and particle fracture and displacement were first determined. These forces were then summed to determine the total forces in the cutting and thrust directions (Equation (5.16)), which were then compared with experimental forces[4] obtained under identical conditions. Additionally, for the same conditions, cutting forces[5] were calculated using the model described by Kishawy *et al.* [45]. Predictions made from this model together with those from the model in Pramanik *et al.* [29] and experimental results are given in Figures 5.21–5.23. Based on the information in [66], the maximum tensile stress of SiC was taken to be 245 MPa.

Figure 5.21 compares the predicted and experimental forces with varying feed. In this figure, the experimental force results are represented by symbols and the predicted results by lines. The solid lines represent predictions made using the model in [29] while the dashed line represents those obtain using the energy based model in [45]. As expected, the experimental cutting and thrust forces can be seen

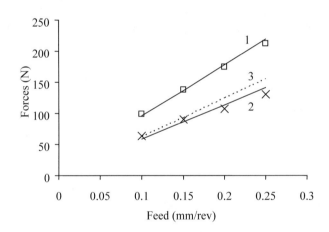

	Experimental symbol	Predicted lines for model	
		Ref. [29]	Ref. [45]
Cutting force	□	1	3
Thrust force	×	2	Unable to predict

Figure 5.21. Comparison of predicted and experimental forces with varying feed (from Pramanik *et al.* [29])

---

[4] In line with the equivalent cutting edge concept used in predicting the forces, the experimental thrust force was taken as the resultant force of experimental feed and radial forces.

[5] Note that the model in [45] allows the determination of the cutting force only.

to increase approximately linearly with the increase in feed. The rate of increase in cutting force is higher than that of the thrust force. The predicted results from the model in [29] show the same trends as the experimental ones with excellent quantitative agreement. On the other hand, the model in [45] can be seen to underestimate the cutting force considerably.

The comparison between the predicted and experimental force results with varying depths of cut is presented in Figure 5.22. Similar to the force variations seen in Figure 5.21, the experimental and predicted force results increase approximately linearly with depth of cut. Once again, a greater rate of increase in cutting force than thrust force as well as excellent qualitative and quantitative agreement between predicted (from the model in Pramanik *et al.* [29]) and experimental results has been obtained. However, the model in [45] underestimates the cutting force within the tested range of depths.

The comparison between the predicted and experimental force results for varying cutting speed is presented in Figure 5.23. Unlike the variations of forces with feed and depth discussed above, in this case cutting and thrust forces can be seen to decrease approximately linearly with increasing speed with somewhat similar rates of decrease. Once again, excellent qualitative and quantitative agreement between predicted (from the model in Pramanik *et al.* [29]) and experimental results seems to exist. On the other hand, the predictions from the energy-based model [45] are not influenced by the cutting speed (hence the horizontal line) and the measured cutting forces are higher than the predictions for all speeds.

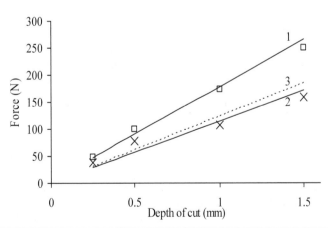

	Experimental symbol	Predicted lines for model	
		Ref. [29]	Ref. [45]
Cutting force	□	1	3
Thrust force	×	2	Unable to predict

Figure 5.22. Comparison of predicted and experimental forces with varying depth of cut (from Pramanik *et al.* [29])

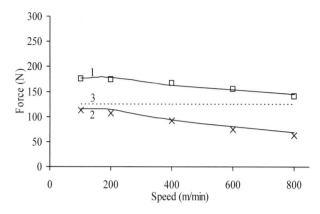

	Experimen-tal symbol	Predicted lines for model	
		Ref. [29]	Ref. [45]
Cutting force	□	1	3
Thrust force	×	2	Unable to predict

Figure 5.23. Comparison of predicted and experimental forces with varying cutting speed (from Pramanik *et al.* [29])

From the above comparisons between predictions (from the model in [29] and that in [45]) and experimental results, it is clear that the predictions made from the model in [29] show excellent quantitative/qualitative agreement with experimental results. However, the model in [45] can be seen to underestimate the predicted cutting force considerably. This underestimation is possibly due to the consideration of only the matrix alloy when estimating the specific energies due to plastic deformation in the primary and secondary shear zones. The influence of reinforcement particles was accounted for by considering the specific energy due to particle debonding. For more accurate predictions from this model, it may be necessary to consider the MMC (instead of the matrix alloy) when estimating the specific energies due to plastic deformation in the primary/secondary shear zones.

In an attempt to further verify their predictive method, Pramanik *et al.* [29] compared their predictions with the experimental results published in the literature. It was found that few experimental force results were available. Notably, Davim *et al.* [7], Chambers [6] and El-Gallab *et al.* [46] measured the forces when machining MMCs with PCD tools for a range of cutting speeds. However, none of these papers gave data in sufficient detail for comparison. The parameters used or applicable for the studies in [7, 6, 46] that were used for further verifying the model are listed in Table 5.1. Note that some of the parameters, which were not given in the above papers, have been assumed in accordance with general machining practice or taken from [29].

For the experimental conditions of Davim *et al.*, Chambers and El-Gallab *et al.*, all the components of cutting and thrust forces for different mechanisms were calculated and added according to the procedure described in the forgoing sections.

Table 5.1. Machining parameters used in the investigations in [7, 6, 46] (from Pramanik *et al.* [29])

Parameters	Davim *et al.* [7]	Chambers [6]	El-Gallab *et al.* [46]
Tool material	PCD	PCD	PCD
Nose radius, r_ε (mm)	0.8	1.6 (assumed)	1.6
Rake angle, γ (degrees)	0	0	0
Approach angle (degrees)	85 (assumed)	85 (assumed)	85
Cutting speeds (m/min)	250–700	50–300	670, 894
Feed, f (mm/rev)	0.1	0.2	0.25
Depth of cut, a (mm)	1	1	1.5
Workpiece material	A356-20%SiC-T6	A356-15%SiC	A356-20%SiC- T71
Yield strength of matrix (MPa)	138	138	124
Average particle diameter,d_p (μm)	20	22.5	12

Figure 5.24 compares the predicted cutting and thrust force results with experimental results obtained from [7, 6, 46]. The number on the abscissa shows the force component and reference from which the experimental forces were obtained. The model predictions are shown by bars while the experimental measurements are denoted by circles with variation ranges.

It can be seen that the agreement between the predicted and experimental cutting/thrust forces is very good for the conditions used by Davim *et al.* [7] and Chambers [6]. However, the model seems to have largely overestimated the cutting force for the conditions used by El-Gallab *et al.* [46]. This is most surprising when one considers the excellent agreement seen between the experimental and predicted results for the experimental conditions in [29] and those in [7, 6]. It is likely that there is an error either in the equipment used by El-Gallab *et al.* or in the experimental conditions or force results given in [46]. Attempts made to confirm this with the authors of [46] were not successful.

In spite of the simplifications made in developing the model in [29], the good agreement seen between the predictions and experimental results indicates that chip formation, ploughing and particle fracture and displacement are indeed the main factors that contribute to the cutting force generation. To investigate the variations of the force components (due to these factors) with feed and speed, cutting/thrust forces due to chip formation, particle fracture/debonding and ploughing were calculated for the MMC using the force data in Figure 5.16. The percentages of these forces in the cutting and thrust directions are presented against the feed and speed in Figures 5.25 and 5.26, respectively. It is found that the percentage of chip formation force is much higher (80–97%) compared to particle fracture/debonding (1.5–20%) and ploughing (0.25–2%) forces. The per-

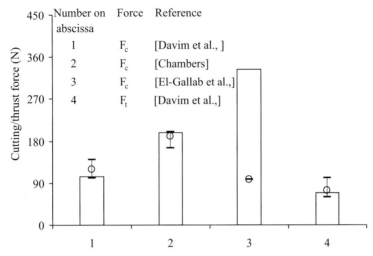

Figure 5.24. Comparison between predicted and experimentally measured forces from literature (from Pramanik *et al.* [29])

centages of the particle fracture/debonding and ploughing forces in the cutting direction decrease and the chip formation force increases with increasing feed (Figure 5.25(a)). The percentages of the particle fracture/debonding and ploughing forces are lower and higher, respectively, in the thrust direction compared to those in the cutting direction (Figure 5.25(b)). No significant change of the percentages of forces is noted with variations of the feed in the thrust direction. With variation of the speed, the percentages of the different forces in the cutting and thrust directions do not appear to vary (Figures 5.26(a and b)). The percentages of the particle fracture/debonding forces in the thrust direction are low compared to those in the cutting direction.

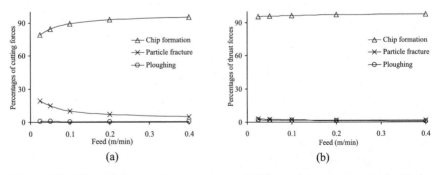

Figure 5.25. Effect of feed on the percentages of different force components in (a) the cutting and (b) the thrust directions (at a speed of 400 m/min and a depth of cut of 1 mm)

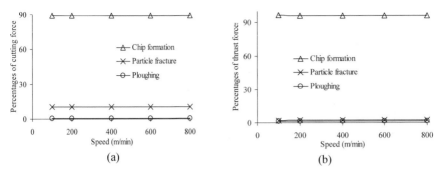

(a)   (b)

Figure 5.26. Effect of speed on the percentages of different force components in (a) the cutting and (b) the thrust directions (at a feed of 0.1 mm/rev and a depth of cut of 1 mm)

5.3.2   Tool–Particle Interaction

A detailed investigation of tool–particle interactions and their consequences are very important for a deeper understanding of process outcomes such as tool wear, surface finish, residual stresses, cutting forces, *etc*. Since analytical and/or experimental methods cannot be used for this type of investigation, Pramanik *et al*. [16] used the finite element method to simulate machining of an MMC. In this work, the development of stress/strain fields was explored for various tool–particle orientations and analysed for possible particle fracture, debonding, *etc*., to provide an insight into the mechanism of MMC machining. Their finite element model is given in Figure 5.27.

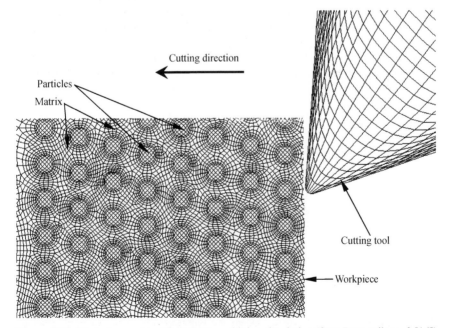

Figure 5.27. Workpiece and tool for MMC machining simulation (from Pramanik *et al*. [16])

This investigation revealed that the magnitude and distribution of stresses/strains in the MMC material and the interaction of particles with the cutting tool are the main reasons for particle fracture and debonding during machining of MMC (Figure 5.28). Additionally, the newly generated surfaces were found to be under compressive residual stress. These surfaces were damaged due to cavities left by the pull-out of particles. These cavities were formed when particles located at the lower part of the cutting edge interacted with the cutting tool. The indentation of particles (located immediately below the cutting edge) due to their interaction with the tool caused localized hardening of machined MMC surface. In these regions, the matrix was seen to deform plastically to a greater depth. High tool wear during machining of MMCs was considered to be due to the sliding of debonded particles over the cutting edge and tool faces that will scratch these contact surfaces (Figure 5.29).

As discussed in this section, during machining of MMCs, the resultant cutting force can be considered to consist of components due to chip formation, ploughing and particle fracture, and displacement. The calculation of these force components based on Merchant's shear plane analysis, slip-line field theory and Griffith theory, respectively, gave excellent predictions. They revealed that, the force due to chip formation is much higher than those due to ploughing and particle fracture. For both the MMC and the aluminium alloy matrix material, forces in turning increased with increasing feed. However, the speed did not affect the forces significantly for the MMC. On the other hand, forces for the matrix material initially

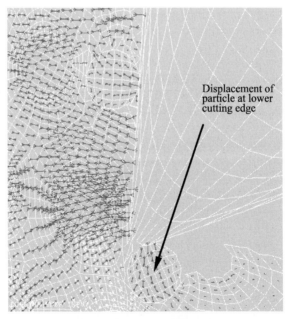

Figure 5.28. Evolution of stress fields for particles along the cutting path during machining of MMC. Compressive and tensile stresses are represented by >—< and <—> symbols, respectively. Their lengths represent comparative magnitudes (from Pramanik *et al.* [16])

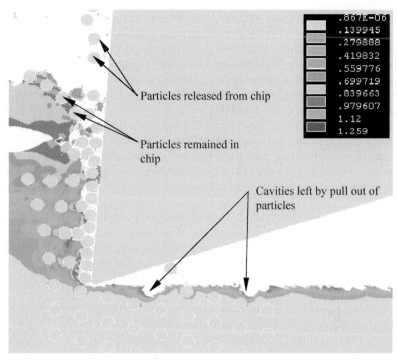

Figure 5.29. Distribution of von Mises strain during machining of MMC (from Pramanik *et al.* [16])

increased with speed but after a certain speed started to decrease. An FEM investigation revealed that the magnitude and distribution of stresses/strains in the MMC material and interaction of particles with the cutting tool are the main reasons for particle fracture and debonding during machining. The indentation of particles (located immediately below the cutting edge) due to their interaction with the tool causes localized hardening of the machined MMC surface. In these regions, the matrix can be seen to deform plastically to a greater depth.

5.4   Tool Wear

Comparatively high tool wear is one of the major concerns during processing of MMCs. Interaction of MMCs' very hard ceramic reinforcement particles with tool appears to be the major contributor to tool wear. High tool wear makes MMC processing very expensive and less efficient.

5.4.1 Performance of Cutting Tools

Performance of cutting tools depends on a number of factors such as tool material, tool geometry, machining conditions and workpiece material. Hence, choosing the right tool for machining any material to minimize the machining cost is a critical issue. It is common practice to perform experiments and select cutting tools and conditions for a particular application based mainly on tool wear and surface finish. For relatively new MMC materials, the above procedure was adopted for selecting suitable tools and machining conditions. Since hard ceramic reinforcements in the MMC have hardness comparable to some cutting tool materials, it is natural that MMCs are difficult to machine.

Comparatively high tool wear, due to tool–particle interactions, is affected by cutting parameters and the size and volume percentage of reinforcements. At high speed and low feed, tool–particle interactions become more frequent and substantial, which accelerates tool wear [4, 48, 49]. Tool wear is lower for small size and low volume percentage of reinforcement and high-hardness cutting tools [67, 68].

Researchers have tested almost all available tool materials to machine different types of MMCs. PCD shows comparatively better performance over any other tool material. However, because of the high cost of PCD tools, use of PCBN [69], CVD diamond-coated [69], TiCN/TiN-coated carbide [5], triple-layer TiC/Al2O3/TiCN-coated carbide [71] and uncoated carbide [5, 23, 72, 73] tools for machining of MMCs has been investigated.

It has been argued that the dominant wear mechanisms during machining of MMC are two- and three-body abrasions and that these are due to the presence of hard reinforced particles and debonded tool material grains [3, 4, 6, 7, 48, 67, 74, 76]. These cause tool pitting, chipping, microcracking and fatigue [75, 78]. Some researchers [74, 77] have argued that adhesion is also a cause of tool wear in addition to abrasion, as thin films of the workpiece material were found to be adhered to the worn areas. They considered that the abrasion was associated with micromechanical damage rather than microcutting. Chemical wear during machining of MMC has not been reported for any tool material. This is because the constituents of MMC, i.e., matrix material and reinforced particles (e.g., SiC and Al_2O_3) are chemically inert to almost all cutting tools (e.g., PCD, CVD-coated, carbide tools) under almost any conditions.

For coated/uncoated tungsten carbide tools, Arsecularatne et al. [79] pointed out that, when machining MMC work materials, tool wear is due to mechanical abrasion.

The increased thickness of diamond CVD coating increase the performance of the tool and the thick (500 μm) CVD diamond coating may be considered a competitor to PCD [77, 78]. Bonding between the substrate and the coating is critical to this tool performance. Coating failure is the dominant wear mechanism for diamond CVD-coated tools during machining of MMCs [70]. Coating failure accelerates at high temperature due to different thermal expansions between the coating and the substrate.

Most of the published research has investigated the performance of PCD tools during machining of MMCs. Wear mechanisms such as abrasion, adhesion and microcracking and fatigue have been used to explain tool wear [79]. This indicates that tool wear mechanism when machining of MMCs is not yet fully understood.

5.4.2 Modelling of Tool Wear

A number of attempts to model tool wear and correlate it with the machining parameters can be found. Most researchers developed a tool wear model assuming MMC to be a monolithic metal, *i.e.*, the influence of the particles was not considered explicitly in the model. They either extended Taylor's equation [80–82] based on experimential analysis or developed empirical relations [20, 83] of tool wear with machining parameters from experimental data. To date only four models are available where effects of particles are considered on tool wear. These assume abrasion to be the dominant wear mechanism and are described below.

Li *et al.* [68] developed a model to calculate the critical weight percentage of reinforcement for tool wear acceleration after a comprehensive study of the tool wear mechanism in cutting MMCs. Similar to Davim *et al.* [7] they argued that wear was due to two- and three-body abrasions, and that wear was accelerated by interference between the reinforcement particles that was associated with a critical weight percentage. Based on these observations, an analytical model was developed to predict the critical reinforcement weight ratio as a function of the densities of the reinforcement and the matrix and of the radius of the reinforcement particles and the cutting tool edge. They assumed that particles were spherical and that the maximum particle displacement was the radius (or function of radius) of the impacted reinforced particle, which occurs when the impact point is near the particle's centroid. A comparision between predictions and their experimental results showed reasonable agreement. The main disadvantage of this model is that it could not correlate tool wear with cutting conditions and tool material properties.

Kishawy *et al.* [84] also developed a model to predict flank wear rate considering two- and three-body abrasion between the MMC and tool, which were formulated using modified Archard's and Rabinowicz's relations, respectively. The rate of volume loss of the cutting tool was calculated by incorporating the contributions of the two types of abrasions by taking the root-mean-square value. The rate of volume loss was correlated with length of flank wear land by considering the geometry of the cutting tool. A comparison between predictions and their experimental results showed reasonable agreement.

Kannan *et al.* [85] proposed a tool wear model to predict flank wear based on Holm–Archard theory [86] by correlating hardness and energy consumption. The energy consumed for machining was determined in terms of the power required by the cutting system, tool volume loss, tool hardness and the properties of the ceramic particulate reinforcement. The power required by the cutting system was calculated as the product of the energy per unit volume of metal removed and the material removal rate. From these expressions the tool volume loss was calculated, which was correlated with the flank wear land width using a two-dimensional geometry of cutting tool.

Pedersen *et al.* [87] proposed an empirical wear model to predict flank wear based on the probability of four possible tool–particle interactions depending on the tool–particle orientations. Based on the probability of tool–particle interaction, it was assumed that only two orientations of these four take part in tool wear significantly. They accounted for workpiece and cutting tool properties in addition to the traditionally used cutting conditions. This model is an expansion of the traditionally extended Taylor's tool life equation based on Rabinowicz's wear equation [86]. A major disadvantage of this model is the large number of empirical constants/exponents which have to be determined experimentally, necessitating time-consuming and costly machining tests.

As discussed in this section, almost all available tool materials (*e.g.*, coated/uncoated tungsten carbide, PCBN and PCD) have been used to machine MMCs. Available experimental results show that the presence of reinforcement particles significantly influences tool wear. Wear of coated/uncoated tungsten carbide tools has been attributed to mechanical abrasion. On the other hand, wear mechanisms such as abrasion, adhesion, and microcracking and fatigue have been used to explain wear of PCD tools. This indicates that wear mechanism(s) of PCD tools are not yet fully understood. Some models are also available in the literature for predicting tool wear. Comparisons between model predictions and investigators' experimental results seem to show good agreement. However, a comprehensive comparison of predicted and experimental results including those available in the literature has yet to be carried out.

Acknowledgements

The authors wish to thank the Australian Research Council for financial assistance. A. P. has been supported by IPRS and IPA.

References

[1]   Rashad RM, El-Hossainy TM (2006) Machinability of 7116 structural aluminum alloy. Mater Manuf Process 21: 23–27

[2]   Durante S, Rutelli G, Rabezzana F (1997) Aluminum-based MMC machining with diamond-coated cutting tools. Surf Coatings Technol 94–95: 632–640

[3]   Heath PJ (2001) Developments in the applications of PCD tooling. J Mater Process Technol 116: 31–38

[4]   Davim JP (2002) Diamond tool performance in machining metal–matrix composites. J Mater Process Technol 128: 100–105

[5]   Pedersen W, Ramulu M (2006) Facing SiCp/Mg metal matrix composites with carbide tools. J Mater Process Technol 172: 417–423

[6]   Chambers AR (1996) The machinability of light alloy MMCs. Composites: Part A 27: 143–147

[7]   Davim JP, Baptista AM (2000) Relationship between cutting force and PCD cutting tool wear in machining silicon carbide reinforced aluminium. J Mater Process Technol 103: 417–423

[8]   Hung NP, Loh NL, Venkatesh VC (1999) Machining of metal matrix composites, in Machining of Ceramics and Composites: ed by S. Jahanmir, M. Ramulu and P. Koshy, 1999, Marcel Dekker, New York. Basel, ISBN 082470178X

[9]   Wang Z, Chen T, Lloyd DJ (1993) Stress distribution in particulate-reinforced metal-matrix composites subjected to external load. Metall Trans 24(A): 197–207

[10]   Huda D, El-Baradie MA, Hashmi MSJ (1994) Analytical study for the stress analysis of metal matix composites. J Mater Process Technol 45: 429–434

[11] Shi N, Arsenault RJ (1991) Analytical evaluation of the thermal residual stresses in Si/Al composites. JSME Int J 34(2): 143–155

[12] Zhao B, Liu CS, Zhu XS, Xu KW (2002) Research on the vibration cutting performance of particle reinforced metallic matrix composites SiCp/Al. J Mater Process Technol 129(1–3): 380–384

[13] Brun MK, Lee M, Gorsler F (1985) Wear characteristics of various hard materials for machining SiC-reinforced aluminum alloy. Wear 104(1): 21–29

[14] Narutaki N (1996) Machining of MMCs. VDI Berichte NR 1276: 359–370

[15] Clyne TW, Withers PJ (1993) An introduction to metal matrix composites, 1st edition. Cambridge University Press, ISBN 0-521-41808-9

[16] Pramanik A, Zhang LC, Arsecularatne JA (2007) An FEM investigation into the behaviour of metal matrix composites: tool-particle interaction during orthogonal cutting. Int J Mach Tools Manuf 47: 1497–1506

[17] Mussert KM, Vellinga WP, Bakker A, Zwaag SVD (2002) A nano-indentation study on the mechanical behaviour of the matrix material in an AA6061 – Al$_2$O$_3$ MMC. J Mater Sci 37(4): 789–794

[18] Leggoe JW (2004) Determination of the elastic modulus of microscale ceramic particles via nanoindentation. J Mater Res 19(8): 2437–2447

[19] Pramanik A, Zhang LC, Arsecularatne JA (2007) Micro-indentation of metal matrix composites – an FEM analysis. Key Eng Mater 340–341: 563–570

[20] Lin JT, Bhattacharyya D, Lane C (1995) Machinability of a silicon carbide reinforced aluminium metal matrix composite. Wear 181–183: 883–888

[21] Pramanik A, Zhang LC, Arsecularatne J. Machining of metal matrix composites: effect of ceramic particles on residual stress, surface roughness and chip formation. Int J Mach Tools Manuf (under review)

[22] Lin JT (1997) Bhattacharyya, D. and Ferguson, W.G., Chip formation in the machining of SiC particle reinforced aluminium matrix composites. Compos Sci Technol 58: 285–291

[23] Karthikeyan R, Ganesan G, Nagarazan RS, Pai BC (2001) A critical study on machining of Al/SiC composites. Mater Manuf Process 16(1): 47–60

[24] Joshi SS, Ramakrishnan N, Ramakrishnan P (2001) Micro-structural analysis of chip formation during orthogonal machining of Al/SiCp composites. J Eng Mater Technol 123: 315–321

[25] Hung NP, Yeo SH, Lee KK, Ng KJ (1998) Chip formation in machining particle-reinforced metal matrix composites. Mater Manuf Process 13(1): 85–100

[26] Ozcatalbas Y (2003a) Investigation of the machinability behaviour of Al$_4$C$_3$ reinforced Al-based composite produced by mechanical alloying technique. Compos Sci Technol 63(1): 53–61

[27] Ozcatalbas Y (2003b) Chip and built-up edge formation in the machining of in situ Al$_4$C$_3$–Al composite. Mater Des 24(3): 215–221

[28] Quan, YM, Zhou, ZH and Ye, BY (1999), Cutting process and chip appearance of aluminium matrix composites reinforced by SiC particle, J Mater Process Technol, 91(1), 231–235

[29] Pramanik A, Zhang LC, Arsecularatne JA (2006) Prediction of cutting forces in machining of Metal Matrix Composites. Int J Mach Tools Manuf 46: 1795–1803

[30] Dieter GE (1988) Mechanical Metallurgy, SI Metric Edition, McGraw-Hill, UK

[31] Clausen AH, Borvik T, Hopperstad OS, Benallal A (2004) Flow and fracture characteristics of aluminium alloy AA5083–H116 as function of strain rate, temperature and triaxiality. Mater Sci Eng A364: 260–272

[32] Wulf GL (1978) The high strain rate compression of 7039 aluminium. Int J Mech Sci 20(9): 609–615

[33] Smerd R, Winkler S, Salisbury C, Worswick M, Lloyd D, Finn M (2005) High strain rate tensile testing of automotive aluminium alloy sheet. Int J Impact Eng 32: 541–560

[34] Davim JP (2007) Application of Merchant theory in machining particulate metal matrix composites. Mater Des 10: 2684–2687

[35] Oxley PLB (1989) The mechanics of machining: an analytical approach to assessing machinability. Ellis Horwood, Chichester

[36] Jaspers SPFC, Dautzenberg JH (2002) Material behaviour in metal cutting: strains, strain rates and temperatures in chip formation. J Mater Process Technol 121: 123–135

[37] Li Y, Ramesha KT, Chin ESC (2004) The mechanical response of an A359/SiCp MMC and the A359 aluminium matrix to dynamic shearing deformations. Mater Sci Eng A 382: 162–170

[38] Li Y, Ramesh KT (1998) Influence of particle volume fraction, shape, and aspect ratio on the behaviour of particle-reinforced metal matrix composites at high rates of strain. Acta Mater 46(16): 5633–5646

[39] Li Y, Ramesh KT, Chin ESC (2000) The compressive viscoplastic response of an A359/SiCp metal-matrix composite and of the A359 aluminium alloy matrix. Int J Solids Struct 37:7547–7562

[40] Zhang ZF, Zhang LC, Mai YW (1995) Particle effects on friction and wear of aluminium matrix composites. J Mater Sci 30(23): 5999–6004

[41] Chichili DR, Ramesh KT (1995) Dynamic failure mechanisms in a 6061-T6 Al/Al$_2$O$_3$ metal-matrix composite. Int J Solids Struct 32(17–18): 2609–2626

[42] Yadav S, Chichili DR, Ramesh KT (1995) Mechanical response of a 6061-T6 Al/Al$_2$O$_3$ metal matrix composite at high rates of deformation. Acta Metall Materialia 43(12):4453

[43] Oxley PLB (1962) Shear angle solutions in orthogonal machining. Int J Mach Tool Des Res 2(3): 219–229

[44] Ng E, Aspinwall DK (2002) The effect of workpiece hardness and cutting speed on the machinability of AISI H13 hot work die steel when using PCBN tooling. Trans ASME 124:588–594

[45] Kishawy HA, Kannan S, Balazinski M (2004) An energy based analytical force model for orthogonal cutting of metal matrix composites. Ann CIRP 53(1): 91–94

[46] El-Gallab M, Sklad M (1998) Machining of Al/SiC particulate metal-matrix composites Part II: Workpiece surface integrity. J Mater Process Technol 83: 277–285

[47] Hong SY, Ding Y, Ekkens RG (1999) Improving low carbon steel chip breakability by cryogenic chip cooling. Int J Mach Tools Manuf 39: 1065–1085

[48] Ding X, Liew WYH, Liu XD (2005) Evaluation of machining performance of MMC with PCBN and PCD tools. Wear 259: 1225–1234

[49] El-Gallab M, Sklad M (1998b) Machining of Al/SiC particulate metal matrix composites. Part II: work surface integrity. J Mater Process Technol 83: 277–285

[50] Cheung CF, Chan KC, To S, Lee WB (2002) Effect of reinforcement in ultra-precision machining of Al6061/SiC metal matrix composites. Scripta Mater 4: 77–82

[51] Sahin Y, Kok M, Celik H (2002) Tool wear and surface roughness of Al$_2$O$_3$ particle-reinforced aluminium alloy composites. J Mater Process Technol 128: 280–291

[52] Monaghan J (1998b) Factors affecting the machinability of Al/SiC metal matrix composites. Key Eng Mater 138–140: 545–574

[53] Lin CB, Hung YW, Liu W-C, Kang S-W (2001) Machining and Fuidity of 356Al/SiC(p) composites. J Mater Process Technol 110: 152–159

[54] Capello E (2005) Residual stress in turning Part I: Influence of process parameters. J Mater Process Technol 160: 221–228

[55] El-Axir MH (2002) A method of modeling residual stress distribution in turning for different materials. Int J Mach Tools Manuf 42: 1055–1063

[56] Brinksmeier E, Cammett JT, Koenig W, Leskovar P, Peters J, Toenshoff HK (1982) Residual stresses – measurement and causes in machining processes. Ann CIRP 31(2): 491–510

[57] Kannan S, Kishawy HA (2006) Surface characteristics of machined aluminium metal matrix composites. Int J Mach Tools Manuf 46: 2017–2025

[58] Davim JP (2001) Turning particulate metal matrix composites: experimental study of the evolution of the cutting forces, tool wear and workpiece surface roughness with the cutting time. J Eng Manuf Proc Inst Mech Eng 215 B: 371–376

[59] Davim JP, Silva J, Baptista AM (2007b) Experimental cutting model of metal matrix composites (MMCs). J Mater Process Technol 183(2–3): 358–362

[60] Merchant ME (1944) Mechanics of the metal cutting process. I. Orthogonal cutting and type 2 chip. J Appl Phys 16: 267–275

[61] Lee EH, Shaffer BW (1951) The theory of plasticity applied to a problem of machining. J Appl Mech December: 15–20

[62] Kobayashi S, Thomsen EG (1959) Some observations on shearing process in metal cutting. Transactions ASME. J Eng Ind August: 251–261

[63] Pugh HLD (1958) Mechanics of cutting process, Proceedings of Conference on Technology of Engineering Manufacture, The Institute of Mechanical Engineers, p. 237–254

[64] Waldorf DJ (2004) A simplified model for ploughing forces in turning. Trans NAMRI SME 32: 447–454

[65] Yan C, Zhang LC (1995) Single-point scratching of 6061 Al Alloy reinforced by different ceramic particles. Appl Compos Mater 1:431–447

[66] Muller HK, Nau B (1998) Fluid Sealing Technology Marcel Dekker, Chapter 14, p. 294

[67] Yanming Q, Zehua Z (2000) Tool wear and its mechanism for cutting SiC particle-reinforced aluminium matrix composites. J Mater Process Technol 100(1): 194–199

[68] Li X, Seah WKH (2001) Tool wear acceleration in relation to workpiece reinforcement percentage in cutting of metal matrix composites. Wear 247(2): 161–171

[69] Sreejith PS (2006) Tool wear of binderless PCBN tool during machining of particulate reinforced MMC. Tribol Lett 22 (3): 265–269

[70] Chou YK, Liu J (2005) CVD diamond tool performance in metal matrix composite machining. Surf Coatings Technol 200: 1872–1878

[71] Ciftci I, Turker M, Ulvi S (2004) Evaluation of tool wear when machining SiC_p-reinforced Al-2014 alloy matrix composites. Mater Des 25: 251–255

[72] Varadarajana YS, Vijayaraghavan L, Krishnamurthy R (2006) Performance enhancement through microwave irradiation of K20 carbide tool machining Al/SiC metal matrix composite. J Mater Process Technol 173(2):185–193

[73] Kilickap E, Cakir O, Aksoy M, Inan A (2005) Study of tool wear and surface roughness in machining of homogenised SiC-p reinforced aluminium metal matrix composite. J Mater Process Technol 164–165: 862–867

[74] Hooper RM, Henshall JL, Klopfer A (1999) The wear of polycrystalline diamond tools used in the cutting of metal matrix composites. Int J Refractory Metals Hard Mater 17: 103–109

[75] El-Gallab M, Sklad M (2000) Machining of Al/SiC particulate metal matrix composites part III: comprehensive tool wear models comprehensive tool wear models. J Mater Process Technol 101:10–20

[76] Weinert K (1993) A consideration of tool wear mechanism when machining metal matrix composite (MMC). Ann CIRP 42: 95–98

[77] Andrewes CJE, Feng H-Y, Lau WM (2000) Machining of an aluminum/SiC composite using diamond inserts. J Mater Process Technol 102: 25–29

[78] D'Errico GE, Calzavarini R (2001) Turning of metal matrix composites. J Mater Process Technol 119: 257–260

[79] Arsecularatne JA, Zhang LC, Montross C (2006) Wear and tool life of tungsten carbide, PCBN and PCD cutting tools. Int J Mach Tools Manuf 46: 482–491

[80] Masounave J, Litwin J, Hamelin D (1994) Prediction of tool life in turning aluminium matrix composites. Mater Des15(5): 287–293

[81] Hung NP, Boey FYC, Khor KA, Phua YS, Lee HF (1996a) Machinability of aluminum alloys reinforced with silicon carbide particulates. J Mater Process Technol 56: 966–977

[82] Hung NP, Zhong CH (1996b) Cumulative tool wear in machining metal matrix composites Part I: Modelling. J Mater Process Technol 58: 109–1 13

[83] Davim JP (2003) Design of optimization of cutting parameters for turning metal matrix composites based on the orthogonal arrays. J Mater Process Technol 132(1–3): 340–344

[84] Kishawy HA, Kannan S, Balazinski M (2005) Analytical modeling of tool wear progression during turning particulate reinforced metal matrix composites. Ann CIRP, 54(1): 55–58

[85] Kannan S, Kishawy HA, Deiab IM, Surappa MK (2005) Modeling of tool flank wear progression during orthogonal machining of metal matrix composites, Trans North Am Manuf Res Inst SME NAMRC 33: 605–612

[86] Rabinowicz E (1995) Friction and Wear of Materials. Wiley Interscience, 2nd edn, pp. 169–170

[87] Pedersen WE, Ramulu M (2005) Proposed tool wear model for machining particle reinforced metal matrix composites. Trans North Am Manuf Res Inst SME Papers Presented at NAMRC 33: 549–556

6

Drilling Polymeric Matrix Composites

Edoardo Capello[1], Antonio Langella[2], Luigi Nele[2], Alfonso Paoletti[3], Loredana Santo[4], Vincenzo Tagliaferri[4]

[1]Dipartimento di Meccanica, Politecnico di Milano, piazza Leonardo da Vinci 32, 20133 Milano, Italy.
E-mail: edoardo.capello@polimi.it
[2]Department of Materials and Engineering Production, University of Naples Federico II, Piazzale Tecchio 32, 80125 Napoli, Italy.
E-mail: antgella@unina.it, luigi.nele@unina.it
[3]Università di L'Aquila, Piazza V. Rivera 1, 67100 L'Aquila, Italy.
E-mail: alpao@ing.univaq.it
[4]Dipartimento di Ingegneria Meccanica, Università di Roma "Tor Vergata", via del Politecnico 1, 00133 Roma, Italy.
E-mail: loredana.santo@uniroma2.it, tagliaferri@mec.uniroma2.it

This chapter presents the basics of drilling of polymeric matrix composites (PMCs). PMCs are becoming widely used in the manufacturing of products where a high mechanical strength must be accompanied by a low weight. However, the machining of PMCs implies coping with problems that are not encountered when machining other materials. Drilling is a particularly critical operation for PMCs laminates because the large concentrated forces generated can lead to widespread damage. This damage causes aesthetic problems but, more importantly, may compromise the mechanical properties of the finished part.

6.1 Introduction

6.1.1 What Are Polymeric Matrix Composites?

Composite materials are made up of two or more constituents [1–8]. The principle behind the study of composite materials and their applications is based on the possibility of using materials with specific characteristics which, when joined together in an appropriate manner, create a single *composite* material with final properties that are not found in any of the raw materials.

One of the constituents, called the *matrix*, usually determines the characteristics of the composite material primarily according to the field of application, the temperature of use and the weight. Its role is to contain another constituent, called the *reinforcement*, which determines the principal mechanical characteristics of the composite material, tensile modulus and mechanical resistance.

In nature, it is possible to find various examples of natural composite materials, such as bone or wood. An artificial composite material can be obtained using various types of matrix and with various types of reinforcement. The reinforcement may be in the form of filaments, called fibres (long or short), or in the form of particles. In theory, all materials can be used for the production of fibres, but in practice only a limited number of materials can be selected, and even fewer of these have those mechanical characteristics that are of interest in engineering applications.

Both matrix and fibres can be metallic, ceramic or polymeric, and the composite materials, classified according to matrix type, can be divided as follows:

- polymeric composites
- metallic composites
- ceramic composites

Polymeric matrix composite materials (PMCs) have a matrix which is made of a thermoset or a thermoplastic polymeric resin. Polymeric matrices are made up of rigid polymeric chains obtained from liquid monomers by means of a process of chemical transformation known as *polymerization*.

Thermoplastic matrices are polymers that have polymeric chains without transverse bonds and thus, after polymerization, an increase in temperature leads to a reduction in viscosity and to a melting of the matrix. The melting from a solid state develops through a gummy state, in a manner which is reversible.

Thermoset matrices are, on the other hand, characterized by the presence of strong transverse bonds which are created during polymerization, and which make the process not reversible. Indeed, an increase in temperature after polymerization does not initially generate substantial variations in the matrix properties, while beyond a certain limit it causes the total degradation of the matrix.

With the use of thermoplastic matrices, the component manufacturer uses a semi-finished product in which the polymerization reactions have already taken place and are guaranteed by the producer of the matrix. These semi-finished products need only to be modelled to obtain the required shape for the final component. As opposed to thermoset matrices, the application of thermoplastic matrices in composite materials has, in the past, been hindered by a low softening temperature, a lower tensile modulus and by reduced adhesion between fibres and matrix. Some types of thermoplastic matrices with properties similar to those of thermoset matrices are currently available, but with the prospect of lower transformation costs. The main thermoplastic matrices used in polymeric composite materials are: poly-ether-ether-ketones (PEEKs), poly-phenylene-sulphide (PPS), and poly-ether-imide (PEI).

With the use of thermoset matrices, the polymerization process must be carried out during the manufacturing cycle of the component, and this requires a rigid, more careful control of the process. The fact that thermoset matrices are so widely used is due to their strong mechanical properties and to the low cost of the matrices. Moreover, manual and semi-manual processing can be easily performed.

The most commonly used types of resin are polyester and epoxy. Unsaturated polyester resins have good mechanical characteristics, are easily workable and polymerize at room temperature. Unsaturated polyester resins can be classified according to the constitutional groups that appear in the chain structure. Among the most common are the ortophtalic polyesters, the isophtalics, the bisphenols and the dicyclopentadienyls.

Bicomponent or monocomponent epoxy resins have better mechanical properties compared to polyester resins, although they have a higher cost. Consequently, their application is generally limited to advanced applications in the aerospace industry.

The most commonly used types of reinforcement are glass, carbon and aramidic fibres.

The most common commercial formats of reinforcements for polymeric matrix composites are:

- **Mat**: The fibres are arranged on a plane in a random manner and are lightly compressed and treated with an adhesive. Alternatively, the unifilo mat can be used, in which a single thread is arranged on a plane in a casual manner. The mats are used to create quasi-isotropic composites with moderate mechanical characteristics;
- **Roving**: Long fibres are wound around bobbins, from which they are unwound in bundles and used as continuous fibres or cut to obtain short fibres;
- **Woven – fabric**: Continuous long fibres are braided together to form a bidimensional or tridimensional material;
- **Pre-pregs**: These are semi-finished products widely used in advanced applications, in which fibres have been impregnated with the right amount of thermoset resin and kept at low temperature, away from the light.

The processes used in the manufacture of polymeric composite materials allow the fibre and matrix volumes to be selected in such a way that particular requirements can be met. The quantity of fibres present usually reaches 60% of the available volume, while the matrix occupies the remaining 40%. The properties of the composite material obviously depend on the properties of the reinforcement and the matrix (Tables 6.1 and 6.2), and on the quantity of these.

A simple rule to predict in a first approximation a property of the composite is the so-called mixture rule which states that:

$$P_c = P_f \cdot V_f + P_m \cdot V_m \qquad (6.1)$$

in which P_c is the resulting property of the composite to be determined, P_m and P_f are the properties of the matrix and the reinforcement and V_m and V_f are the volumetric fractions of the matrix and reinforcement present, respectively.

The reinforcement and the matrix are arranged on a plane, forming a so-called *lamina* in which the fibres (short or continuous) are suspended in the matrix (Figure 6.1). The fibres may be placed casually or they may be aligned. The study of interaction between fibres and matrix is known as micromechanics.

Table 6.1. Properties of matrices commonly used from polymeric composite materials

Matrix material	Tensile modulus (E) (GPa)	Tensile strength (σ_u) (GPa)	Density (ρ) (g/cm^3)	Ultimate elongation (%)
Epoxy	2.75–4.10	55–130	1.2–1.3	4–8
Polyester	2.80–3.50	20–80	1.1–1.4	1,4–4,0
PEEK	3.2	100	130–1.32	50
PPS	3.3	83	1.36	4
PEI	3	105	1.27	60

Table 6.2. Properties of fibres commonly used from polymeric composite materials

Fibre material	Tensile modulus (E) (GPa)	Tensile strength (σ_u) (GPa)	Density (ρ) (g/cm^3)	Ultimate elongation (%)
Glass E	80 – 81	3.1–3.8	2.62	4.6
Glass S	88– 91	4.4–4.6	2,48	5.4–5.8
Carbon, low modulus [*]	170–241	1.4–3.1	1.90	0.9
Carbon, high modulus [*]	380–620	1.9–2.8	2.00	0.5
Aramidic (Kevlar 29)	83	3.6	1.44	4.0
Aramidic (Kevlar 49)	131	3.6–4.1	1.44	2.8

[*] Mesophase pitch-based carbon fibre

A lamina is assumed to behave orthotropically. The overlapping of several laminae with various fibre orientations makes up a laminate (Figure 6.2). The sequence of fibre orientations in the various laminae must be precisely defined as it influences the mechanical behaviour.

In fact, the overlapping of several orthotropic laminae can create a laminate with orthotropic or anisotropic behaviour, depending on the various orientations and the different positions within the thickness of the laminate. The study of the behaviour of laminates is known as macromechanics.

Figure 6.1. Fibre disposition in a lamina of a fibre-reinforced composite

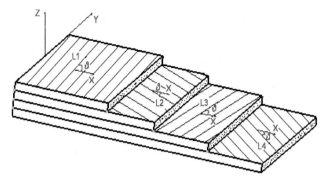

Figure 6.2. Overlapping of several laminae to form a laminate

Table 6.3. Properties of conventional structural materials and fibre composites with fibre volume fraction of 60%

Material	Tensile modulus (E) (GPa)	Tensile strength (σ_u) (GPa)	Density (ρ) (g/cm3)	Specific Modulus (E/ρ)	Specific strength (σ_u/ρ)	Ultimate elonga-tion (%)
Mild steel	210	0.45–0.83	7.8	26.9	0.058–0.106	38–50
Aluminium	70	0.26–0.41	2.7	25.5–27.0	0.096–0.152	30–40
E-Glass epoxy	21.5	0.57	1.97	10.9	0.26	2.5
E-Glass polyester	22.1	0.38	1.61	13.7	0,23	3.4
Kevlar 49–epoxy	40	0.65	1.4	29.0	0.46	1.8
Carbon fibre–epoxy	90	0.38	1.54	58.4	0.25	1.0

6.1.2   The Importance of Drilling

In most cases a composite part needs to be assembled to other parts, either in composite or in a different material (steel, aluminium alloys, wood, *etc.*). Since composite materials cannot be welded, and glueing is quite complex (and cannot be disassembled), mechanical joining (rods, pins, fasteners, rivets, *etc.*) is the solution commonly adopted to assemble a composite part to other parts (Figure 6.3).

The holes required by mechanical joining are generally drilled in the semi-finished composite part. This procedure is generally preferred to the one where a core is placed in the mould during the curing phase. Only large-diameter or complex contour holes are manufactured using this technique.

Since a composite part should perfectly match with the other parts during the assembly phase, holes must be placed in the exact position required, and must have the correct diameter. Moreover, due to their load transfer commitment once in use, holes generally undergo intense localised stress. Consequently, holes are generally subjected to specifications both in geometrical and mechanical terms.

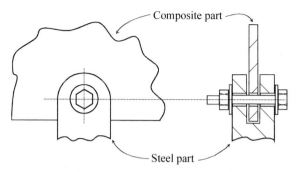

Figure 6.3. Mechanical joining of a composite part

Geometric specifications are the same as the usual given on holes drilled in other materials. They generally consist of specifications on dimension (diameter), position (of the hole centre) and shape (roundness, cylindricity and straightness). Microgeometry (*i.e.*, roughness) con be subject to a specification too, and is generally given in terms of the roughness of the cylindrical surface of the hole. Roughness, cylindricity and straightness specifications are seldom used on thin composite laminates. All these specifications must be reported on the mechanical drawings following the active ISO standards.

The mechanical properties that the material presents around the drilled hole are generally different from (*i.e.*, lower than) the ones that can be found far from the hole. This is because drilling is an invasive process and the material around the hole undergoes structural damage that will be described later. The term *residual mechanical properties* is generally used to describe the mechanical properties that the material around the hole presents after the drilling process.

The residual mechanical properties strongly depend on *how* the hole has been drilled. Abusive parameters can deeply damage the material around the hole, thereby leading to limited residual mechanical properties and a hole that cannot withstand the required mechanical load and will fail once in use.

Mechanical specifications are more seldom used in composite manufacturing, and no standards exist for their expression on mechanical drawings. Generally speaking, the material around a hole must withstand a sufficient static load and have an adequate fatigue limit. These characteristics must be present both at the end of the manufacturing phase and during the whole life of the part.

To summarise, it can be stated that drilling conditions and parameters strongly influence the quality of the drilling process from both the mechanical and geometrical point of view. Poor hole quality was estimated to cause 60% of composite part rejection during quality control in the composite manufacturing industry. This rejection is particularly severe from the economic point of view since it occurs in the last phases of production when the manufacturing cost of the composite part has already been faced.

Consequently, a carefully designed drilling process is the first step to obtaining an economic composite part that can be assembled and is fail safe.

6.2   Drilling Technology of Polymeric Matrix Composites

6.2.1   Conventional Drilling Process

6.2.1.1 The Twist Drill

The most commonly used tool in the conventional drilling of composite materials is the twist drill (Figure 6.4), generally obtained in high-speed steel (HSS). The twist drill [19–21] is made up of a cylindrical shank into which two opposite helical grooves have been cut, forming two cutting lips at the end surface (AB and CE). A central chisel edge (AC) is present near the drill axis to connect the two cutting lips.

The flank (α) and rake (γ) angles (described in Chapter 2) play an important role in PMCs drilling, as they influence the hole quality.

In fact in drilling the angles α_f and γ_f observed in the feed plane of the tool-in-hand reference system (Figure 2.4) may differ substantially from the ones (α and γ) observed in the tool-in-use reference system.

Considering Figure 6.5, it can be observed that a generic point on the cutting lip rotates with a tangential speed of v_t [m/s]:

$$v_t = \omega \cdot r = \frac{2\pi}{60 \cdot 1000} n \cdot r \qquad (6.2)$$

where ω [rad/s] is the rotation velocity, n is the revolution per minute [rpm] and r [mm] is the distance of the considered point from the drill axis. Therefore, it can be observed that in drilling the cutting speed varies along the cutting lip: it is at its maximum value on the periphery of the drill and decreases to zero on the axis.

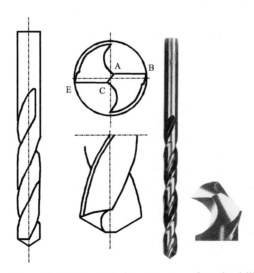

Figure 6.4. Characteristic parameters of a twist drill

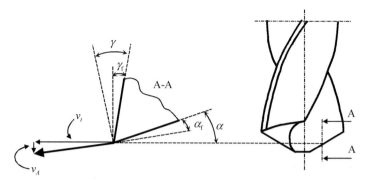

Figure 6.5. Relief and rake angles of a twist drill cutting lip

The same lip point is fed along the drill axis at a speed v_A [m/min] given by:

$$v_A = \frac{1}{60 \cdot 1000} f \cdot n \qquad (6.2)$$

where f [mm/rev] is the feed rate.

It can be easily observed that that the tool-in-use angles are related to the tool-in-hand feed rate plane angles through the following relationship (Figure 6.5):

$$\gamma = \gamma_f + tan\left(\frac{v_A}{v_t}\right) = \gamma_f + tan\left(\frac{f}{2\pi r}\right) \qquad (6.3)$$

$$\alpha = \alpha_f - tan\left(\frac{v_A}{v_t}\right) = \alpha_f - tan\left(\frac{f}{2\pi r}\right) \qquad (6.4)$$

As the considered point on the cutting lip is moved closer to the drill axis, r becomes small, and special care must be given to avoid negative values of the flank angle α. If the flank angle becomes negative there is no longer a cutting action by the lips but the twist drill acts like a punch. In PMCs drilling with punching action is particularly dangerous as it leads to the damage of the material, which dramatically reduces the structural integrity of the material around the hole.

Negative values are generally avoided by adopting a proper tool geometry (α_f increases from the periphery to the axis while γ_f decreases) and by selecting low feed rates.

6.2.1.2 Wear of the Twist Drill

One of the main limitations when drilling PMCs with the conventional HSS twist drill is the excessive wear experienced by the tool. In fact, while a HSS twist drill can be used to drill hundreds of holes in carbon steel before being worn out, in PMCs drilling the same drill may last for fewer than ten holes.

This rapid wear is due to the abrasive nature of the reinforcement fibres, and is present regardless of the shape or length of the fibres. The wear generally increases with the fibre volume fraction and hardness.

Tool wear has a significant effect on hole damage as the thrust force increases as tool wear proceeds.

6.2.1.3 Thrust Force

During drilling, a vertical force, that is, a thrust force, is generated. This thrust force can be considered as the sum of several components, each one rising either from the cutting process or from the friction between material and cutting tool (Figure 6.6).

The cutting process occurs along the cutting lips and at the chisel edge. The cutting process along the lips generates a force on each lip that has a component $F_{cl,A}$ parallel to the axis of the drill, that is, the feed direction. Moreover, the chisel edge generates a vertical penetration force, called F_{ce}.

The friction forces arise from two components. The first is related to the friction between the side surface of the tool and the generated hole surface, which leads to the vertical force F_{ss}. The second component is related to the friction between the chip flow along the helical grooves, which generates the vertical force F_{hg}.

The total axial thrust force acting on the drill is therefore:

$$F_A = 2F_{cl,A} + F_{ce} + F_{ss} + F_{hg} \tag{6.5}$$

as $F_{cl,A}$ is generated on both cutting lips.

The thrust force observed during drilling not only depends on the geometry of the drill and on the type of material and laminate being worked upon, but also on the relationship between the feed rate and the cutting speed, as well as on the degree of wear of the drill [22–24].

In Figure 6.7 a qualitative trend of thrust force F_A as a function of the drilling time t is shown. As can be seen, most of the time the thrust force is positive, that is, a pushing action is exerted by the drill on the workpiece. In the first period the thrust force continues to increase as an increasing part of the cutting lips is engaged in the material; in the second phase the thrust force remains at an almost

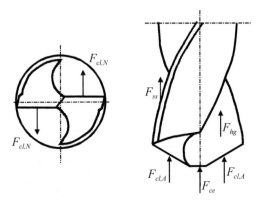

Figure 6.6. Thrust forces during drilling

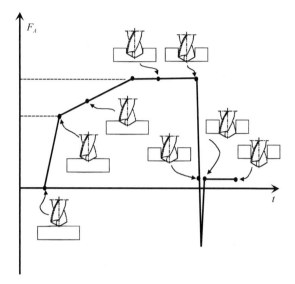

Figure 6.7. General trend of the thrust force as a function of drilling time

constant value as the drill sinks into the workpiece. In the third phase the thrust force rapidly decreases when the twist drill exits, sometimes causing a negative thrust force, that is, a pulling force.

The actual level of the force depends on the material being drilled, on tool geometry, material and wear, and on process parameters. An experimental measure of the thrust force as a function of feed rate is reported in Figure 6.8 [25, 26].

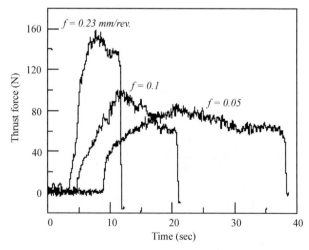

Figure 6.8. Experimental trend of the thrust force as a function of drilling time and feed rate [26]

6.2.1.4 Torque

During drilling also a torque T is generated, once again being due to the cutting process and to friction.

The cutting torque is due to the horizontal component $F_{cl,N}$ (normal to the cutting lip, see Figure 6.6) of the cutting force. This force is applied somewhere along the cutting lip; as a first approximation it can be supposed to be applied at half the radius. Therefore, the cutting torque is:

$$T_c = 2\frac{r}{2}F_{cl,N} = r \cdot F_{cl,N} \tag{6.6}$$

The frictional torques are due to the torque generated by the friction between the chisel edge and the workpiece (T_{ce}), the friction between the side surface of the drill and the inside hole surface (T_{ss}) and to the friction of the chip flow along the two helical grooves (T_{hg}).

The total torque value is:

$$T = T_c + T_{ce} + T_{ss} + T_{hg} \tag{6.7}$$

A typical trend of torque as a function of drilling time is shown in Figure 6.9.

As can be seen, the torque initially increases rapidly in a linear manner up to the value T_i [25, 26]. This is due to the fact that an increasing part of the cutting lips is involved in the process This is the part of the torque which is effectively due to the cutting operation.

Subsequently, a further but less steep linear increase in torque can be noted up to value T_{max}. This increase is mainly due to the torque which is generated by friction between the lateral surface of the drill and the inside surface of the hole.

In the final phase of the process, when the drill breaks through the lower surface of the workpiece, the only remaining torque is related to friction between the

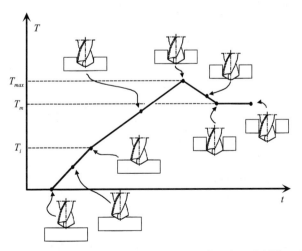

Figure 6.9. Trend of torque during drilling as a function of drilling time

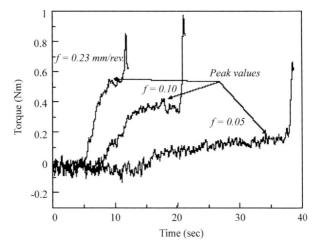

Figure 6.10. Experimental measure of torque during drilling as a function of drilling time and feed rate [26]

lateral surface of the drill and the inside surface of the hole. Consequently, the torque decreases to a value of T_m and remains constant.

As for the thrust force, the actual torque depends on the material being drilled, on several process parameters, on the tool shape, material and wear, and on the fixture system. A torque measurement obtained at different feed levels is reported in Figure 6.10.

6.2.2 Unconventional Drilling Processes

The main unconventional processes that can be used in composite drilling are laser cutting, water jet (WJ) cutting and abrasive water jet (AWJ) cutting.

In laser cutting, CO_2 and Nd:YAG resonators are generally used (at wavelengths of 10.6 μm and 1.064 μm, respectively), even though the CO_2 resonators are far more frequently applied, as the 10.6 μm wavelength is very well absorbed by polymer-based materials. Typical power ranges from 50 W to 1 kW. The generated laser beam is focussed on the surface with a spot of about 0.2 mm diameter, therefore yielding a power density (power per unit of exposed surface) as high as 10^2–10^4 W/cm². This power density very rapidly heats the exposed material, which generally vaporises in less than ten milliseconds. Therefore, laser cutting is basically a thermal process.

If the diameter of the hole to be drilled is small (say from 0.1 mm up to 2 mm), the process can be performed without relative movement between the beam and the workpiece. For larger hole diameters a hole is first pierced in the material and then the beam contours the circular profile of the hole. This is actually a cutting action of the beam.

Although interesting for the absence of contact between tool and workpiece, which leads to the absence of cutting forces or tool wear, laser cutting presents some drawbacks that must be carefully considered.

The main limit of laser processing is that the maximum thickness that can be cut with reasonable quality is about 5–10 mm. Above this thickness an evident taper can be observed. Moreover, since laser cutting is a thermal process, thermal damage can be experienced around the drilled hole. Furthermore, since composites are actually made of two different materials that present different thermal properties (matrix and reinforcement), poor hole quality can be generally observed.

In WJ cutting a highly pressurized jet (up to 500 MPa) is formed through a sapphire orifice (diameter 0.05–0.3 mm) and can travel at close to twice the speed of sound in air (about 500 m/s). The formed jet directly impinges on the composite surface and removes material through a set of complex physical mechanisms, where cavitation plays the main role.

The hole pierced by WJ cutting is generally too small for many mechanical fasteners, as the diameter of the hole is the same size as the orifice diameter. Therefore, holes are generally machined using a contour technique.

The main advantages of WJ cutting derive from the absence of tool wear and from the very limited force exerted by the jet on the workpiece. On the other hand, some delamination may occur, especially when piercing a composite laminate.

AWJ is similar to WJ, but the water jet is mixed with a flow of natural or synthetic abrasive particles (mass flow rate 100–800 g/min). The abrasive particles are then accelerated inside a carbide nozzle (diameter about 1 mm) by the water jet, and the speed of the particles is believed to reach a speed slightly below the speed of sound (about 250 m/s). The result is a jet of water and abrasive particles that can cut the composite through an erosion mechanism (it can actually cut almost any material, ranging, for example, from steel to concrete).

The maximum thickness that can be drilled and cut can be much higher than in laser or in WJ cutting (up to say 200 mm) and very limited delamination is observed even in piercing.

Apart from drilling, these energy beam processes can also be used to trim the composite contour after curing or to cut slots in the manufactured part (such as, the visor window in a composite motorcycle helmet).

6.3   Modelling of Conventional Drilling

6.3.1   The Need for Modelling

As will be discussed later, several types of damages that can be observed in composite drilling can be directly related to cutting force and torque. In particular, delamination can be related to the thrust force during drilling. Therefore, it is particularly important to model the cutting action and to derive an analytical model that predicts the thrust force as a function of process parameters.

The cutting action of a twist drill is a complex process of oblique three-dimensional cutting. The cutting speed and the rake and relief angles vary with the radial distance ρ along the cutting lips of the drill, therefore the process conditions vary noticeably along the cutting lips.

6.3.2 Cutting Force Modelling

Many studies are available in literature dealing with the modelling of the metal drilling process, and reference to Merchant's shear plane model [9] is present in all of them. Unfortunately, all models developed for metals have proved to be unsuitable for composites, as the cutting mechanism is different [10–15]. Consequently, a tailored set of model for drilling PMCs had to be developed.

Since it was experimentally observed that $2F_{cl,A} > F_{ce} \gg F_{ss}$; F_{hg} , only the action of the cutting lips and of the chisel edge is generally considered in modelling, and the other two forces are neglected.

Since the cutting process varies along the cutting lip, each cutting lip can be described as being composed of several infinitesimal elements aside one other. On each element a orthogonal cutting process occurs, which is characterised by different process parameters and different forces.

The cutting force per unit of lip length generated in the orthogonal cutting of composite materials can be divided into two components, one parallel to the cutting velocity (δF_{vt}), the other (δF_{vn}) perpendicular to it (Figure 6.11). These two components can be predicted using the following experimental relationships [16–18]:

$$\delta F_{vn} = B \cdot 10^{-a\gamma} \cdot t_1^{0.5} \tag{6.8}$$

$$\delta F_{vt} = A + B \cdot 10^{-b\gamma} \cdot t_1 \tag{6.9}$$

where t_1 is the thickness of the uncut chip and A, B, a and b are four experimental coefficients that mainly depend on the machined material.

The $F_{cl,N}$ force, acting normal to the cutting lip, and the $F_{cl,A}$ force, acting parallel to the twist drill axis, can be obtained by adding together the contributions of the unitary forces δF_{vt} and δF_{vt}, using an integration operation along the whole cutting lip [15,16] (Figure 6.12):

$$F_A = 2 \cdot \int_{\tau}^{1} \delta F_{vn} \left(1 - \frac{w^2 \sin^2(\varepsilon/2)}{2\rho^2 R^2} \right) R \sin(\varepsilon/2) d\rho$$
$$= 2 \cdot \int_{\tau}^{1} B \cdot 10^{-a\gamma} (f/2)^{0.5} \left(1 - \frac{w^2 \sin^2(\varepsilon/2)}{2\rho^2 R^2} \right) R \sin(\varepsilon/2) d\rho \tag{6.10}$$

$$T = 2 \cdot \int_{\tau}^{1} \delta F_{vt} \left(1 - \frac{w^2 \sin^2(\varepsilon/2)}{2\rho^2 R^2} \right) \rho R^2 d\rho$$
$$= 2 \cdot \int_{\tau}^{1} [A + B \cdot 10^{-b\gamma} (f/2)] \left(1 - \frac{w^2 \sin^2(\varepsilon/2)}{2\rho^2 R^2} \right) \rho R^2 d\rho \tag{6.11}$$

where the integration variable is $\rho = r/R$ and the integration limits can be obtained from Figure 6.13.

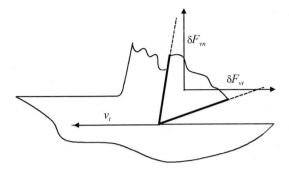

Figure 6.11. Orthogonal cutting condition and forces per unit of cutting lip length

The actual rake angle, γ, is given by the sum of two addends as described by Equation (6.3). Since the closed analytical solution of Equations (6.10) and (6.11) is too complex, it is possible to adopt an average rake angle, γ_m, expressed as:

$$\gamma_m = \frac{\int_{\tau}^{1}\left(\gamma_f + tan^{-1}\left(\frac{f}{2\pi\rho R}\right)\right)d\rho}{\int_{\tau}^{1}d\rho} \tag{6.12}$$

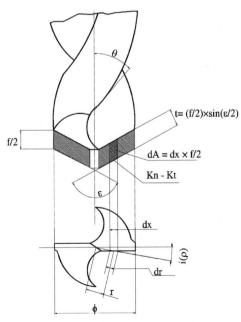

Figure 6.12. Twist drill and relevant parameters

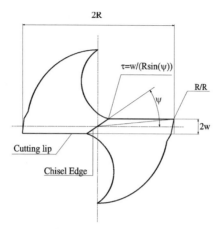

Figure 6.13. Twist drill seen from below

Thus, the following quantities con be determined:

$$G = \int_\tau^1 \left[1 - \frac{w^2 \sin^2(\varepsilon/2)}{2\rho^2 R}\right] R^2 \rho d\rho = \frac{1}{2}\left[(1-\tau^2)R^2 + w^2 \sin^2\left(\frac{\varepsilon}{2}\right) \ln\tau\right] \quad (6.13)$$

$$G' = \int_\tau^1 \left[1 - \frac{w^2 \sin^2(\varepsilon/2)}{2\rho^2 R}\right] R\sin(\varepsilon/2)d\rho = \frac{\sin\left(\dfrac{\varepsilon}{2}\right)(1-\tau)\left[2\tau R^2 - w^2 \sin\left(\dfrac{\varepsilon}{2}\right)\right]}{2\tau R}$$

$$(6.14)$$

The value of the thrust force and of the torque can be calculated using the simplified relations:

$$T' = 2 \cdot B \cdot 10^{-b\gamma_m} \cdot (f/2)^{0.5} \cdot G' \quad (6.15)$$

$$M' = 2 \cdot \left[A + B \cdot 10^{-b\gamma_m} \cdot (f/2)\right] \cdot G \quad (6.16)$$

In a similar way it is possible to evaluate the contribution of the chisel edge to the thrust force (the contribution to the torque is negligible) [16]:

$$T = C \cdot 10^{-c \cdot \gamma_{chisel}} \cdot f^{0.5} \cdot 2w \quad (6.17)$$

where C and c are two experimental coefficients.

The identified model was verified for glass fiber reinforced plastic (GFRP) laminates and theoretical values have shown a good correspondence to the experimental ones, as can be seen in Figure 6.14 [16].

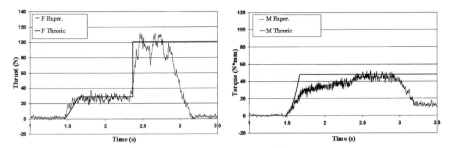

Figure 6.14. Typical run of the theoretical–experimental comparison of thrust and torque – $n = 1250$ rpm; $f = 0.25$ mm/rev; $\varepsilon = 140°$; $\theta = 30°$

6.4 Damage Generated During Drilling and Residual Mechanical Properties

6.4.1 Structural Damage

The main types of structural damage generated during drilling of polymeric matrix composites are delamination, microcracks, fibre–matrix debonding, matrix cratering and thermal damage[1]. The presence and extension of the different kinds of damage depend on the composite material characteristics, tool geometry and material, and the process parameters [27–36]. The damage is particularly detrimental to the residual mechanical properties and significantly reduces the composite performance in use. Consequently, special care should be given to avoid the generation of defects during drilling.

6.4.1.1 Damage Evaluation Methods

Different techniques and parameters can be used to assess the damage caused by drilling [27, 33, 39, 40]. Sophisticated non-destructive techniques can be employed in addition to low-magnification microscopy in order to identify internal defects, while destructive techniques are rarely employed. Moreover, a scanning electron microscope can be used to observe the cut surface morphology [28,29].

There are various methods for delamination evaluation. On semi-transparent composites, a coloured liquid can be used to penetrate the material through the cut surface. The extent of the damage around the hole is highlighted by the contrasting colours and can be measured using an optical microscope. In [39] an ultrasonic C-scan technique is used, but X-ray computerized tomography (CT) can also be successfully used, as mentioned by the author.

In [33] the principal parameters used to characterize delamination after drilling are discussed. One group of authors tends to use dimensional parameters, such as delamination area or length, while another proposes non-dimensional parameters

---

[1] Unfortunately there is no consistency in the terminology used in the literature, and care should be taken to compare the damage described by different authors.

(ratio of damage area to the hole area, ratio of the drill radius to the delamination radius, *etc.*).

6.4.1.2 Delamination

Delamination is the damage which is most evident after drilling composite materials. It consists of local debonding around the drilled hole of one or more plies (Figure 6.15).

Delamination is commonly classified as peel-up delamination at the twist drill entrance and push-down delamination at the twist drill exit (Figure 6.16).

Peel-up delamination at drill entry is not always present. Push-down delamination is generally more extensive and is consequently considered to be the most dangerous. The last plies of composite tend to open up as the drill pierces through the laminate while generating a hole. This happens because the last lamina undergo a push-down action, thus debonding the plies interface.

Several phenomena contribute to the delamination mechanism. It is assumed that delamination mainly depends on the thrust force exerted by the drill point. The process parameters play a main role in determining the thrust force, and consequently the extent of delamination.

The influence of the drilling parameters and tool geometry has been assessed for different fibre-reinforced composites.

Caprino and Tagliaferri [29] assert that the damages observed after drilling GFRP with an HSS drill are strongly affected mainly by the feed rate. Furthermore, Tagliaferri *et al.* [28] indicate that the GFRP damage depends on the ratio between rotation speed and feed rate, irrespective of reinforcement form, resin type and fabrication method.

Capello and Tagliaferri [32] show that the degree of peel-up delamination depends on the feed rate and on the helix angle of the twist drill. Push-down delamination is mainly affected by the feed rate, by the presence of a support beneath the specimen and by the twist drill temperature.

In aramid composite drilling, a linear relationship exists between feed rate, thrust force and push down delamination, as proposed by Veniali *et al.* [31]. With regard to the tool geometry, the chisel edge width seems to be the most important factor that contributes to the thrust force and hence to the occurrence of delamination, as shown by Jain and Yang [36]. The point angle is of minor significance.

Figure 6.15. Delamination around a drilled hole: bottom view (left) and cross section (right)

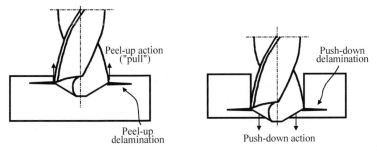

Figure 6.16. Delamination at the twist drill entrance (left) and exit (right) when drilling PMCs laminate

Davim and Reis [45] studied the effect of two distinct geometries of cemented tungsten carbide drills on carbon-fibre-reinforced plastic (CFRP) laminates. The authors concluded that delamination at the drill entry and exit are affected by distinct parameters, *i.e.*, at drill entry feed rate was the most significant factor affecting delamination whereas at the drill exit, delamination was primarily affected by cutting speed.

6.4.1.3 Modelling Delamination

A model predicting push-down delamination was identified using the classical plate-bending theory and linear elastic fracture mechanics [40–44].

The model describes the last stage of drilling as schematically reported in Figure 6.12. At this point the twist drill exerts a force (thrust force F_A) on the last lamina, which can be considered as a circular plate subject to a point central load.

When the drill is fed downwards, the delamination propagates. In energetic terms, the work done by the thrust force is used to bend the free part of the last lamina (similar to a circular plate) and to widen the intra-lamina crack, that is, the push-down delamination.

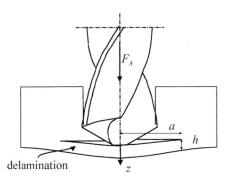

Figure 6.17. Circular plate model for delamination analysis

Therefore, the energy balance equation can be written as:

$$G_{IC}dA = F_A dz - dU \qquad (6.18)$$

where z is the vertical displacement, dU is the infinitesimal strain energy, dA is the increase in the area of the delamination crack and G_{IC} the critical crack propagation energy per unit area in mode I.

If an isotropic behaviour and a pure bending of the laminate are assumed, for a circular plate clamped at its periphery the stored strain energy U is [44]

$$U = \frac{Eh^3}{12(1-v^2)}\frac{8\pi z^2}{a^2} = M\frac{8\pi z^2}{a^2} \qquad (6.19)$$

where E is the modulus of elasticity, v is Poisson's ratio and M is the flexural rigidity of the plate.

The thrust force at the onset of crack propagation can be calculated as:

$$F_{A,th} = \pi\left[\frac{8G_{IC}Eh^3}{3(1-v^2)}\right] = \pi\sqrt{32G_{IC}M} \qquad (6.20)$$

where h is the uncut thickness of the last lamina.

In order to avoid delamination, the applied thrust force should not exceed this threshold value, which is a function of the material properties and the uncut thickness. If the thrust force exceeds the $F_{A,th}$ value, delamination occurs and propagates.

This model was later extended to various twist drill types, such as saw drill, candle stick drill, core drill and step drill [48, 49], where the load exerted by the drill is distributed in a different way.

6.4.1.4 Other Types of Damage

Cratering and thermal alteration of the matrix, pull-out and fuzzing of the fibres and intra-lamina cracks are reported in the literature as other types of damage occurring in a composite material subjected to drilling [29, 30, 32].

The thermal alterations in both fibre and matrix, generally limited to a small volume around the hole, are related to the energy converted into heat by frictional forces. The extent of this thermal damage depends on the thermal conductivity of fibre and matrix. For instance, low thermal conductivity in aramide/epoxy laminates promotes the increase of temperature and possible thermal damage of the material around the hole [30].

Fibre pull-out and fuzzing can be present along the entire hole surface and mainly depend on the fibre orientation and on the feed rate, as well as on the tool geometry.

6.4.2   Residual Mechanical Properties

The damage generated during drilling in a PMC laminate is particularly detrimental to the residual mechanical properties.

In order to obtain a fail-safe composite part the relationship between the drilling parameters and the different kinds of damage must be understood, and the influence of these kinds of damage on the residual mechanical behaviour must be identified.

Some mechanical tests both under static and cyclic load conditions are proposed in the literature to evaluate these relationships [28, 38, 49].

6.4.2.1 Evaluation Methods of Residual Mechanical Properties

The most commonly used method for the evaluation of the residual mechanical properties is the tensile test, *i.e.,* the static and cyclic bearing load tests and the fatigue test.

There are two ways of testing the residual mechanical properties of a drilled composite. One way, referred to as a load test, is to pull a rectangular specimen which has a hole drilled in the centre, at both short ends. The other method, generally referred to as a bearing load test, is to place a pin in the drilled hole and to apply a pulling force to the pin and to one of the short ends (Figure 6.18). In the bearing load test the hole undergoes a localized stress similar to the stress it would probably undergo in use. The result of this test is more sensitive to the level of damage generated during drilling than the results of the conventional load test.

Furthermore, in both tests, the load can be either constant or cyclic. If the load is constant (static bearing load (SBL)), the specimen is elongated until breakage occurs or a given displacement has been imposed. If the load is cyclic (cyclic bearing load (CBL)), it is varied, generally following a sine wave, again until breakage occurs or a given displacement has been imposed.

6.4.2.2 Static and Fatigue Resistance

Several data are available concerning the static and fatigue resistance of GFRP laminate. The results discussed in [28] show that static tensile strength of specimens with a hole is not affected by drilling conditions and consequently by damage, whereas a strong decrease in bearing strength could be observed as the damage around is increased.

In [38] the results of tests with static and cyclic bearing load show that the main cause of mechanical failure is the microdamage generated at the inner part of the hole surface, the influence of delamination being less evident. This underlines the

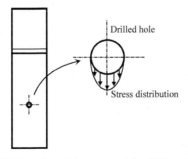

Figure 6.18. Experimental setup used for SBL and CBL tests

role played by the feed rate, which is the most important process parameter that promotes the generation of microcracks along the whole hole section.

The presence of a support under the material positively affects only the SBL but has a mild effect on the CBL mechanical behaviour. The twist drill preheating, even if it reduces the size of delamination [32], promotes thermal alteration uniformly distributed in the hole section, and therefore has a negative effect in both SBL and CBL.

In [49] the influence of different drilling tool on the fatigue behaviour is discussed. In particular the S–N curves are influenced by the used tool and then show a direct correlation between the damage and the fatigue behaviour of the material around the hole.

6.5   Damage Suppression Methods

6.5.1   Introduction

Damage reduction or suppression can be achieved through:

- a careful selection of process parameter
- an enhancement of drilling conditions
- the use of a special tool.

A careful selection of process parameters is the first step to reducing drilling damage, as this can be done without any particular effort. Drilling conditions may also be varied, but this is not always feasible. It may not be possible to place a predrilled support under the workpiece and align it perfectly with the twist drill axis. Lastly, the standard HSS twist drill tool might be substituted by tools in different materials and with optimised geometries, but these are generally more expensive [34, 51, 52].

6.5.2   Process Parameters Selection

The two parameters that significantly influence the drilling process, and, in particular, the quality of the obtained holes, are the cutting speed v_t and the feed rate f [60, 61].

An increase in cutting speed results in lower thrust force and torque due to the higher temperatures reached by the tool and the machined material. On the other hand, higher twist drill temperatures negatively influence the internal quality of the hole and the residual mechanical properties.

Large values of feed rate are associated the failure modes typical of impact damage, with step-like delamination, intra-lamina cracks and high-density microfailure zones [24, 30, 50, 54]. With intermediate values of feed rate the failure consists essentially in push-down delamination, generated when the chisel edge and the inner portion of the lips have already left the work material [29]. Even lower values of feed rate may lead to a damage-free hole, at least for the supported drilling condition.

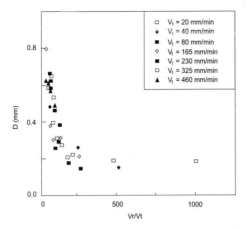

Figure 6.19. Size of delamination zone as a function of the ratio between cutting speed and feed rate [28]

It should be noticed that the type and extension of the damage is strongly influenced by the ratio between the cutting speed v_t and feed rate f, rather then on the two separately. For large values of the vt/f ratio the internal surface of the hole is smooth and regular with little delamination, both in the case of supported or unsupported drilling; even though in the latter case, the push-out delamination is sensibly wider.

For small values of the vt/f ratio the hole is significantly damaged: the push-out delamination is wider and involves several layers, internal cracks and debonding of the fibres occur. Figure 6.19 reports the size of delamination (in terms of average diameter of the delaminated area) as a function of the ratio between cutting speed and feed rate [28].

The effect of cutting parameters on surface roughness is quite difficult to generalise, as the result is influenced by composite characteristics and, in particular, by fibre volume fraction.

For glass fibre reinforced/epoxy composites the hole surface finish can be improved by increasing cutting speed and fibre volume fraction. Holes drilled with low cutting speed and feed rates exhibit a large roughness. On the other hand, composites with high fibre volume fraction exhibit a contrary behaviour [26].

6.5.3   Drilling Conditions

The selection of drilling conditions is concerned with the application of particular cutting methods, such as the use of a pre-heated tool, the presence of a support under the workpiece, the vibration-assisted drilling and the employment of a damper for unsupported drilling.

The use of a pre-heated tool drastically reduces the thrust force, yielding a narrower push-out delamination and a lower energy required to drill the hole. Unfortunately, the thermal damage induced in the material around the hole is stronger and a strong reduction in residual mechanical properties can be observed [40]. Precooled tools, on the other hand, present an increase in drill thrust force (and in the probability of delamination), but this drawback is compensated by the improved surface finish, hole quality and far superior tool life [57].

The presence of the support beneath the workpiece significantly reduces the level of the push-out delamination because the threshold thrust force at the onset of delamination is increased, but mildly affects the internal damage. Consequently, a commonly followed practice in the industry is the use of a support on the back to prevent deformations leading to exit delamination [37].

Better quality holes and high efficiency can be obtained by the vibration-assisted cutting technique [38, 55, 58]. Conventional drilling is a continuous cutting process, whereas vibration-assisted drilling is a pulsed intermittent cutting process using a piezoelectric crystal oscillator [59].

In the case of unsupported drilling, delamination can be sensibly reduced by using a damper alongside the twist drill. In fact, the damper greatly reduces the workpiece dynamics, avoiding the fast release movement of the workpiece when the twist drill exits the laminate. Thus the actual feed rate in this particular final phase is reduced and delamination almost eliminated [37].

6.5.4   Special Tools

Several geometries of drill tools have been investigated, aiming mainly at the reduction of thrust force and delamination. The most investigated geometries are the candlestick drill, saw drill, core drill and step drill (Figure 6.20) [33, 39, 53, 56].

All these tools were developed with the aim of reducing push-out delamination by reducing the thrust force. In Figure 6.21 the correlation between thrust force and feed rate for various drills is shown.

As can be seen, the drilling thrust of the twist drill is the highest, followed by the saw drill and core drill, while the candlestick drill and step drill are the lowest. This behaviour means that the twist drill is more susceptible to cause delamination damage due to the larger thrust force it generates during drilling.

The peel effect of plies along the edge of the major drill cutting edges takes place when the tool tip encounters the first ply. This defect increases with the rake angle and tends to produce ply detachment [44]. To reduce or eliminate this en-

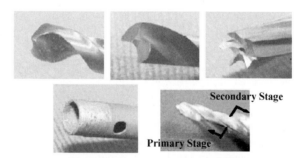

Figure 6.20. Examples of different drilling tool geometries (from left to right): the traditional twist drill, the candle stick drill, the saw drill, the core drill and the step drill

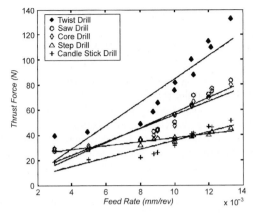

Figure 6.21. Correlation between thrust force and feed rate for special drills [39]

trance defect, a small rake angle prevents the first plate ply from lifting up and tearing off. A rake angle lower than 6° is usually recommended.

References

[1]   Crivelli Visconti I (1974) Materiali Compositi, Tamburini, Italia
[2]   Agarwal BD, Broutman LJ (1980) Analysis and Performance of Fiber Composites. Wiley, New York, NY
[3]   Schwartz, MM (1984) Composite Materials Handbook. McGraw-Hill, New York, NY
[4]   Ecklod G (1994) Design and Manufacturing of Composite Structures. Woodhead, Cambridge, England
[5]   Hollaway L (1994) Handbook of Polymer Composites for Engineers. Woodhead, Cambridge, England
[6]   Gürdal Z, Haftka RF, Hajela P (1999) Design and Optimization of Laminated Composite Materials. Wiley, New York, NY
[7]   Staab GH (1999) Laminar Composites. Butterworth-Heinemann, Boston, MA
[8]   ASM Handbook, Volume 21 Composites (2001) ASM International. Material Park, OH
[9]   Merchant ME (1944) Basic mechanics of the metal-cutting process. ASME J Appl Mech 11: 168–175
[10]   Oxford CJ (1955) On the drilling of metals-I. Basic mechanics of the process. Trans ASME 77: 103–114
[11]   Mauch CA, Lauderbaugh LK (1990) Modeling the drilling process – An analytical model to predict thrust force and torque, computer modeling and simulation of manufacturing processes. ASME Prod Eng Div 48: 59–65
[12]   Armarego EJA, Wright JD (1984) Predictive models for drilling thrust and torque – a comparison of three flank configurations. Ann CIRP 33: 5–10
[13]   Stephenson DA, Agapiou JS (1992) Calculation of main cutting edge forces and torque for drills with arbitrary point geometries. Int J Mach Tools Manuf 132(4): 521–538

[14] Ehmann KF, Kapoor SG, DeVor RE, Lazoglu I (1997) Machining process modeling: a review. J Manuf Sci Eng 119: 655–659

[15] Chandrasekharan V, Kapoor SG, DeVor RE (1995) A mechanistic approach to predicting the cutting forces in drilling: with application to fiber-reinforced composite materials. J Eng Ind 117: 559–570

[16] Langella A, Nele L, Maio A (2005) A torque and thrust prediction model for drilling composite materials. Compos Pt A 36: 83–93

[17] Caprino G, Nele L (1996) Cutting forces in orthogonal cutting of unidirectional GFRP composites. J Eng Mater Technol 118: 419–425

[18] Caprino G, Nele L, Santo L (1996) On the origin of cutting forces in machining unidirectional composite materials. ASME International. Reprinted from: PD-Vol 75, ESDA Engineering System Design and Analysis Conference, 83–89

[19] Arshinov V, Alekseev G (1973) Metal Cutting Theory and Cutting Tool Design. MIR, Moscow

[20] Spur G, Stöferle T (1987) Enciclopedia delle Lavorazioni Meccaniche, vol. 3 asportazione di truciolo. Tecniche Nuove, Milano

[21] Santochi M, Giusti F (2000) Tecnologia Meccanica e studi di fabbricazione, Casa Editrice Ambrosiana, Milano

[22] Jain S, Yang DCH (1994) Delamination – free drilling of composite laminates. J Eng Ind 116: 475–481

[23] Di Ilio A, Paoletti A, Veniali F (1996) Tool wear in drilling thermosets and thermoplastic matrix composites.Eng Syst Des Anal ASME 3: 41–46

[24] Khashaba UA (2004) Delamination in drilling GFR-thermoset composites. Compos Struct 63: 313–327

[25] Abrate S, Walton DA (1992) Machining of composite material: Part I: Traditional methods. Compos Manuf 2: 75–83

[26] El-Sonbaty I, Khashaba UA, Machaly T (2004) Factors affecting the machinability of GFR/epoxy composites. Compos Struct 63: 329–338

[27] Davim JP, Rubio JC, Abrao AM (2007) A novel approach based on digital image analysis to evaluate the delamination factor after drilling composite laminates. Compos Sci Technol 67:1939–1945

[28] Tagliaferri V, Caprino G, Diterlizzi A (1990) Effect of drilling parameters on the finish and mechanical properties of GFRP composites. Int J Mach Tools Manuf 30 (1):77–84

[29] Caprino G, Tagliaferri V (1995) Damage development in drilling glass fibre reinforced plastics. Int J Mach Tools Manuf 35 (6): 817–829

[30] Di Ilio A, Tagliaferri V, Veniali F (1991) Cutting mechanisms in drilling of aramid composites. Int J Mach Tools Manuf 31 (2): 155–165

[31] Veniali F, Di Ilio A, Tagliaferri V (1995) An experimental study of the drilling of aramid composites. J Energy Resour Technol 117:271–278

[32] Capello E, Tagliaferri V (2001) Drilling damage of GFRP and residual mechanical behavior-part 1: drilling damage generation. J Compos Technol Res 23 (2):122–130

[33] Abrao A M, Faria PE, Campos Rubio J., C., Reis P, Paulo Davim J (2007) Drilling of fiber reinforced plastics: A review. J Mater Process Technol 186:1–7

[34] Jain S, Yang DCH (1993) Effects of federate and chisel edge on delamination in composite drilling. Trans ASME J Eng Ind 115: 398–405

[35] Mathew J, Ramakrishnan N, Naik NK (1999) Investigations into the effect of geometry of a trepanning tool on thrust and torque during drilling of GFRP composites. J Mater Process Technol 91:1–11

[36] Jain S, Yang DCH (1994) Delamination-free drilling of composite laminates. Trans ASME J Eng Ind 116 (4):475–481

[37] Capello E (2004) Workpiece damping and its effect on delamination damage in drilling thin composite laminates. J Mater Process Technol 148:186–195

[38] Capello E, Tagliaferri V (2001) Drilling damage of GFRP and residual mechanical behavior – part 2: static and cyclic bearing loads. J Compos Technol Res 23 (2):131–137

[39] Hocheng H, Tsao CC (2006) Effects of special drill bits on drilling-induced delamination of composite materials. Int J Mach Tools Manuf 46:1403–1416

[40] Tsao CC, Hocheng H (2003) The effect of chisel length and associated pilot hole on delamination when drilling composite materials. Int J Mach Tools Manuf 43:1087–1092

[41] Timoshenko S, Woinowsky-Keiger S (1959) Theory of plates and shells. McGraw-Hill, New York, NY

[42] Tsao CC, Hocheng H. (2007) Effect of tool wear on delamination in drilling composite materials. Int J Mech Sci 49:983–988

[43] Tsao CC (2006) The effect of pilot hole on delamination when core drill drilling composite materials. Int J Mach Tools Manuf 46:1653–1661

[44] Hocheng H, Dharan CKH (1990) Delamination during drilling in composite laminates. J Eng Ind 112:236–239

[45] Davim JP, Reis P (2003) Drilling carbon fiber reinforced plastics manufactured by autoclave-experimental and statistical study. Mater Des 24 (5):315–324

[46] Won M S, Dharan CK, H (2002) Chisel edge and pilot hole effects in drilling composite laminates. Trans ASME J Manuf Sci Eng 124:242–247

[47] Cantwell WJ, Morton J (1989) Geometrical effects in the low velocity impact response of CFRP. Compos Struct 12:39–60

[48] Hocheng H, Tsao CC (2003) Comprehensive analysis of delamination in drilling of composite materials with various drill bits. J Mater Process Technol 140:335–339

[49] Persson E, Eriksson I, Zackrisson L (1997) Effects of hole machining defects on strength and fatigue life of composite laminates. Compos Pt A, 28: 141–151

[50] Konig W, Wulf C, Grass P, Willerscheid (1985) Machining of fiber reinforced plastics. Ann CIRP 34 (2): 537–548

[51] Velayudham A, Krishnamurthy R (2007) Effect of point geometry and their influence on thrust and delamination in drilling of polymeric composites. J Mater Process Technol 185: 204–209

[52] Tsao CC, Hocheng H (2007) Parametric study on thrust force of core drill. J Mater Process Technol 192: 37–40

[53] Piquet R, Ferret B, Lachaud F, Swider P (2000) Experimental analysis of drilling damage in thin carbon/epoxy plate using special drills. Compos Pt A 31: 1107–1115

[54] Konig W, Grab P, Heintze A, Okcu F, Schmitz-Justen Cl (1985) New developments in drilling and contouring composites containing kevlar aramid fiber, Fabrication of Composite Materials, Source Book edited by M. M. Schwartz, ASM, Metal Park, Ohio, 236

[55] Wang X, Wang LJ, Tao JP (2004) Investigation on thrust in vibration drilling of fiber-reinforced plastics. J Mater Process Technol 148: 239–244

[56] Tsao CC, Hocheng H (2005) Effect of eccentricity of twist drill and candle stick drill on delamination in drilling composite materials. Int J Mach Tools Manuf 45: 125–130

[57] Bhattacharyya, Horrigan DP (1998) A study of hole drilling in kevlar composites. Compos Sci Technol 58, 267–283

[58] Arul S, Vijayaraghavana L, Malhotrab SK, Krishnamurthy R (2006) The effect of vibratory drilling on hole quality in polymeric composites. Int J Mach Tools Manuf 46: 252–259

[59]   Zhang LB, Wang LJ, Liu XY, Zhao HW, Wang X, Luo HY (2001) Mechanical model for predicting thrust and torque in vibration drilling fibre-reinforced composite materials, Int J Mach Tools Manuf 41: 641–657
[60]   Chen WC (1997) Some experimental investigations in the drilling of carbon fiber-reinforced plastic (CFRP) composite laminates. Int J Mach Tools Manuf 37 (8): 1097–1108
[61]   Stone R, Krishnamurthy K (1996) A neural network thrust force controller to minimise delamination during drilling of graphite-epoxy laminates. Int J Mach Tools Manuf 36 (9): 985–1003

7

Ecological Machining: Near-dry Machining

Viktor P. Astakhov

General Motors Business Unit of PSMI, 1255 Beach Ct., Saline MI 48176, USA.
E-mail: astvik@gmail.com

This chapter points out major ways to reduce the ecological and health impacts of metal working fluids (MWFs). In the continuous quest for dry machining, only one process can offer a near-term solution for practical applications. This process uses a minimum quantity of lubrication and is referred to as "near-dry". In near-dry machining (NDM), an air–oil mixture called an aerosol is fed onto the machining zone. Compared to dry machining, NDM substantially enhances cutting performance in terms of increasing tool life and improving the quality of the machined parts. This chapter presents a classification of NDM methods, discussing their advantages and drawbacks. Analyzing the available information on the performance of NDM, the chapter offers a physically attractive explanation of why NDM works. It considers the essential components of the whole NDM system, arguing that a 360° vision approach is the key to successful implementation of NDM.

7.1   Introduction

The basic functions of a metal working (cutting) fluid (MWF, or simply the coolant) are to provide cooling and lubrication and thus reducing the severity of the contact processes at the cutting tool–chip and cutting tool–workpiece interfaces. A MWF may significantly affect the tribological conditions at these interfaces by changing the contact temperature, normal and shear stresses and their distributions along the interfaces, the type and/or mechanism of tool wear, machined surface integrity and machining residual stresses induced in the machined parts, *etc*. In some applications, it is expected that MWF fluid should also provide secondary service actions, for example, washing of the machined part or chip transportation in deep-hole drilling, in which the MWF transports the chip over significant distances [1].

Historically, until the 19th century, water was used for centuries as a cooling medium to assist various metalworking operations. Taylor [2] was probably the first to prove the practical value of using liquids to aid in metal cutting. In 1883, he demonstrated that a heavy stream of water flooding the cutting zone increased the allowable cutting speed by 30–40%. It was found however that, although water is an excellent coolant due to its high thermal capacity and availability [3], the use of water as a coolant had the drawbacks of corrosion of parts and machines and poor lubrication. Further developments followed quickly.

Mineral oils were developed at this time as they had much higher lubricity. However, their lower cooling ability and high costs restricted their use to low-cutting-speed machining operations. Finally, between 1910 and 1920 soluble oils were initially developed to improve the cooling properties and fire resistance of straight oils [3]. Other substances were also added to these to control problems such as foaming, bacteria and fungi. Oils as lubricants for machining were also developed by adding extreme-pressure (EP) additives. Today, these two types of MWFs (coolants) are known as water emulsifiable oils and straight cutting oils. Additionally, semi-synthetic and synthetic MWFs were developed to improve the performance of many machining operations [4]. Today, MWFs play a significant role in manufacturing processes, supporting their high productivity and efficiency.

7.2   Amount and Cost

Although the significance of MWFs in machining is widely recognised, cooling lubricants are often regarded as supporting media that are necessary but not important. In many cases, the design or selection of the MWF supply system is based on the assumption that, the greater the amount of lubricant used, the better the support for the cutting process. As a result, the machining zone is often flooded with MWF without taking into account the requirements and specifics of an operation. Moreover, the selection of the type of the MWF for a particular machining operation is often based upon recommendations of sales representatives from MWF suppliers without clearly understanding the nature of this operation and the clear objectives of MWF application. The brochures and websites of MWF suppliers are of little help in such a selection. The technique of MWF application, which includes the MWF pressure, flow rate, nozzle design and location with respect to the machining zone, filtration, temperature, *etc.*, are often left to the discretion of machine and tool designers. Moreover, operators of manual and semi-automatic machines are often those who decide the point of application and flow rate of the MWF for each particular cutting operation.

MWFs also represent a significant part of manufacturing costs. Just two decades ago, MWFs accounted for less than 3% of the cost of most machining processes. These fluids were so cheap that few machine shops gave them much thought. Times have changed and today MWFs account for a significant portion of direct manufacturing costs. Moreover, MWFs, especially those containing oil, have become a huge liability. Not only does the Environmental Protection Agency (EPA) regulate the disposal of such mixtures, but many states and localities have also classified them as hazardous wastes. Although the cost of MWF application and disposal is high, it varies from one industry to the next and from one country (state, province or region) to another, depending upon many factors. After one European automotive plant reported that the contribution of MWF was up to 17% of manufacturing cost, while the tool cost was up to 8% [5], this was generally accepted as a kind of general reference without understanding the particular conditions and regulations involved. Based on these data, the conclusion was drawn that MWF costs are more than twice tool-related costs, although the main attention of researchers, engineers and managers is focussed on improving cutting tools. At the other extreme, a cost of 0.9% has also been reported by a large manufacturer [5].

Our survey shows that the share of the cost due to the MWF depends highly on the selection, maintenance and disposal practices adopted for a particular plant. When properly selected and maintained and when a plant has its own MWFs disposal facility (common for the automotive industry), the share of the manufacturing cost due to MWF is 8–10%. As such, the share of the tool cost varies depending upon the complexity of the product. For example, for advanced powertrain plants it is 12–15% due to the high cost of gear manufacturing, whereas for simpler automotive plants dealing mainly with turning and milling (shaft production), the cost contribution of direct tooling is 5–7%.

The cost of using MWFs is increasing as the number and extent of environmental protection laws and regulations increase. The cost of using MWFs is about US $48 billion a year in the USA [6], about one billion marks in Germany [7] and about ¥71 billion in Japan [8].

7.3   Health and Environmental Aspects

Ecological and health aspects of metalworking fluids' manufacture, use and disposal have become very important due to new stricter legislation, notably in Europe [9]. In the European Union, MWFs and lubricants are included in the voluntary Eco-label initiative, which is dedicated to stimulating the supply and demand of products with reduced environmental impact. The number of country-specific standards for environmental labeling of metalworking fluids and other chemicals is ever increasing, including Blue Angel certification in Germany, the SS 155434 standard in Sweden, and VAMIL regulation in The Netherlands. In the UK, the Health and Safety Executive is developing more stringent mandatory workplace exposure limits to airborne mist from metalworking fluids.

It is estimated that MWF consumption is more than 100 million gallons per year in the USA [8]. A typical large automobile metal processing facility utilizes more than 600,000 gallons (>2.28 million litres) of MWF concentrates per year and more than 300,000 gallons (>1.14 million litres) of straight oil per year. In Germany, coolant consumption is about 75,500 tons a year [10]. In Japan, MWF consumption is 100,000 kilolitres of water-immiscible (disposal cost ¥35–50 per litre), 50,000 kilolitres of water-soluble coolant without chlorine (disposal cost ¥300 per litre) and 10,000 kilolitres of water-soluble coolant with chlorine (disposal cost ¥2250 per litre) [8].

According to the National Institute for Occupational Safety and Health (NIOSH), over 1 million workers in the USA are exposed to MWFs. Machinists, machinery mechanics, metalworkers and other machine operators and setters have the greatest contact with these fluids. However, workers performing assembly operations can also be exposed if MWFs remain on the machined product. Workers other than machinists may also be exposed to MWF mists if ventilation systems are poorly designed or inadequate. Workers may be exposed by skin contact, inhaling (breathing in), or ingesting (swallowing) particles, mists and aerosols.

Although recent changes in MWF formulations have resulted in safer products, it is important to realize that MWFs may cause a variety of health problems. Skin exposure to MWFs can result in conditions such as dermatitis or folliculitis. Repeated inhalation of MWF mists has been shown to decrease lung function over the course of a workshift. MWF mists may also cause several respiratory diseases, including asthma, bronchitis and hypersensitivity pneumonitis [11–16]. Exposure to some MWFs and/or their additives may cause cancer. The health hazards of MWFs depend upon the type of fluid used, as well as the additives and contaminants that may be present in the fluid. MWFs have received increased attention recently because, based on multiple evidences, it was concluded that workers may be suffering from asthma and possibly other lung diseases at the current occupational exposure limits. Consequently, NIOSH and the Occupational Safety and Health Administration (OSHA) are considering lowering the occupational exposure limit. NIOSH recommends that occupational safety and health programs that include medical monitoring should be established at workplaces using MWFs. OSHA has convened a committee of national experts to develop a proposal for a new standard to lower workplace exposures and increase worker protection.

The implementation of multiple available regulation and recommendations are not that simple as risk assessments are difficult due to numerous variables. Facilities utilizing MWFs have complex exposure patterns, as MWF formulations are complex chemical mixtures. As a result, the intensity of the drive for improvement in MWFs is determined mainly by economics rather than by safety and environmental causes.

7.4   Principal Directions in the Reduction of MWF Economical, Ecological and Helth Impacts

There are several principal ways to reduce both ecological and economical impacts of MWFs:

1. **Balanced selection of MWF.** Although the selection of MWF is a complicated process that includes various facets of metal-machining business [1], the priorities in this selection are shifting towards ecological and health aspects. Different metal working processes have varying cooling and lubrication needs. MWFs have many properties that increase their efficiency, extend their life, and minimize the potential to damage tools and products. Therefore, when choosing an MWF, many factors need to be

considered, including: ramp oil rejection; ability to settle out solids; bacterial resistance; corrosion and rust resistance; emulsification capability; foaming nature and resistance; optimal MWF life; longevity of MWF as measured against current industrial standards; cost; chemical restrictions and reactivity; lubricity; biodegradability; recyclability; capacity of recycled MWF to prevent galvanic attack; water compatibility and requirements: pH, deionized water requirements, mineral content and hardness; history of dermatitis and others.

Utilization of environmentally responsible metalworking fluids is definitely gaining momentum globally. In the private sector, a growing interest in renewable-resource-based metalworking fluids is evidenced by the fact that ISO 15380 includes environmentally friendly fluids, and that a major automotive manufacturer has endorsed bio-based metalworking fluids. In addition, the Canadian Auto Workers Union has recommended vegetable-oil-based alternatives. As the cost of crude oil rises, vegetable-oil-based fluids are becoming more attractive alternatives.

2. **Proper application of MWF**. In a huge number of cases, maybe the majority, MWF is applied improperly, so that a large amount of MWF is used when much better results would be achieved by precisely using much smaller amounts [17]. Applying MWFs properly and consistently is critical to grinding and machining operations. It reduces the usage of MWF, the amount of hazard mist generated and the exposure of the operator, decreases scrap and reduces cycle time and tool wear. All this means significant cost reduction. Areas to consider are [17]: nozzle design, placement, and conditions (including blockage and wear); discharge coefficient, coherence of MWF flow including laminar flow, air dams and air barriers.

    MWFs can be applied through manual, flood or high-pressure through-tool application methods. In manual application, an operator uses an oilcan to apply cutting fluid. Although this is apparently the simplest cheapest method, it has limited use in any but the simplest applications and is very inconsistent. In flood cooling, MWF is directed under relatively low pressure to the work area. To reduce the inconsistency of this technique due to nozzle location relative to the tool, large amounts of MWF have to be delivered. Advanced MWF flood application techniques use a movable nozzle whose position is controlled by the CNC for optimum MWF delivery at the same place as the tool and/or the workpiece is moving during machining. High-pressure through-tool MWF application utilizes high-pressure MWF pumped through the spindle and then through internal channels made in the tool directly to the machining zone. This is probably the best application technique available today as it significantly improves the effectiveness of MWF, limits is usage and reduces hazard exposure. Although its application requires a number of additional units such as as high-pressure pumps, rotary units to introduce MWF to the rotating spindle, special tool holders and special tools, the benefits of its application significantly offset these additional costs.

3. **Meticulous management of MWF**. Although this aspect appears to have the greatest impact on the ecological, health and cost aspect of MWFs, it is still the most neglected one. The objective of MWF management is two-fold: firstly to extend MWF life as much as possible and secondly to dispose of unusable MWFs at the lowest cost. An MWF management plan normally includes: continuously removing metal chips and tramp oil; thorough cleaning of the MWFs system according to the preventive management (PM) schedule; pumping MWFs from the sump; removing all metal chips and fines; cleaning any oily residues that remain on any surface; filling the sump with a good cleaner using clean water and circulating the cleaner through the coolant system for several hours; application of a cleaning solution to machine surfaces that are not contacted by the MWF during machine operation; pumping cleaning solution from the sump; wiping cleaning solution residues from the sump; rinsing the entire coolant system with clean water; rinsing the system again if necessary to remove all residues; and recharging the system with reclaimed or new coolant immediately to protect metal surfaces against corrosion. Skerlos discussed the realization of particular steps in this plan with modern equipment [18].

4. **Gradual reduction of MWF usage** by increasing the use of near-dry and dry machining. At present, many efforts are being undertaken to develop advanced machining processes using less or no MWFs. Machining without the use of MWFs has become a popular avenue for eliminating the problems associated with the MWF management [10].

Dry machining has its advantages and associated drawbacks. The advantages of dry machining are obvious: cleaner parts, no waste generation, reduced cost of machining, reduced cost of chip recycling (no residual oil), *etc*. However, these advantages do come at a cost. The most prohibitive part of switching to dry machining is the large capital expenditure required to start a dry machining operation. Machines and tools designed for MWFs cannot be readily adapted for dry cutting [10].

New, more powerful machines must be purchased, and special tooling is often needed to withstand the high temperatures generated in dry cutting. The quality of machined parts may be affected significantly as the properties of the machined surface are significantly altered by dry machining in terms of its metallurgical properties and residual machining stresses. High cutting forces and temperatures in dry machining may cause the distortion of parts during machining. Moreover, parts are often rather hot after dry machining so their handling, inspection gauging, *etc.*, may present a number of problems.

Near-dry machining (NDM) formerly known as minimum quantity lubrication (MQL) machining, was developed to provide at least partial solutions to the listed problems with dry machining. This following sections aim to present some important aspects of NDM.

7.5   Nearly Dry Machining (NDM)

7.5.1   How NDM Operates

General speaking, near-dry machining (NDM), also known as minimal quantity lubrication (MQL) machining, supplies very small quantities of lubricant to the machining zone. It was developed as an alternative to flood and internal high-pressure coolant supply to reduce MWFs consumption.

In NDM, the cooling media is supplied as a mixture of air and an oil in the form of an aerosol (often referred to as the *mist*). An aerosol is a gaseous suspension (hanging) into air of solid or liquid particles. In NDM, aerosols are oil droplets dispersed in a jet of air. An idealized picture of NDM is shown in Figure 7.1: small oil droplets carried by the air fly directly to the tool working zone, providing the needed cooling and lubricating actions.

Aerosols are generated using a process called atomization, which is the conversion of bulk liquid into a spray or mist (*i.e.*, collection of tiny droplets), often by passing the liquid through a nozzle. An atomizer is an atomization apparatus; carburetors, airbrushes, misters, and spray bottles are only a few examples of the atomizers used ubiquitously. In internal combustion engines, fine-grained fuel atomization is instrumental to efficient combustion. Despite the name, it does not usually imply that the particles are reduced to atomic sizes. Rather, droplets of 1–5 µm are generated. Because MWF cannot be seen in the working zone, and because the chips look and feel dry, this application of minimum-quantity lubricant is called near-dry machining.

An atomizer is an ejector in which the energy of compressed gas, usually air taken from the plant supply, is used to atomize oil. Oil is then conveyed by the air in a low-pressure distribution system to the machining zone. The principle of the atomizer is shown in Figure 7.2. As the compressed air flows through the Venturi path, the narrow throat around the discharge nozzle creates a Venturi effect in the mixing chamber, *i.e.*, a zone where the static pressure is below the atmospheric pressure (often referred to as a partial vacuum) [19,20]. This partial vacuum draws the oil up from the oil reservoir where the oil is maintained under a constant

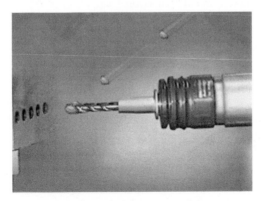

Figure 7.1. Idealized image of NDM

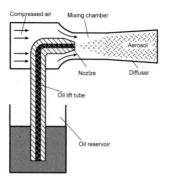

Figure 7.2. Model of a simple atomizer

hydraulic head. The air rushing through the mixing chamber atomizes the oil stream into an aerosol of micron-sized particles.

The design of the atomizer is critical in NDM as it determines the concentration of the aerosol and the size of droplets. Unfortunately, this is one of the most neglected aspect in casual application of NDM. Several machine tool companies have patented their designs (for example, US patents 6 923 604 and 6 602 031). In such designs, the position of the discharge nozzle is controlled by the machine controller so the parameters of the aerosol are changed depending upon the machining conditions.

7.5.2   Classification of NDM

Unfortunately, there is no accepted classifications of NDM so it is very difficult for a practical engineer or plant manager to make the proper choice about the regimes of NDM and equipment needed.

7.5.2.1 Classification by Aerosol Supply

The first level of NDM classification includes a way by which aerosol is supplied into the machining zone:

1. NDM 1: NDM with external aerosol supply. In NDM 1, the aerosol is supplied by an external nozzle placed in the machine similar to a nozzle for flood MWF supply.
2. NDM 2: NDM with internal (through-tool) aerosol supply. In NDM 2, the aerosol is supplied through the tool similar to the high-pressure method of internal MWFs supply.

As the name implies, NDM with an external aerosol supply (NDM 1) includes the external nozzle that supplies the aerosol. There are two options in NDM 1, which are shown in Figure 7.3:

1.1. NDM 1 with an ejector nozzle. The oil and the compressed air are supplied to the ejector nozzle and the aerosol is formed just after the nozzle, as shown in Figure 7.3. One of the possible designs of ejector nozzle is shown in Figure 7.4. As can be seen, it has two air passages. The first one is ex-

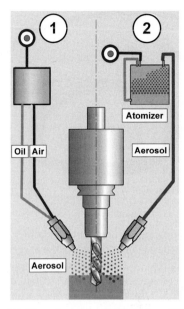

Figure 7.3. The principles of NDM 1.1 and NDM 1.2

ternal and creates the air envelope that served as the mixing chamber. The second one provides the atomizing air supply. The oil to be atomized is supplied through the central passage.

1.2. NDM 1 with a conventional nozzle. The aerosol is prepared in an external atomizer and then supplied to a conventional nozzle, as shown in Figure 7.3. The nozzle deign is similar to that used in flood MWF supply.

NDM 1.1 is probably the cheapest and simplest method. For example, the Spra-Kool Midget unit shown in Figure 7.5 is advertised as an economical method of applying an MWF spray for machining. The Spra-Kool unit works on an air pressure of 0.2–1.0 MPa, which should be adjusted on the compressor. Attaching the ball-check fitting to the air supply, dropping the suction tube into an oil container, and locating the nozzle by means of the spring-wire attaching clip, one can get NDM for a cost of US $30. The soft wire in the nose can be bent to direct the spray to the work. It is designed for easy transfer from one machine to another.

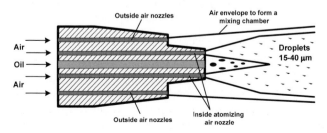

Figure 7.4. Nozzle design for NDM 1.1

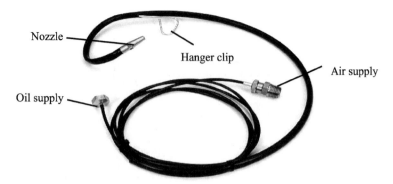

Figure 7.5. Spra-Kool Midget

In reality, however, adjustments are not that simple. If no special precautions are taken, the unit generates a dense mist that covers everything in the shop, including the operator's lungs. To prevent this from happening and to gain the full control on the parameters of aerosol, one needs to have (design or buy) a hydraulic unit similar to that shown in Figure 7.6. When such a unit is used, the parameters of the aerosol can be adjusted in a wide range in terms of droplet size and oil flow rate by setting appropriate air and oil flow rates and by adjusting the pressure in the oil reservoir. Moreover, such a device prevents oil spills as it shuts down the oil supply line when the air supply is not available.

NDM 1.2 is probably the simplest method. An external atomizer required is an off-the-shelf product, such as that shown in Figure 7.7. As can be seen, the same nozzle as is used for flood MWF is used, so that NDM 1.2 can be used with the same sets of nozzles installed in the machine.

NDM 1 has the following advantages:

- Inexpensive and simple retrofitting of the existing machines
- The same cutting tools used for flood MWF will work
- Easy to use and maintain equipment
- The equipment can be moved from one machine to another
- Relative flexibility of NDM 1.2 as the position of nozzle can be adjusted for the convenience of operator. As such, the parameters of the aerosol do not depend on the particular nozzle location.
- The equipment can be moved from one machine to another.
- Various standard and special nozzle designs are available with NDM 1.2 to suite most common metal machining operations and tool designs. It is proven to be particularly effective in face milling and sawing

NDM 1 has the following disadvantages:

- Both NDM 1.1 and 1.2 do not work well with drills and boring tools as an aerosol cannot penetrate into the hole being machined
- A critical aspect for NDM 1.1 and an important aspect for NDM 1.2 is the location of the nozzle relative to the working part of the tool. For both

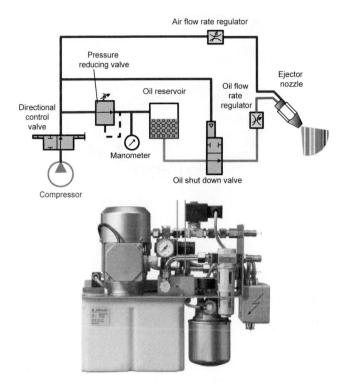

Figure 7.6. Aerosol control unit

methods, this location must be fixed, *i.e.*, should not change as the tool moves. Note that this issue is not that important in flood MWF supply where gravity and the energy of MWF flow cover a much wider range of possible nozzle locations

- The parameters of the aerosol should be adjusted for each particular metal machining operation and the work material. This makes NDM 1 less attractive option in a job-shop environment

As the name implies, NDM with internal aerosol supply (NDM 2) includes internal passages for aerosol supply. There are two options in NDM 2: NDM 2.1 with an external atomizer as shown in Figure 7.8 and NDM 2.2 with an internal atomizer located in the spindle of the machine as shown in Figure 7.9.

In NDM 2.1, the aerosol is prepared in an external atomizer and then supplied trough the spindle and the internal channels made in the tool. When NDM 2.1 is used on machining centres or manufacturing cells, the aerosol supply unit has to react to the frequent tool changes that nowadays take only 1 or 2 s, setting the proper aerosol parameters for each given tool/operation. If the aerosol unit is shut down every time a tool change takes place then it requires some time to fill the whole system with aerosol again. VOGER, a NDM equipment suppler, has developed the bypass principle illustrated in Figure 7.8. The aerosol is produced con-

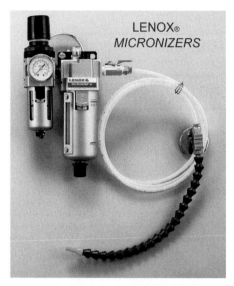

Figure 7.7. Accessories with the external atomizer by Lenox Co. for NDM 1.2

tinuously and supplied to the directional control valve, which allows aerosol into the spindle as soon as a tool change is over.

NDM 2.1 has the following advantages:

- Lowest initial cost
- The possibility of keeping two cooling system on the machine: flood and NDM
- Relatively simple installation and control
- Accurate control of the aerosol parameters so that they can be easily adjusted by the machine controller for a given tool or even operation with the same tool

NDM 2.1 has the following drawbacks:

- Spindle rotation creates a centrifugal force field that coats the wall of the aerosol delivery channel with oil that must be removed periodically. For a high-volume production manufacturing factory (plant, shop, line, cell), this downtime may be intolerable and costly. This additional cost can easily offset the savings on NDM.
- Special care should be taken of the position of the flexible line that connects the external aerosol unit with the spindle. Firstly, it should be as close to the machine as possible. Secondly, the radius of curvature of this line should not be less that 200 mm to prevent aerosol decomposition.
- The parameters of the aerosol that enters the machining zone can be different than those controlled in the outside unit depending upon the spindle speed and the conditions of the supply channels.

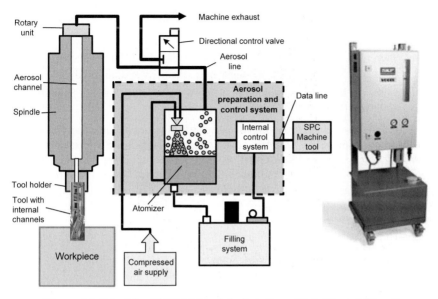

Figure 7.8. Principle of NDM 2.1 and the LubriLean DigitalSuper1 (Vogel)

Generally, NDM 2.2 is a more attractive concept: to mix the air and oil as close as possible to the tool in a well-designed mixing chamber. To do this, the oil is supplied through the spindle through a central tube within the surrounding annular air duct (Figure 7.9). The air and oil are mixed in the mixing chamber close to the tool. Because the air–oil aerosol is influenced only by the spindle rotation for a short distance, the discharge response from the tool tip is said to be improved. As discussed above, the design of the mixing chamber and ejector nozzle as well as the controlling of the oil discharge are critical for the application, and is normally patented by the machine tool manufactures.

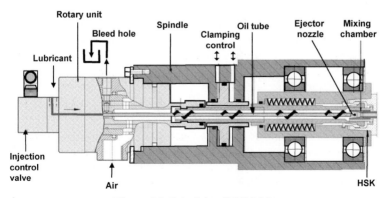

Figure 7.9. Principle of NDM 2.2

In implementing this NDM 2.2 approach, two important issues need to be addressed:

- Two nested rotary couplings must be provided for the air and oil connections, which raises questions regarding reliability and durability.
- The system plumbing must be prevented from shifting and changing the spindle's residual unbalance.

As the spindle speeds increase, both of these design issues become more of a problem. Moreover, to gain maximum advantage of NDM 2.2, a special computer-controlled solenoid must be installed on the machine to adjust the parameters of the aerosol for each particular tool/operation. All of these factors make retrofitting of an existing machine with NDM 2.2 rather cumbersome and the economic gain becomes uncertain.

7.5.2.2 Classification by Aerosol Composition

In the simplest cases of NDM 1 and 2, the aerosol is an air–oil mixture. The discharge of the oil in this mixture is selected to be in the range 30–600 ml/h depending upon the design of the NDM system, the nature of the machining operation, the work material and many other factors. Unfortunately, not many recommendations of the mixture composition are available. Advanced NDM (ANDM) uses aerosol that includes not only oil but also some other components. This Section considers two examples of ANDM: oil on water droplet (OoW NDM) and advanced minimum quantity cooling lubrication NDM (AMQCL NDM).

OoW NDM includes the supply of water droplets covered with a thin oil film [21–23]. As claimed by its authors, this method possesses both great cooling and lubricating abilities. The former is due to water properties (high specific heat capacity, density and thermal conductivity compared to air) and its evaporation. The latter is due to the specific droplet configuration.

The concept of OoW NDM is shown in Figure 7.10, which shows an ideal OoW droplet moving towards a hot surface. When the droplet reaches the tool or hot workpiece surface, the lubricant oil spreads over the surface in advance of water spreading. The water droplets are expected to perform three tasks: carrying the lubricant, spreading the lubricant effectively over the surface due to inertia and

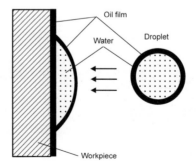

Figure 7.10. The concept of the oil-on-water NDM (NDM 3)

cooling the surface due to its high specific heat and evaporation. To make this concept practical, *i.e.*, to generate OoW droplets, a specially designed discharge nozzle is needed.

Figure 7.12 presents one of the designs [23] that, according to its authors, provides reliable continues supply of oil-on-water droplets in a controlled fashion. Rated by the flow meter, compressed air is fed through the central channel of the nozzle. Another air flow goes into the control unit where it is mixed by the control valve with oil in a known measurable proportion. Then the resulting mixture is fed into the first array of the ejector nozzles, through which the mixture is ejected and atomized by the main air flow. Water is supplied through a flow meters to the second arrays of the ejector nozzles as shown in Figure 7.11. Passing through these nozzles, the water is also atomized. According to the authors, when two atomized flows meet after the second nozzles, adhesion of oil to the surface of water droplets takes place as the average size of water droplets is greater than that of the oil droplets due to the difference between the surface tensions of water and oil. Selecting a particular oil that has favorable spreading properties over a water surface, one can achieve automatic generation of OoW droplets. So far, all known OoW nozzles work using this principle. However, the nozzle shown in Figure 7.11 is designed with third and fourths array of ejector nozzle to increase opportunities for the collision of water and oil particles to form OoW particles. This allows an increase of the feasible range of oil-to-water ratios. Moreover, these additional nozzles prevent oil particles from agglomerating and accumulation inside the mixing chamber and the discharge nozzle.

The size of OoW droplets when they hit the surface of the workpiece or tool is about 100–200 μm, which is much greater than that with the aforementioned traditional NDM approaches. This is considered a plausible explanation for the lower cutting forces in aluminium machining compared to traditional NDM with oil aerosol or even to the usual MWF flood machining. Improvements in surface finish and tool life with this method are also reported.

The advanced minimum quantity cooling lubrication (AMQCL) approach and the corresponding system developed by CoolTool Co. (Eagan, MN; Valencia, CA, USA) combines a source of propellant gas (*i.e.*, compressed air), lubrication additives (*i.e.*, soy oil) and solid and/or gaseous CO_2 (*i.e.*, coolant) in various concentrations to form a widely adjustable aerosol. A schematic of the AMQLC method is shown

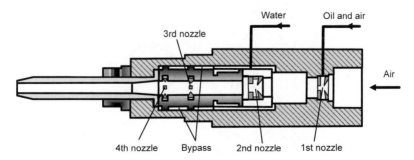

Figure 7.11. One of the developed nozzles for aerosol containing air, oil and water

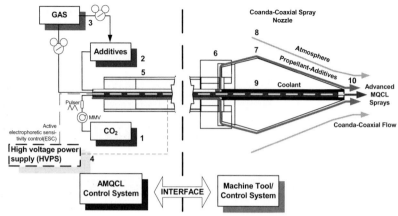

1. CO_2 coolant generator (CoolPulse™)
2. Lubrication additives (gases, liquids, solids)
3. Propellant gases (clean air, clean dry air, gaseous nitrogen, carbon dioxide)
4. Active ES charging unit (Optional)
5. Triaxial (CO_2/Additive/Propellant) delivery line
6. Triaxial-to-Coanda stream separator
7. Coanda propellant-additive stream flow
8. Coanda-induced airstream flow
9. Coaxial coolant stream flow
10. Injection, cooling, charging and spray formation

Figure 7.12. Schematic of the AMQCL system (Courtesy of Mr. D. Jackson, ToolCool Co.)

in Figure 7.12. The AMQCL system employs a novel Coanda-coaxial injector and spray applicator. The applicator employs a passive electrostatic charging mechanism to enhance droplet uniformity, spray force and machined surface deposition. Alternatively, an active electrostatic charging system may be employed to provide combination spray charging capability. The AMQCL system is interfaced both mechanically and electronically with a machining operation.

AMQCL technology exploits the Coanda effect for both additive injection and spray flow pattern control. A mixture of propellant gas and additive move along the Coanda profile as a thin film, powered by both propellant gas and atmospheric pressure shear stresses (Figure 7.13). The higher-pressure (and warmer) propellant-additive film sheds into and mixes with the lower-pressure (and cooler) coax-

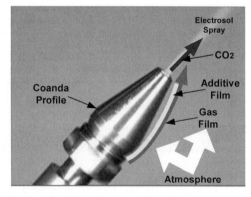

Figure 7.13. Coanda effect (Courtesy of Mr. D. Jackson, ToolCool Co.)

Advanced MQCL

MQCL - Minimum Quantity Cooling Lubrication

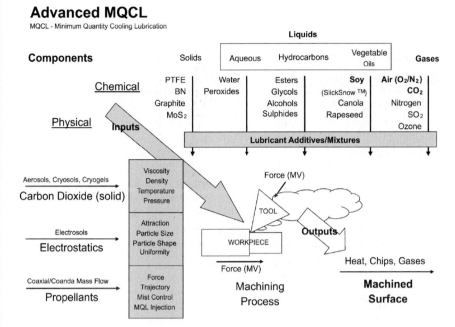

Figure 7.14. AMQCL chemistries

ial CO_2 stream. This is analogous to a waterfall, resulting in rapid vortex mixing and electrostatic charging of the cooling lubricant aerosol (referred to as an *Electrosol* by ToolCool Co.). Moreover, the Coanda effect creates an atmospheric-pressure tunnel which extends for a long distance from the nozzle tip towards the machining operation. The Coanda tunnel confines and shields the charged cooling lubrication spray as it moves from the nozzle to the cutting zone.

AMQCL aerosol chemistries are formed and delivered on the fly by combining one or more chemical and physical cooling lubricant ingredients. The composition may include liquids, extreme pressure solid additives, and reactive gases which are combined with a propellant gas and injected into a metered flow of charged CO_2 gas–solid aerosol (Figure 7.14). Each ingredient provides or controls a physical and/or chemical dimension of the resulting machining fluid chemistry,including cooling capacity, penetration power, boundary layer reactivity, lubricity, viscosity, spray particle size and uniformity. AMQCL sprays have variable geometry, including adjustable cooling lubricant characteristic from dry to wet composition, room temperature to cryogenic temperature, and spray pressures ranging from 0.07 to 1 MPa, or much higher if desired.

The combinational use of renewable bio-based feedstocks such as CO_2 and vegetable lubricants (SlickSnow^{TM}) provides an environmentally sound and worker-friendly alternative to petroleum-based cooling lubricants. Vegetable-based lubricants offer significant advantages such as superior tool lubricity, better surface finish, higher feed rates and safer chemistry. Using a bio-based lubricant in a minimum quantity application eliminates the costs associated with treatment, environmental compliance and disposal, among other operational cost factors. Moreover, vegetable oils such as soybean oil exhibit natural rust inhibition on metal surfaces, a very important characteristic for ferrous machining applications.

Compared to petroleum-based fluids, bio-based lubricating fluids perform as well as or better when machining steel and aluminium, cooling and lubricating the cutting surface as they remove small metal chips and enabling faster, more accurate machining. Bio-based lubricating fluids are biodegradable, non-toxic, have low volatile organic compounds (VOCs) emissions, high flash point and no offensive odour.

According to CoolTech Co, bio-based oils have been demonstrated to provide superior metal machining characteristics due to their somewhat polar nature. Unlike non-polar lubricating films, polar boundary layer films have a natural affinity for the workpiece surface, aligning in a highly ordered orientation (like magnets) with respect to the surface. This type of boundary layer lubrication is much stronger (*i.e.*, higher shear), producing better finishes, reducing tool chatter and increasing the longevity of cutting tool edge surfaces.

Used in combination with CO_2, spray mixtures produced are non-flammable and non-corrosive even at extremely high temperatures with sparking. Moreover, articles being processed are blanketed in a fire-quenching media (CO_2), which prevents chips from igniting or burning and oil smoking.

Test results provided by CoolTech Co show that, unlike simple gases such as nitrogen and air, CO_2 exhibits very strong hydrocarbon solubility (*i.e.*, CO_2 gas exhibits $> 600\%$ higher solubility in oils compared with compressed air). Due to this unique physico-chemical properties and cohesion energy, the CO_2 gas modifies the lubricant and coolant additive properties to produce mixtures with lower surface tension and lower viscosity, which aids in penetration into chip–tool capillary interfaces. Moreover, CO_2 itself behaves as a reactive boundary layer lubricant, forming carboxylic acid functional groups during tribochemical reactions.

AMQCL technology provides widely adjustable and fractional cooling–lubricant compositions of CO_2 coolant (Fc), propellant gases (Fp), and minimum quantities of lubrication additives (Fa). Adjustable spray pressure, temperature, coolant particle size and lubricant additive concentration allow a machinist to customize a cooling lubricant composition for any application. One or more individually controllable and flexible Coanda spray applicators may be employed to provide an optimum cooling, lubricating, and cleaning pattern. Moreover, the CO_2 coolant spray may be pulsed rapidly (CoolPulse^{TM}) to provide enhanced spray penetration and cooling in the cutting zone in certain applications, but without interrupting or compromising lubrication delivery.

7.5.3 Why NDM Works

7.5.3.1 Some Reported Results

To understand why NDM machining works, a great body of the reported results on NDM have to be classified in a systemic fashion and analyzed using the fundamentals of metal cutting tribology. The lack of information on the experimental conditions, including the parameters of the aerosol, prevents any reasonable systematization of the work done. Wu and Chien evaluated NDM of three different steel materials [24]. They found that the machining performance is affected by the lubrication type and its flow rate, the nozzle design, the distance between nozzle and tool tip, and the workpiece material. All these parameters are found to be dependent on the work material and process conditions. They concluded that only when the appropriate oil quantity and appropriate distance between the nozzle and tool tip are selected properly does NDM provides the optimum process condition. Unfortunately, many of these important parameters are not reported in many research documents and papers on NDM. As a result, a process or manufacturing engineer who wants to implement NDM is not equipped with sufficient recommendations to make proper choices of the equipment, parameters and regimes of NDM for a particular machining operation. Although professional magazines have published a number of articles on NDM (for example [25–29]), these articles do not present complete and systematic information on NDM even at the application level, limiting their scope to rather promotional aspects.

Some research work aimed to answer the question about how good NDM is compared with dry cutting. Ueda *et al.* [30] found that temperature reduction in oil-mist turning is approximately 5%, while in oil-mist end milling it is 10–15% and in oil-mist drilling it is 20–25% compared to the temperature in dry cutting. Khan and Dhar [31] found that NDM with vegetable oil reduced the cutting forces by about 5–15%. The axial force decreased more predominantly than the power force. They attributed this reduction as well as the improved tool life and finish of the machined surface to reduction of the cutting zone temperature as the major reason for the improved performance of machining operations. Similar results were obtained in machining 1040 steel [32]. Li and Liang [33] found cutting forces in machining 1045 steel lower in NDM compared with dry cutting. They also attributed this reduction to the cutting temperature difference.

Other groups of researches have compared NDM with wet machining. Dhar *et al.* studied the effect of NDM in tuning of 4340 steel using external nozzle and aerosol supply to the tool [34]. They found that the temperature at the tool–chip interface reduced by 5–10% (depending upon the particular combination of the cutting speed and feed) in NDM compared to wet machining. As a result, tool life and finish of the machined surface improved by 15–20%. Interestingly, the authors found that the tool life is the same for dry and wet machining, which is in direct contradiction with common shop practice. Filipovic and Stephenson [35] did not find any difference in tool life in gundrilling and cross-hole drilling of crankshafts between wet machining and NDM. Yoshimura *et al.* [23] found that, in machining of aluminium, the cutting force is lower and the surface finish is better with OoW NDM compared with dry, traditional NDM and wet machining.

Obikawa *et al.* [36] evaluated the performance of NDM in high-speed grooving of a 0.45%C carbon steel with a carbide tool coated with TiC/TiCN/TiN triple coating layers. Studying tool life in grooving, they found that a vegetable oil supplied at constant rate of 7 ml/h reduced the corner and flank wears more effectively than water-soluble oil at high cutting speeds of 4 and 5 m/s.

Using a minimum quantity of lubricant (MQL) and a diamond-coated tool in the drilling of aluminium–silicon alloys, Braga *et al.* [37] showed that the performance of the NDM process (in terms of forces, tool wear and quality of ma-

chined holes) was very similar to that obtained when using a large amount of water-soluble oil, with both coated and uncoated drills.

Studying turning of brasses, Davim *et al.* [38] concluded that, with proper selection of the NDM system, results similar to flood lubricant condition can be
achieved.

The result of a comprehensive and well-grounded study of turning of AISI
1045 steel of HB 163 hardness by Li and Liang [39] showed that the cutting force
and temperatures are always lower for wet machining with a water-soluble oil
compared with NDM and dry machining.

7.5.3.2   Explanations Provided

Surprisingly, very few explanations are provided to clarify why NDM works. The
known explanations of the efficiency of NDM came from the oil mist principle
that was developed by a bearing manufacturer in Europe during the 1930s. Today,
this method of lubricating is applied to all types of other machine tools, web and
sheet processing equipment, belt and chain conveyors rolling mills, vibrators,
crushers, centrifuges, kilns, pulverizers, ball mills, dryers and liquid processing
pumps. Boundary lubricating layer of extreme-pressure (EP) additives and accurate delivery of the lubricant to the contact surface just before they engaged into a
contact plush cooling by oil droplets evaporation were borrowed to explain the
effectiveness of NDM in metal cutting.

Obikawa *et al.*, trying to explain the effectiveness of NDM, suggested that the
air supply plays an important role in transporting the oil mist to the interface between the flank wear land and machined surface (the tool–workpiece interface)
[26], *i.e.*, it was assumed that the oil can penetrate the tool–workpiece interface
due to the high pressure of the compressed air that serves as a vehicle for the oil
droplets. No comparison between the contact pressure at this interface and that of
compressed air was made.

Liao and Lin [40] argued that MQL can provide extra oxygen to promote the
formation of a protective oxide layer that that might form between the chip and the
rake face over the tool–chip interface. This layer is basically quaternary compound
oxides of Fe, Mn, Si and Al, and has proved to act effectively as a diffusion bar-
rier. Hence, the strength and wear resistance of a cutting tool can be retained,
which leads to a significant improvement in tool life. It was found that there exists
an optimal cutting speed at which a stable protective oxide layer can be formed.
Unfortunately, no direct conclusive evidence of the existence of such a layer were
provided.

López de Lacalle *et al.* [41] suggested that the flow of air with oil drops acts in
three different ways:

- It removes the heat generated during cutting by the convection of the injected air and the evaporation of part of the oil.
- It decreases friction between chip and the rake face of the tool, since oil
  drops are small enough to get into the chip–tool interface.
- It removes the chip of the working area due to the action of compressed
  air.

Unfortunately, no direct supporting evidence was given for these suggestions, the validity of which was justified by the reduced cutting force and tool wear. The boundary conditions selected for their finite element analysis as well as the parameters of the jet of compressed air and parameters of the coolant were not based on the known experimental data. This results in a conclusion that machining with any kind of nozzle arrangement of their NDM resulted in lower tool flank wear compared with that found in wet machining with a water-soluble oil as the coolant. Although this computational result is in direct contradiction with everyday machining practice, no explanation for this was provided.

7.5.3.3 Feasible Explanation of the Effectiveness of NDM

To understand the subject, one should ask himself a few simple questions: Why does NDM work at all? Why does it show somehow better results than flood or even high-pressure MWF supply? In other words, how is this *physically* possible, as the flood and high-pressure MWF approaches definitely remove much more thermal energy? Why does an aerosol containing an oil and air mixture have (according to all NDM research and promotional papers) greater cooling ability than water-soluble MWFs even through the heat capacity of water is ten times greater than that of oil and much greater than that of air, bearing in mind that heat removal due to the evaporation of oil droplets is negligible small because of the inherent properties of oil and tiny oil flow rate (30–60 ml/h)?

A thorough consideration of the possible action of the aerosol in NDM [1] allows the following conclusions to be drawn:

- Aerosols do not act as lubricants or boundary lubricants in metal cutting. It has been conclusively proven that, even when cutting fluids are applied, a high contact pressure exists between the chip and tool rake face, particularly along the plastic part of the tool–chip contact length, which combined with high temperature along the actual contact area and surrounding zone, precludes any fluid access to the tool–chip and tool–workpiece interfaces [1]. This is the case even if a jet of MWF of extremely high pressure (15 MPa as maximum available on commercial high-pressure MWF supply equipment) is used. As the maximum pressure of the aerosol cannot exceed that of compressed air (normally available at a maximum of 0.7 MPa and then reduced due to hydraulic losses to 0.5 MPa), the penetration ability of the aerosol is far too low compared even with MWF supplied through the tool at a high pressure.
- Aerosols used in NDM always have lower cooling ability than a properly selected water-soluble MWF. There are two possible cooling actions of the aerosol in NDM, namely cooling due to droplet evaporation and due to forced convection. As the oil flow rate in NDM is very small (normally 30–60 ml/h), the cooling action due to droplet evaporation is also small. Astakhov introduced a parameter called the cooling intensity K_h to account for forced convection of the MWF in metal cutting [1]

$$K_h = v_{cf}^{0.65} k_{cf}^{0.67} C_{P-cf}^{0.33} \gamma_{cf}^{0.33} v_{cf}^{-0.32} \qquad (7.1)$$

As can be seen, the cooling ability of a cooling medium depends upon the velocity of this medium, v_{cf}, its thermal conductivity k_{cf}, its specific heat C_{P-cf}, its specific gravity γ_{cf} and its viscosity v_{cf}. A simple comparison of the listed parameters in the case of a high-pressure water-soluble MWF with those in NDM shows that the former has much greater cooling ability [1].

- Aerosols cannot change the shape of the forming chip due to their direct mechanical action regardless of the nozzle location. To comprehend this, one should consider their momentum. For a moving body, the momentum is just the product of the mass of the body and its speed, Momentum is traditionally labeled by the letter p, and the definition of momentum is therefore $p = mv$ for a body with mass m moving at speed v. As the mass of the aerosol is small compared to that of a jet of a high-pressure water-soluble MWF, its action on the forming chip is negligibly small compared with that of a high-pressure water-soluble MWF.

Having read and analyzed these conclusions, one may ask a logical question: Why does NDM work as reported by many researchers and practitioners? As the proper answer to this question is not yet known, trial-and-error methods are used to determine the parameters of NDM.

One of the feasible and physics-based explanations is the so-called embrittlement action of the cutting fluid, which reduces the strain at facture of the work material. This action is based on the Rebinder effect [1]. Rebinder's studies were directly concerned with the metal cutting process. Conducting a great number of cutting tests under different cutting conditions and with different cutting fluids, he observed the formation and healing of microcracks. The latter was particularly pronounced when machining ductile materials where large plastic deformation of the layer being removed was observed. The results of Rebinder's study showed that absorbed films prevent microcracks from closing (healing due to plastic deformation of the work material). Because each microcrack in the machining zone serves as a stress concentrator, a lower energy was required for cutting. Pursuing this direction, Epifanov [1] found that the penetration of the foreign atoms (from the decomposition of the cutting fluid) produced an embrittlement effect in a manner similar to hydrogen embrittlement. He concluded that plasticity of the work materials reduces. Our current understanding of the Rebinder effect is as the alternation of the mechanical and physical properties of materials due to the influence of various physiochemical processes on the surface energy [1].

As shown in Chapter 1, the work of plastic deformation of the layer being removed in metal cutting is the greatest, so it is the prime cause of tool failure due to various mechanisms of wear. Once this work is done, it is too late to mitigate its consequences using inferior (in terms of cooling ability) NDM as it cannot compete against the cooling ability of a flood or a high-pressure though water-soluble tool MWFs. The only feasible way for NDM to work in metal cutting is to increase the embittlement of the layer being removed and thus to reduce the work of plastic deformation done in the transformation of the layer being removed into the chip. Available information about the practice of NDM suggests that an atomized oil possesses great ability to enhance the Rebinder effect. In the author's opinion, this explains the effectiveness of NDM. As the lowering of the surface energy of a solid to alter its mechanical properties can be achieved by adsorption, chemisorption, surface electrical polarization and surface chemical reactions [1], the research, selection of designs and parameters and testing of the effectiveness of NDM should be driven in this direction.

7.5.4   Consideration of the NDM System Components

Experience of implementing of NDM shows that this technology can be successful if and only if the whole NDM system of components is considered and special attention is paid to each and every component individually while maintaining the coherence of the system. Therefore, a 360° analysis of all components listed in Figure 7.15 is a prerequisite before any practical implementation of NDM. Unfortunately, this is not the case in practice, which in the author's opinion is the major hurdle in implementing this technology.

Regardless of its scope, this analysis should always start from the NDM setup. At this stage, a particular NDM type (see above), aerosol components and parameters should be selected and tested in terms of their suitability for the considered project. In other word, the feasibility of NDM should be verified and data for the efficiency analysis should be collected for critical machining operations, tools and work materials to be involved with NDM.

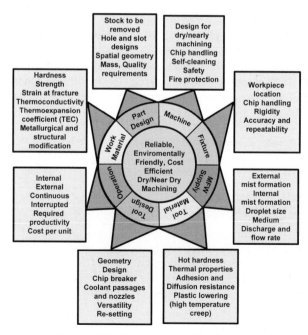

Figure 7.15. Components of the NDM system

A detailed cost analysis and the entire scope of the manufacturing system have to be studied prior to implementation of NDM. As such, the depth and extent of analysis depend on the scope of the project, starting with implementing NDM for a single stand-alone machine and finishing with a NDM manufacturing system. The wider the scope, the larger the system that should be included in such an analysis.

7.5.4.1 Cutting Tool

The next important issue to be considered is the cutting tool. The implementation of NDM requires specially designed cutting tools if the real advantages of NDM are to be achieved.

Unfortunately, a common notion is that with external NDM (NDM 1.1 and NDM 1.2) the same cutting tool used for flood coolant will work [27]. Although this is true in some particular cases, the maximum advantage of NDM is achieved if the cutting tool is specially designed for NDM, including specific tool geometry and tool material (including coating). Cutting tools and tool holders for NDM 2 should or modified be designed specially for NDM.

Round-ended cutting tools such as drills and reamers should be designed with special channels having a maximum allowable cross-sectional area and extremely smooth configurations without dead ends. A dedicated line should go to every contact (cutting, rubbing, locating, *etc*) area. Void and trapped areas in the aerosol flow path should be eliminated. Exit holes and nozzles must be positioned to provide aerosol directly to the chip-forming zones and special consideration should be give to changing position of these holes with tool re-sharpenings. In multi-stage reamers, the cross-sectional area of these holes should be selected to provide balanced aerosol supply to all stages or, better, balanced in proportion to the amount of work material cut by various stages. An example of channel design is shown in Figure 7.16.

The design of non-round tools requires special attention. The flow of aerosol coming through the centre of the spindle and a central hole in the tool holder has to change direction to reach the machining zone. The best the designer can do is to make the bend of the aerosol channel as gradual as possible, avoiding sharp turns

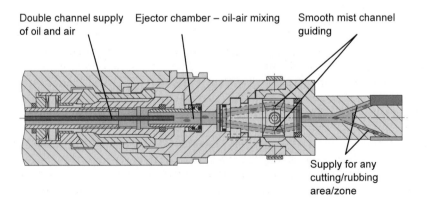

Figure 7.16. MQL internal channel design for a reamer (courtesy of Mapal USA Co.)

at any cost. This present a great challenge for small-sized tools. A number of design concepts should be considered and discussed before making the final decision on the proper tool design, as tool manufactures have very little experience with the design of tools for NDM.

The foregoing consideration implies that standard inexpensive tools cannot be used in NDM, as tools for NDM should substantially re-designed or, at least, modified, which makes them special. Therefore, one should be prepared to bear greater initial tooling costs and these costs should be a part of the analysis made prior to considering an NDM application. As these new tools are not off-the-shelf products stored by manufacturers and distributors, the lead time to deliver, modify and repair this tool is much longer than with standard tools.

New procedures in tool setting should also be introduced. One of the most critical of these is the verification of the aerosol flow behaviour after each setting or re-setting of the cutting tool [42]. To do this, special equipment that matches the production machinery (the same aerosol system, oil, oil flow rate, operating parameters, *etc.*) should be available for spray pattern testing of the tool–tool holder assembly. Naturally, this adds time and thus cost that should also be included in the cost analysis.

An additional important tooling issue that should be addressed in NDM is the modification of standard tool holders to prevent any clearance in the line that introduces aerosol to the tool holder and any cavity in the tool holder [42].

7.5.4.2 Chip Management

MWFs perform two important technological functions: (a) cooling the workpiece so that it does not distort excessively and does not get excessively hot. As a result, its proper in-process and post-process gauging as well as interoperation handling (loading/unloading and transportation) do not present problems, (b) flushing away chips so the tool, the workpiece, the fixture and the machine are not damaged. The challenge for NDM is to provide substitutes for these critical functions. Although since the late 1990s a number of research programs and pilot projects have been conducted to establish and verify the fundamentals of dry cutting [10], it is not yet obvious whether such substitutions can be achieved economically in industrial applications.

Smooth, continuous chip evacuation is a key feature in any high-productive machining operation, particularly when high-penetration-rate tools, *i.e.*, in the automotive industry in machining of aluminium alloys, are considered. High-pressure MWF is effective in ejecting chips from the drills, reamers, taps *etc.* In NDM, air replaces the liquid, which presents a problem as air cannot produce the ejection force that a liquid at a high flow rate can exert. A feasible solution should be sought in the location of aerosol exit holes, increased and optimized chip flutes and adjusted speeds and feeds.

Even if the chips are removed from the hole being machined, they must still be removed from the tool and fixtures into a collection hopper to prevent damage to the tool, workpiece and machine tool. To do this, special vacuum hoods have to be designed and installed. Such a hood can be a part of the machine, as shown in Figure 7.17. It is obvious that these hoods make tool change, tool maintenance and setting more difficult and time consuming.

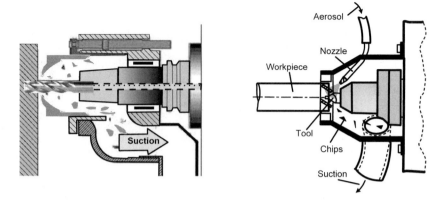

Figure 7.17. Vacuum hoods installed in the machine for MQL2 and MQL1

Vacuum chip-removing hoods can also be a part of the tool assembly as shown in Figure 7.18. Each tool should be equipped with such a hood, which increases the cost of the tool and its setting.

The worktable mechanism offers one approach to moving chips to the collection mechanism. In some designs, the table mechanism flips the workpiece from an upright position for loading to an upside-down position for machining. The flipping actions throws off the stray chips collected on the fixture and worktable, and the majority of these chips fall down onto a chip conveyor belt built into the base of the machine.

Another design necessity for reducing machine downtime for clean-up is to design the walls of all interior surfaces surrounding the working area to be steep enough that chips will slide down and drop onto the chip conveyor belt. As this is not always possible, some machine designs offer blowers that can blast loose chips that have adhered to the walls. Properly located blowers enable the machine to be cleaned automatically without opening the machine hood, avoiding significant downtime.

Figure 7.18. Vacuum hoods on the tool (courtesy of Mapal USA Co.)

7.5.4.3 Concluding Remarks

From the steps that must be taken to implement NDM, it is apparent that, to make this technology reliable, environmentally friendly and cost efficient, the whole picture has to be considered. First of all, a machine tool (manufacturing cell, production line *etc.*) and cutting tools should be designed specifically for NDM. Although some machine tools have been retrofitted for NDM, retrofitting an existing machine does not appear to be an attractive alternative.

Ideally, the implementation of NDM starts from the part design, which should make NDM easier in terms of chip removal and evacuation. For example, deep blind holes are to be avoided giving preference to core holes; threads should be designed to make them suitable for form rather than cutting threading taps; preference should be given to open flat faces suitable for face and interpolating milling; parts (particularly aluminium parts in the automotive industry) should be designed to be less susceptible to thermal distortion due to heat generated in machining with the aid of FEM thermal distortion analysis *etc.* The choice of the part material and its metallurgical state should also be made, accounting for machinability and susceptibility to thermal distortion. Chip-breaking problems should be one of the prime concerns as, if this problem is not resolved, it can easily make the whole NDM project impractical.

As discussed above, the cutting tools should be designed for NDM. Not only should the aerosol-supply channels be suitable, but also the tool geometry, tool materials and design of the tool body (back tapers, reliefs, undercuts and supporting elements) should be optimized for NDM. Additional procedures in tool setting and maintenance as well as additional equipment for aerosol verification must be implemented and followed.

The success of NDM can be achieved if all components of machining or manufacturing system are suitable for this technology.

References

[1]   Astakhov VP (2006) Tribology of Metal Cutting. Elsevier, London
[2]   Taylor FW (1907) On the art of cutting metals. Trans ASME 28: 70–350
[3]   Childers JC (1994) The chemistry of metalworking fluids. In Byers JP (ed.) Metalworking Fluids. Marcel Dekker, New York, NY, pp. 165–189
[4]   Mariani G (1990) The selection and use of semi-synthetic coolants. SME Paper MF90-321: 1–10
[5]   Graham D (2000) Dry out. Cutting Tool Eng 52: 1–8
[6]   Narutaki N, Yamane Y, Tashima S, Kuroki H (1997) A new advanced ceramic for dry machining. Ann CIRP 16(1): 43–48
[7]   Granger C (1994) Dry machining's double benefit. Mach Prod Eng 54: 14–20
[8]   Feng FS, Hattori M (2000) Process Cost and Information Modeling for Dry Machining NIST publication
      http://www.mel.nist.gov/msidlibrary/doc/cost_process.pdf (accessed May 17, 2008).
[9]   Balulescu M, Herdan JM (1997) Ecological and health aspects of metalworking fluids manufacture and use. J Synth Lubr 15(1): 35–45
[10] Klocke F, Eisenblaetter G (1997) Dry cutting. Ann CIRP 46(2): 519–526

[11] Balulescu M, Herdan JM. (1997) Ecological and health aspects of metalworking fluids' manufacture and use. J Synth Lubr 14: 35–45

[12] Piacitelli GM, Sieber WK, O'Brien DM (2001) Metalworking fluid exposures in small machine shops: an overview. AIHA J 62: 356–70

[13] Simpson AT, Stear M, Groves JA, Piney M, Bradley SD, Stagg S, Crook B (2003) Occupational exposure to metalworking fluid mist and sump fluid contaminants. Ann Occup Hyg 47(1): 17–30

[14] Sprince NL, Palmer JA, Popendorf W, Sprince NL, Palmer JA, Popendorf W, Thorne PS, Selim MI, Zwerling C, Miller ER. (1996) Dermatitis among automobile production machine operators exposed to metal-working fluids. Am J Ind Med 30: 421–429

[15] Zacharisen MC, Kadambi AR, Schlueter DP, Zacharisen MC, Kadambi AR, Schlueter DP, Kurup VP, Shack JB, Fox JL, Anderson HA, Fink JN. (1998) The spectrum of respiratory disease associated with exposure to metal working fluids. J Occup Environ Med 40: 640–647

[16] Stear M (2005) Metalworking fluids – clearing away the mist? Ann Hyg 49(4): 279–281

[17] Walz TJ. Managing Coolants from Machining and Grinding Operations (on line book http://www.carbideprocessors.com/Coolant/book/)

[18] Skerlos SJ (2007) Prevention of metalworking fluid pollution: environmentally conscious manufacturing at the machine tool. In Kutz M (ed.) Environmentally conscious manufacturing. Walley, Hoboken, NJ, pp. 95–122

[19] Astakhov VP, Frazao J., Osman MOM (1991) On the design of deep-hole drills with non-traditional ejectors. Int J Prod Res 28(11): 2297–2311

[20] Astakhov VP, Subramanya PS, Osman MOM (1996) On the design of ejectors for deep hole machining. Int J Mach Tool Res Manuf 36(2): 155–171

[21] Yoshimura H, Itoigawa F, Nakamura T., Niwa K (2006) study on stabilization of formation of oil film on water droplet cutting fluid. Trans Jpn Soc Mech Eng C 72: 941–946

[22] Itoigawa F, Childs THC, Nakamura T, Belluco W (2006) Effects and mechanisms in minimal quantity lubrication machining of an aluminum alloy. Wear 260(3): 339–344

[23] Yoshimura H, Itoigawa F, Nakamura T, Niwa K (2005) Development of nozzle system for oil-on-water droplet metalworking fluid and its application to practical production line. JSME Int J Series C 48(4): 723–729

[24] Wu C-H, Chien C-H (2007) Influence of lubrication type and process conditions on milling performance. Proc Inst Mech Eng Pt B: J Eng Manuf 221: 835–843

[25] Quaile R (2005) Understanding MQL. Modern Machine Shop 36(1):12–19

[26] Landgraf D (2004) Factors to consider when dry or near-dry machining. Cutting Tool Eng 56(1): 1–5

[27] Woods S (2006) Near dry. Cutting Tool Eng 58(3): 16–17

[28] Canter NM. (2003) The possibilities and limitations of dry machining. Tribol Lubr Technol 59: 30–35

[29] Lorincz J (2007) The right solution for coolant. Manuf Eng 139(2): 89–96

[30] Ueda T, Nozaki R, Hosokawa A (2007) Temperature measurement of cutting edge in drilling – effect of oil mist. Ann CIRP STC C 56(1): 93–96

[31] Khan MMA, Dhar NR (2006) Performance evaluation of minimum quantity lubrication by vegetable oil in terms of cutting force, cutting zone temperature, tool wear, job dimension and surface finish in turning AISI-1060 steel. J Zhejiang Univ Sci A 7(11): 1790–1799

[32] Dhar NR, Ahmed MT, Islam S (2007) An experimental investigation on effect of minimum quantity lubrication in machining AISI 1040 steel. Int J Mach Tools Manuf 47(5) 748–753

[33] Kuan-Ming Li K-M, Liang SY (2007) Modeling of cutting forces in near dry machining under tool wear effect. Int J Mach Tools Manuf 47(7–8): 1292–1301

[34] Dhar NR, Islam S, Kamruzzaman M (2007) Effect of minimum quantity lubrication (MQL) on tool wear, surface roughness and dimensional deviation in turning AISI-4340 steel. GU J Sci 20(2): 23–32

[35] Filipovic A, Stephenson DA (2006) Minimum quantity lubrication (MQL) applications in automotive power-train machining. Mach Sci Technol 10(1): 3–22

[36] Obikawa T, Kamata Y, Shinozuka J (2006) High-speed grooving with applying MQL. Int J Mach Tools Manuf 46(14): 1854–1861

[37] Braga DU, Diniz AE, Miranda GWA, Coppini NL (2002), Using a minimum quantity of lubricant (MQL) and a diamond coated tool in the drilling of aluminum–silicon alloys. J Mater Proces Technol 122(1): 127–138

[38] Davim JP, Sreejith PS, Silva J (2007) Turning of Brasses Using Minimum Quantity of Lubricant (MQL) and Flooded Lubricant Conditions. Mater Manuf Process 22(1) 45–50

[39] Li K-M, Liang S (2006) Performance profiling of minimum quality lubrication in machining. Int J Adv Manuf Technol, DOI 10.1007.s00170-006-0713-1

[40] Liao, YS, Lin HM (2007) Mechanism of minimum quantity lubrication in high-speed milling of hardened steel. Int J Mach Tools Manuf 47(11): 1660–1666

[41] López de Lacalle LN, Angulo C, Lamikiz A, S´anchez JA (2006) Experimental and numerical investigation of the effect of spray cutting fluids in high speed milling. J Mater Process Technol 172: 11–15

[42] Stroll A, Furness R (2007) Near dry machining (MQL) is a key technology for driving the paradigm shift in machining operations. SME Paper TP07PUB22:1–14

8

Sculptured Surface Machining

L. Norberto López de Lacalle and A. Lamikiz

Department of Mechanical Engineering, University of the Basque Country
Escuela Técnica Superior de Ingenieros Industriales, c/Alameda de Urquijo s/n,
E-48013 Bilbao, Spain.
E-mail: norberto.lzlacalle@ehu.es, aitzol.lamikiz@ehu.es

This chapter is related to the machining of complex surfaces, ruled or sculptured, which is very common for forge and stamping dies, injection moulds and aeroengine components. Difficult-to-cut materials and complex part geometry are the two factors involved. The current technology for finishing is high-speed milling in three axes using ball-end milling tools, or in five axes using ball or flat-end mills.

Dies and moulds require precise final forms and good roughness. Modelling the milling operation helps to define optimal tool paths and cutting conditions. Some examples illustrate current machining procedures.

8.1 Introduction

The manufacturing of complex surfaces is common to several sectors. Machining is the most important technology to achieve a precise shape of dies and punches, moulds, or blades.

The challenge is to produce precise free-forms on difficult-to-cut materials, with narrow tolerances and good economical performance. The latter aspect is critical because low-wage countries are new competitors in the dies and moulds market. Therefore, several aspects must be taken into account: the use of three- or five-axis machines, powerful computer-aided manufacture (CAM) systems, high-tech tools and having skilled programmers and machinists.

Within the general group of complex surfaces, two types can be identified: *ruled surfaces*, used in blades and compressor disks, and *sculptured surfaces* (or *free-form surfaces*), typical in moulds and dies. For industrial applications, four classes are defined, whose main features are shown in Table 8.1:

- **Forge dies**: Constitutive materials are treated steels, especially made for hot work, with hardness ranging from 30 to 60 HRC. Tolerances and final roughness are quite large. Narrow and deep zones are not usual.
- **Stamping dies**: Constitutive materials are ductile iron casting. However, in the last 5 years, the use of advanced high-strength steels (AHSS) for car bodies has become more common. Consequently, a higher proportion of harder surfaces on the forming tool, tempered iron, and 60-HRC insert blocks must be machined. In addition, the finishing machining step may take several hours. For this reason, many users prefer less durable but more reliable tool materials (such as carbide) rather than harder but more fragile tool materials (such as PCBN); in the latter case, an uncontrolled allowance may produce an unexpected tool breakage, with catastrophic consequences for the die.
- **Moulds**, for plastic or aluminium alloy injection. The mould is made of steel tempered to 50–55 HRC. Shape accuracy and good finishing are required by the injection process. The complexity of plastic parts leads to mould design with very narrow and deep zones, some on the borderline between cutting technology and electro-discharge machining.
- **Ruled surfaces**: Some energy or aircraft engine parts present this type of surface, including workpiece materials of Ti6Al4V, Inconel 718 (or similar super-alloys) or aluminium alloy. In these applications, high precision in the interface zones is required. A particular type of finishing operation, flank milling [1], tries to make use of ruled geometry, following tool paths in which the tool generatrix (a cylinder or a cone) is tangent to the part rulings. Otherwise, this surface is machined with the same procedures as freeform surfaces.

Table 8.1. Requirements of different sectors involving complex surfaces

Sector	Raw material	HRC	Roughness Rt	Tolerances	Fillet radius
Forge dies	56NiCrMoV7 (L6) X40CrMoV5/11 (H13)	42–60	5–10 µm	0.2 mm	More than 1.5 mm
Plastic injection	X37CrMoSiV5 1 (H11) X40CrMoV5 1 (H13) X30WCrV9 (H21)	48–50	1 µm (0.5 Ra), mirror-like finishing	0.01 mm in interfaces	Sharp edges
Aluminium injection	X40CrMoV5 1 (H13)	50–55	1–2 µm, mirror-like finishing	0.05 mm	Several cases
Stamping dies	GG25 (ASTM 48 Grade 40B) GGG70 (ASTM 100-70-03)	220–270 HBN	20 µm	0.1 mm in free forms	Small
Blades	Ti6Al4V	35	2–5 µm	0.05 mm	3 mm but external

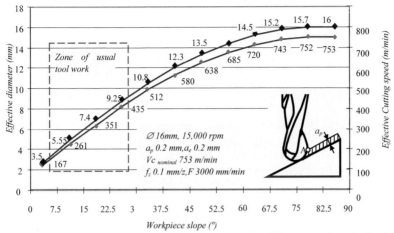

Figure 8.1. Effective tool diameter and cutting speed for different surface inclinations, using a ball-end milling tool. A is the point of maximum cutting speed

In this chapter, we focus on high-speed milling as the basic technology for finishing, the final and high-added-value stage when complex forms are produced. This operation commonly involves the milling of a 0.3-mm allowance [2], which is usually done using ball-end mills with diameters below 20 mm because of the intricate shape details. Considering that slopes commonly found in sculptured forms are from 0° to 90° and that effective cutting speed must be between 300 and 400 m/min (the maximum recommended for TiAlN-coated carbide tools), the spindle rotational speed must be over 15,000 rpm. This requirement necessitates the use of high-speed spindles. The maximum rotational speed of industrial high-speed machines is around 20,000–25,000 rpm, with power ranging from 14–20 kW. For this rotational speed, bearing in mind a recommended feed per tooth of around 0.07–0.1 mm/th, the maximum linear feed is 10–15 m/min. These values can be obtained by typical linear ball screws connected to synchronous motors, or by linear motors. The control of interpolated curves at this feed is not a problem for current numerical controls [3].

In Figure 8.1, values of the maximum cutting speed at the effective diameter are shown for a ∅16-mm integral ball-end milling tool. Integral carbide tools are used more than tools with inserts for finishing.

8.2   The Manufacturing Process

The design of complex moulds, forging dies or stamping dies strongly depends on the final process technology. However, after the design process, the problem is somewhat similar for all: the machining of a complex surface, most often of sculptured (free-form) type. At this stage, the main differences are the tolerance of dimensions, roughness levels and the lead time imposed by the competitive market.

8.2.1 Technologies Involved

Before the generalized use of high-speed milling, the technology employed in mould manufacture was a combination of conventional milling and electro-discharge machining (EDM). Figure 8.2 shows the three processes that have been used in making a mould: (A) the traditional process before the advent of high-speed milling; (B) the process with high velocity of the 1990s; and (C) the process dating from 2000.

Conventional milling (d1, e1) can be applied directly only to moulds made of aluminium or soft steel, before tempering. Because of the slow linear feed, the finishing operation requires a long machining time. EDM (g1) is used with tempered steels. The electrode may be machined in graphite (f1), making use of high-speed milling machines. The graphite electrode is used for finishing the piece to its final dimensions, and the orbital motion of the electrode can be used to enhance the surface finish. Mould accuracy depends directly on electrode accuracy.

From 1997 to 1999, roughing (d2) and semi-finishing (e2) were usually carried out with conventional machines, with mould steel in a soft state before tempering. Subsequently, heat treatment was applied. After that, finishing was performed in high-speed machining centres (f2). There were two reasons for this sequence:

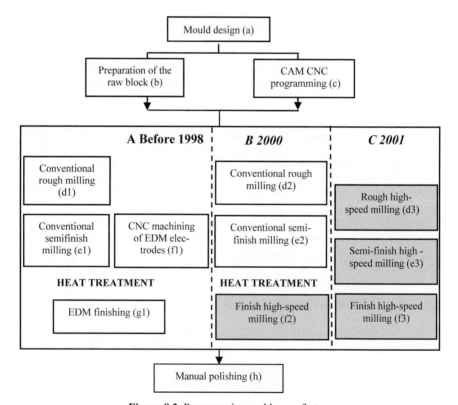

Figure 8.2. Processes in mould manufacture

- Roughing, with its few precision requirements, was done in machines having a per-hour cost a quarter that of high-speed machines. Moreover, tool wear was light because of the low hardness of the workpiece material.
- The typical high-speed spindles available at that time could not deliver sufficient torque below 1,500 rpm, making roughing impossible.

In 2000, improvements made to high-speed spindle control resulted in an improved capacity to deliver enough torque even at low rotational speeds. Roughing in high-speed machines became possible, with an application process similar to that used in the conventional case. Therefore, a new procedure was defined, starting directly from a block that was initially heat treated, with all operations (d3, e3, f3) carried out consecutively in the same machine. The main advantage of this simpler process was the shorter time needed to launch a new mould because less time was required for setup between successive operations. At the same time, accuracy and workpiece reliability also increased because of the elimination of workpiece-zero setups between operations. Two new roughing strategies have been developed in recent years: plunge milling, which involves a big-end mill that removes a large amount of material through successive drilling, and high-feed milling, in which a high fz is applied (even more than 2 mm/th) along with small axial depths of cut. In the latter case, the semi-finishing operation is eliminated, and a low process time is achieved.

At present, economic criteria and required lead times drive the decision of whether to use high-speed machines with tempered raw material, or conventional roughing of untempered steel followed by tempering and high-speed milling. As an illustration, within 10 days of order acceptance, iron-forging firms must launch preliminary runs of a new component to be produced. This short lead time makes the former approach very attractive. In relation to mould production, from our experience, two conclusions can be drawn:

- If more than 7000–8000 cc of material must be removed, roughing time in a high-speed machine may be excessive and too costly. This figure may be the decisive criterion in choosing between the two procedures.
- If less than 1250 cc of material is to be removed, roughing on the same high-speed machine makes production of a finished mould possible in less than 150 minutes, with one single clamping and the same zero reference.

At present, shrinkage EDM is clearly falling out of favour for the manufacturing of moulds, being used only in very deep and low-fillet-radii zones. On the other hand, new applications of wire EDM are being identified at a rapid pace.

8.2.2 Five-axis Milling

In five-axis milling, two additional orientation axes allow the machining of very complex parts that cannot be machined using three-axis machines. For example, in the automotive sector, the five faces of engine blocks can be machined in only one setup and fixturing, as depicted in Figure 8.3 (left).

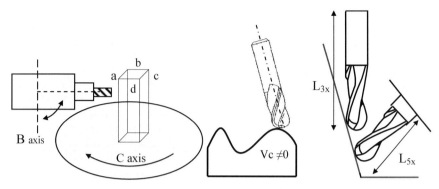

Figure 8.3. The three main advantages of five-axis milling: working on all workpiece sides in one setup, avoiding the tool tip cutting, and using shorter tools ($L_{5x} < L_{3x}$)

On the other hand, cutting speed is zero at the tool tip, making tool cutting very unfavourable. Because of this factor, when ceramics or PCBN tools are used, a typical failure mode is the breakage of the tool tip. With five axes, milling can be performed avoiding the tool tip cutting, as is shown in Figure 8.3 (centre). The machine tool manufacturer Starrag gives another example [4]: the Sturz (also called P-milling®) machining strategy, in which bull-nose tools for the milling of free-form surfaces are used, reducing process time by two-thirds.

Moreover, using five-axis milling, tool overhang can be reduced. Therefore, tool stiffness is greater, which increases machining precision and reduces the risk of tool breakage (Figure 8.3, right). As shown, tool stiffness [5, 6] is directly related with the tool slenderness factor L^3/D^4 (Section 8.4); thus, a tool length reduction dramatically reduces tool deflection and the lack of precision arising from this effect. In the last 5 years, at industrial fairs (EMO in Milano 2003, Hannover 2005 and 2007), many five-axis milling centres have exhibited machining in the $3+2$ operation mode, orienting the tool axis with respect to the target surface and machining only with linear interpolation of the three Cartesian axes of the machine.

Other technological advantages can be highlighted. In the finishing operation of ruled surfaces, the flank milling strategy can be used, machining all inclined side faces with the cylindrical (or conical) part of the tool, applying a large axial depth of cut. In addition, in the machining of inclined planes, with the use of the correct tool-axis orientation, a face milling operation can be carried out instead of a ball-end sculptured milling operation. These approaches reduce processing time and improve surface quality.

However, there are two drawbacks when five-axis machining is applied: those related to the complexity of CAM and tool-path generation, and potential interference during the process from collisions of the tool against the part and workpiece fixtures, and even between the different components of the machine. The geometrical calculation of the position of the tool centre point (TCP) and the tool-axis orientation is carried out directly for all commercial software with five-axis capabilities (Unigraphics™, Catia™, Openmind™, GibbsCam™, Powermill™ and others); therefore, with their use, a skilled CAM user will not experience any problems. The algorithms used in these systems are explained in their theoretical manuals, and abundant technical information about them is also available [6]. There are also CAM software packages specifically focussed on special applications, for example impellers and other turbo-machinery components.

From the CAM user's point of view, the main trouble during tool path generation appears in the post-processing step, when the tool path generated by the CAM is translated into a CNC code for a particular five-axis machine. For example, a machine with the two rotary additional axes in the bed is very different from those with two orientation angles (the twist and tilt angles) in the spindle head. The same part and even the same automatically programmed tool (APT) code obtained from CAM leads to very different CNC codes for each of these.

Another real risk is tool collision during milling. Collisions can damage the weak hybrid bearings of the high-speed spindle (these bearings are composed of steel races with ceramic balls), involving high repair costs and long off-production times. Even if the machine's spindle is not damaged, it must be remembered that the five-axis process is usually applied on complex and high added-value parts, such as impellers made of titanium or super-alloys, or near net shape precision cast parts; therefore, machining errors can also damage the workpiece, wasting a lot of previous machining time and expensive raw material.

As explained in [2] for the three-axis machining of complex surfaces, in five-axis milling, a new approach to the CAM stage must be applied, improving the reliability of the whole process. Here the definition of reliability is achieving good productivity with low risk to parts out of tolerances or with irrecoverable errors. In five-axis milling, the CAM is the centre of gravity of the planning process. Workshop workers can only change the actual values of the cutting speed and feed rate, making use of the machine dials (which modify the actual feed and spindle rotation speed with respect to those programmed in the CNC code); it is not possible for them to change the complex tool path directly in the CNC interface. A new intelligent CAM procedure is presented below. This production scheme includes a scientific model to evaluate the cutting forces.

8.3   The CAM, Centre of Complex Surfaces Production

In Figure 8.4, a reliable scheme for CAM and process planning for three- and five-axis milling is proposed. The commercial CAM software achieves the common requirements for three- and five-axis machining to define tool paths on the machined surfaces. For this purpose, machining strategies, such as zig-zag, z-level, rest milling and by-user definition of the *intol* and *outol* parameters, are provided. However, new concepts are included in the five-axis scheme compared to the three-axis one.

Thus, there is the possibility of using new finishing strategies for applying five-axis milling, which allows improvements in precision and quality of complex surfaces. For example, using bitangencies in corners achieves a better use of the tool cutting edge in the finishing of sharp surfaces. The selection of the tool-axis angle with respect to the surface to be machined and the feed direction with respect to the reference line are parameters that the CAM operator freely chooses. Commercial software allows the user to define different values for these angles, but no recommendation is given about how to do so correctly. In a published analogy [7], a cutting force estimation model is presented as an analysis utility for the best selection of the minimal force tool path in three-axis machining, a problem similar to the climber's choice of the best path for scaling a mountain. A climber will choose a climbing path depending on his or her training, but the ascent will require a great effort, regardless. However, in five-axis milling, the mountaineer (the CAM operator) can knock the mountain down to reach the peak.

Based on the above observations, some recommendations have been introduced into the CAM scheme for prior selection of tool-surface orientation. Tilt and lead angles of tools related to surface and machining direction, respectively, can be pre-selected to reduce cutting forces. This point is elaborated on in Section 8.6 and is shown as a module (f) in Figure 8.4.

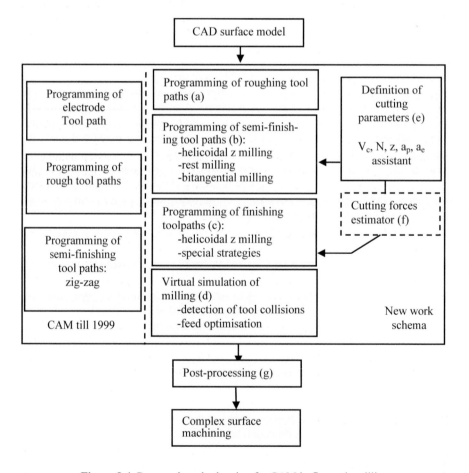

Figure 8.4. Proposed work planning for CAM in five-axis milling

The next step after tool path generation is post-processing, presenting those problems and features explained in the previous section. The post-processor definition is a difficult task needing a great knowledge of the postprocessor generator utility and of the machine tool itself.

After post-processing, the next step is tool path verification. Strongly recommended in three-axis milling, this step becomes essential in the case of five-axis machines. Available software systems (Vericut™, Predator™, *etc.*) allow the user to perform a virtual simulation before actual machining, allowing detection and correction of problems, as follows:

- Collisions and interferences between tool and part, toolholder and part, or even between spindle head and machine bed.
- Problems from the tool gouging into the workpiece, an important aspect of the machining of corners.

The output of the virtual verification is a list of collisions, interferences, and detections of bad cuts (milling under the theoretical CAD surface or uncut overstocks). After the list examination, changes in the CAM tool paths are done by the CAM operator, generating a new CNC program free of interferences. However, the result is far from a true reliable tool path. Virtual simulation takes into account only geometric collisions, and problems from the cutting process itself are not revealed. In spite of these limitations, the virtual simulation is a powerful tool for achieving a good machining process, allowing very fast pre-process error detection. Nowadays, the trend is to include virtual simulation inside the CAM software, or to include a direct link from CAM to other partner software programs for verification.

Finally, the machining of the workpiece will be performed. Even with a correct previous simulation, this step leads to its own problems: wrong cutting parameters or an incorrect tool selection can lead to bad results, even with correct CAM tool paths. CAM operators can select cutting parameters based on the recommendations from tool manufacturers' databases (module e in Figure 8.4), or based on the historical background collected in the company's database. In aluminium and other light alloys, this procedure is enough to define good cutting conditions. However, in the cases of tempered steels (for moulds), heat-resistant alloys (impellers and blades) and titanium alloys (blades and frame components), cutting conditions usually calculated only for good tool life are not optimal. Other aspects such as tool deflection, which directly influences parts precision and roughness, are more important. Tool deflection is directly related to the cutting forces that are going to be calculated and offered as information to the CAM user (Section 8.6).

This CAM scheme has been implemented in several small and medium-size enterprise (SME) companies in the Basque Country of Spain. Waste part reductions of 20% (and even 35% in the best case) have been reported since 2003.

8.4   Workpiece Precision

Dimensional errors on sculptured surfaces depend on the chordal error applied by CAM, the machine-tool kinematic errors, the tool deflection caused by the cutting forces and on the delay error of the CNC along curve interpolation. A complete study of the stiffness chain of the global system has been presented in two publications, one [8] for the conventional scale and one [9] for micromilling (machine, toolholder and tool), giving the following figures for the former scale:

- The machine stiffness, measured as the displacement of the spindle nose with regard to the machine bed due to a force, can be obtained by experimental or finite element tests. Typical values are about 30–62 N/μm in X, 30–40 N/μm in Y, and 67–95 N/μm in Z (vertical machining centres).
- The most common shank/toolholder is the HSK 63 type. Radial stiffness of these shanks ranges from 25 N/μm at 5,000 rpm to 20 N/μm at 25,000 rpm.
- With respect to tools, 0.15 N/μm for 6∅ mm, 4.4 N/μm for 12∅ mm, and 1.6 N/μm for 16∅ mm were measured at the tool tip. Even in the best case (12∅ mm), tool stiffness is approximately four times less than the shank. Usual models for tool deflection consider it as a cylindrical cantilever beam. Then, deflection conforms to the equation:

$$\delta = \frac{64\,F}{3\pi\,E}\frac{L^3}{D^4}$$

(8.1)

where E is the Young's modulus for tool material, and L^3/D^4 is the tool slenderness parameter, where D is the equivalent tool diameter and L is the overhang length. F is the cutting force component perpendicular to the tool

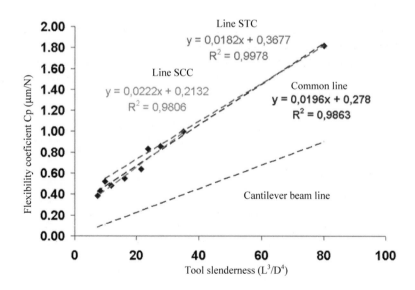

Figure 8.5. Approximate extrapolation lines for stiffness coefficients: lines for shank with cylindrical collet (SCC) and shank with tapered collet (STC), line common to all cases and lines of the cantilever beam case

axis and contained on the plane composed by the tool axis and the normal to surface at the cutting point, called the deflection force hereinafter. The Young's modulus of micrograin-grade carbide tools sintered with 10–12% cobalt is approximately 6×10^5 N/mm² (three times stiffer than steel).

This model is good enough to explain qualitative aspects but overlooks the effects of the machine itself and the toolholder. Some useful values have been proposed for the whole system in [8, 10] (see also Figure 8.5), in which the user can approximate tool tip flexibly for typical tapered or cylindrical collet shanks, and values of L^3/D^4. The flexibility is an input for the deflection estimation of a particular tool.

Natural frequencies of the spindle-tool assembly for three- and five-axis machining centres were measured for obtaining and checking the stability lobes of machining operations. For a typical case, the value of the low frequencies was 436 and 1115 Hz on the X-axis and 430, 720, and 1110 Hz on the Y-axis. These values, along with the small cut depth of finishing, avoid dynamic problems such as chatter in high-speed finishing of moulds, dies and other complex free-forms.

8.4.1 Cutting Forces

Cutting forces are a function of various factors, including tool geometry, work material, workpiece geometry, cutting conditions and the sense of machining with respect to the surface. Forces vary greatly with the machining strategy, both in magnitude and direction. Dimensional error on the surface also depends on the value and direction of the deflection (see Equation (8.1)).

Some research has focussed on the estimation of cutting forces [11–14]. Basically there are three approaches:

- The finite element method: the complexity of the three-dimensional (3D) cutting model requires a long computation time that renders this model type untenable for real-time calculations. These models are interesting for tool design or for hypothesis or results interpretation in research projects about machining.
- A mechanistic model for the cutting forces calculus, along with a solid modeller to define the swept volume at each surface point [15].
- A mechanistic model as in the previous case, but including the effect of the part slope through geometrical considerations. This has been well explained previously [16, 17].

Figure 8.6 provides a brief explanation of this last model. Cutting force can be calculated as a combination of the shear and ploughing cutting force components, based on the expressions presented by Lee and Altintas [18]. The tangential, radial and binormal components are calculated at each edge point as shown in Equation (8.2).

$$\begin{cases} dF_t\left(\theta,z\right) = K_{te}dS + K_{tc} \quad \cdot t_n\left(\varPsi,\theta,\kappa\right) \ db \\ dF_r\left(\theta,z\right) = K_{re}dS + K_{rc} \cdot t_n\left(\varPsi,\theta,\kappa\right) \ db \\ dF_a\left(\theta,z\right) = K_{ae}dS + K_{ac} \cdot t_n\left(\varPsi,\theta,\kappa\right) \ db \end{cases} \tag{8.2}$$

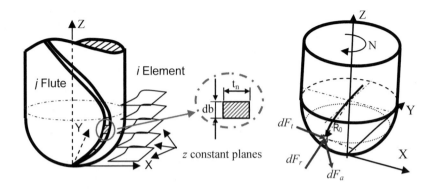

Figure 8.6. The basis of the mechanistic model for cutting force estimation: discretization of the cutting edge, calculus of the chip section, and forces at each discrete edge element

where dF_t, dF_r, dF_a (in N) are the tangential, radial, and axial components; K_{tc}, K_{rc}, and K_{ac} (in N/mm^2) are the shear-specific coefficients; K_{te}, K_{re}, K_{ae} (in N/mm) are the edge-specific coefficients; dS (mm) is the length of each discrete element of the cutting edge; t_n (in mm) is the undeformed chip thickness; and db (mm) is the chip width in each cutting edge discrete element. Therefore, it is necessary to calculate the undeformed chip thickness and the length of each discrete element of the cutting edge, which requires geometrical modelling of the tool.

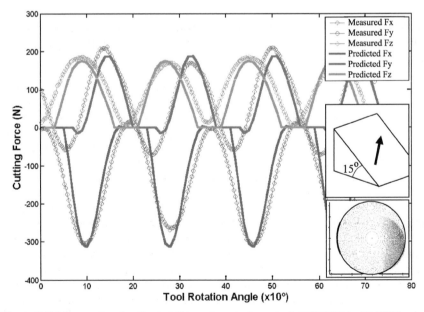

Figure 8.7. Measured and estimated forces for a test on steel AISI H13 to 52 HRC using a 45° upward feed sense strategy on a 15° slope, $a_p = 1$ mm, $f_z = 0.032$ mm. Image of the tool engagement into material

A coordinate transformation is applied to introduce the case of slope milling. Thus, the force over each cutting edge discrete element is obtained, and finally the resulting force is determined by the numerical sum along the edge engaged with the material, simultaneously taking into account the teeth that are cutting. One result for a milling test of a 15° slope is presented in Figure 8.7.

For calculating the specific cutting coefficients, a set of experimental tests has to be carried out. Forces experimentally measured during these tests are used as input data, together with the cutting parameters (a_p, a_e, Vc, fz) of each test, to obtain the model specific coefficients. The calculation is based on an inverse method of the force model used.

There are differences regarding the kind of specific coefficients to be used. The simpler models use constant coefficients, but these are only valid for end mills. A usual approach is to calculate the force coefficients by obtaining a function capable of fitting the force values measured in the set of characterization tests; these are the typical polynomial functions (with different degrees). Polynomials depend on the undeformed chip thickness or the z position of each cutting edge element. For example, Feng and Menq [11] applied cubic polynomial shear coefficients with good results. Regarding models that introduce the ploughing cutting coefficients, only Engin and Altintas [12] have introduced z-dependent quadratic coefficients for inserted milling cutters. In Lamikiz et al. [16], these z-linear shear and constant ploughing coefficients are demonstrated as sufficiently precise.

8.5   Workpiece Roughness

In finishing, not only good accuracy but also good roughness levels must be achieved. Theoretical roughness (Figure 8.8) for a ball-end milling operation conforms to the equation

$$P = \sqrt{(8 \cdot R \cdot R_t - 4\,R_t^2)}\ \ cos\,\alpha \tag{8.3}$$

Here P (named *step*, similar to a_e) is the radial width of cut, R is the tool radius, R_t is the maximum theoretical roughness (R_t is similar to C, the scallop height) and α is the slope of the surface (in the direction in which the step P is increased). However, in practice there are other factors, such as vibration, material plasticity and tool deflection, which affect the values predicted by Equation (8.3), causing roughness that is perhaps 20–30% greater.

The estimation of real topography remains a research issue [19–21]. However, in many industrial cases, the exact prediction of roughness may not be so useful because of the final by-hand polishing usually performed, or alternative finishing operations such as ball burnishing [22] or in some very particular cases laser or chemical polishing. The former has been successfully applied for complex surfaces [22].

The burnishing devices recommended for die and moulds are based on a *hydrostatic spring*, whose main advantage is that the ball load is constant during the process and related to the maximum pressure survey by an external pump. This is a high-pressure pump with low flow, taking coolant up from the machine-tool reservoir.

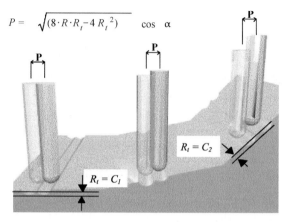

$$P = \sqrt{(8 \cdot R \cdot R_t - 4 R_t{}^2)} \quad \cos \ \alpha$$

Figure 8.8. Influence of slope variation α on scallop height C; with the same P, $C_2 > C_1$

A movement of the ball head up to 10 mm is possible without changes in the force value. The key element is a ceramic ball $\varnothing 6$ mm; this material exhibits low adhesion to steels and cast irons, the constitutive materials of moulds and dies. The ceramic ball is entirely supported by the fluid, freely rotating on the workpiece surface, as shown in Figure 8.9. The ball-burnishing head is placed into an HSK63 shank and manually inserted into the main machine spindle in which rotation has been blocked.

Burnishing is applied in a zig-zag, at the maximum machine linear feed (usually 10–15 m/min). The rolling ball smashes the peaks into the valleys by plastic deformation, creating a new topography. The aspect of the final surface becomes

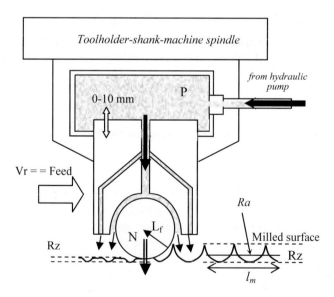

Figure 8.9. Hydrostatic head of the ball roller system [22], with reduction of roughness

Table 8.2 Results of ball burnishing on two mould steels, P20 and H13, for different ball-end milling radial width of cut

AISI P20			Milled surface, radial width of cut a_e (Radial width of burnishing was 0.1 mm)			
Burnishing pressure	Surface hardness	Roughness parameters μm	0.2 mm	0.3 mm	0.4 mm	0.6 mm
Before burnishing	32 HRC	R_a	1.04	1.65	1.83	2.43
		R_z	8.38	12.05	16.42	14.44
20 MPa	38 HRC	R_a	0.13	0.07	0.10	0.13
		R_z	2.33	1.15	0.96	1.01
AISI H13			Milled surface, radial width of cut a_e (Radial width of burnishing was 0.1 mm)			
Burnishing pressure	Surface hardness	Roughness parameters μm	0.4 mm	0.45 mm	0.9 mm	1 mm
Before burnishing	52 HRC	R_a	1.46	1.72	1.97	2.78
		R_z	6.29	7.19	7.31	12.17
20 MPa	58 HRC	R_a	0.2	0.28	0.68	1.24
		R_z	1.28	1.27	2.96	4.15

a combination of the previously machined one and the effect of burnishing. In many cases, the final surface aspect is mirror-like on steels, and somewhat poorer on iron castings because of the graphite particles.

Using a large radial depth of cut in the ball-end milling, together with a small radial depth during burnishing, achieves a very acceptable final roughness. A radial depth of cut typical of semi-finishing ($a_e = 0.6$–0.9 mm) may be sufficient if a precise burnishing is later applied ($a_b = 0.05$–0.1 mm). Burnishing is always applied at the maximum feed (15 m/min), instead of the 1–2 m/min of milling, with a 20–30% time saving.

Some results on two different mould steels are given in Table 8.2. As shown, results on treated steel (P20) are very good, resulting in a near mirror-like finishing. Results of hardened (H13) steels are also good.

Regarding large dies, in ductile cast irons (GGG70, ASTM 100-70-03, hardness 280 HBN), the roughness of the zones milled applying $a_e = 0.2$ mm dramatically decreases after burnishing from 1.45 to 0.27 μm Ra. Reduction of time required for the global process, *i.e.*, milling ($a_e = 0.6$ mm) and burnishing ($a_b = 0.1$ mm), is 25% compared to applying only a good milling finishing ($a_r = 0.2$ mm).

8.6   Tool Path Selection Using Cutting Force Prediction

In upmilling, surface error is created at the tool entrance; in downmilling, error is caused at the tool exit. However, in high-speed ball finishing (in both cases), determination of the exact point where the final surface is produced is complex, requiring more research [23]. In ball-end milling of inclined surfaces, the error depends on tool-axis orientation (and feed sense) with respect to the part surface. A simplification has been proposed to optimise and select milling tool paths to maintain maximum values of the *deflection component* under a certain threshold, or selecting the tool path with minimum deflection force in comparison with other milling strategies. Two approaches are proposed for the three- and five-axis cases, respectively.

8.6.1 Three-axis Case

In three-axis milling, the tool axis (Z axis) is strictly fixed with respect to workpiece surfaces; therefore, only the feed direction can be studied and varied by the CAM user. Figure 8.10 shows the proposed new methodology for the implementation in CAM programming. The user defines a set of control points on the surface, projecting a grid. The finer the grid, the greater the amount of information about cutting forces that can be collected. At each surface point, the excess of material to be machined (theoretically the depth of cut a_p) is calculated from the CAM output file of the previous semi-finishing operation. With this input data, the cutting forces prediction model presented in Section 8.4.1 is applied. Cutting force components are obtained each 15° (24 directions at each control point) for both down-milling and upmilling cases. Two options can be followed at this step:

- Select a global milling direction (zig, zag, or zig-zag) that minimises the mean value of the deflection force. This leads to simple linear tool paths that are easily programmed.
- Define tool paths specially adapted to the minimum force direction at each control point, performing a milling process with tool paths winding on the part surface. CAM programming is supported by the guidelines obtained from the linkage of all the minimum deflection force directions at each of all the control points. Sometimes, this approach can result in complex tool paths that can produce poor roughness results or time-intensive operations; therefore, this method should be applied with these other considerations in mind to achieve a high-quality surface and good machining time.

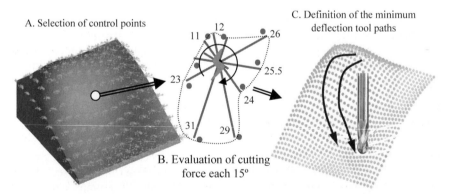

Figure 8.10. Integration of cutting force into three-axis CAM. (A) Selection of control points; (B) estimation along each 15°, upmilling in this case. (C) Selection of the minimum force tool path

We have implemented the cutting forces model in C++ [23] to increase the speed of the calculation process. This module has been implemented into CAM using the application program interface (API) functions of Unigraphics version NX.

8.6.2 Five-axis Case

In the five-axis case, the CAM user is free to define the tool-axis orientation with respect to the part surface. For each orientation, there are multiple choices for the selection of feed direction. Thus, the proposed approach is focussed on achieving optimum values for the *tilt angle* of the tool axis with respect to the normal of the surface, and the *lead angle* with respect to the line reference (usually the maximum slope line in the tool reference system).

Based on the calculation utility for the force deflection component, a procedure scheme has been developed, with several steps for optimal programming:

a) Separation of surface areas taking into account several criteria:
- Division of part surfaces into zones with similar precision requirements and involving only smooth changes in tool orientation, considering several faces as only one case for machining. For example, in car body dies, large surface sets can be formed by adding up adjacent small areas, making preparation of CNC programs much easier.
- Use the same tool for different part faces with similar requirements for roughness. This approach reduces the number of tool changes, with a rationalization of tool paths.

b) Selection of the orientation angle, *tilt angle* α, see Figure 8.11. This angle is restricted by the tool overhang with respect to the depth of the part to be machined, to avoid the tool interference with the inclined walls. The deeper the part to be machined, the narrower the taper angle in which the tool axis can be oriented. Once a possible tool orientation is selected by user, the next step is performed.

c) Selection of the feed direction, *lead angle* δ. A reference line must be previously defined for this selection, the intersection line of the tangent plane at each control point and the plane containing both the tool axis and the surface normal vector at each control point; the line is denoted by *lmp* in Figure 8.11. The lead angle is selected by testing values of the deflection force along each 15° on the surface tangent plane, with the ability to test both the downmilling or upmilling cases.

Therefore, in five-axis milling, there are multiple choices for α and δ, and the CAM user must select a pair of adequate values following other machining optimisation criteria, such as making smoother tool paths, keeping feed speed as high as possible, achieving the maximum length of continuous tool paths, or avoiding idle movements of tools over surfaces.

In this five-axis case, a special toolbox has been developed in C++, which can be executed simultaneously with the CAM software, helping the user select good tool paths for feed sense and tool axis orientation.

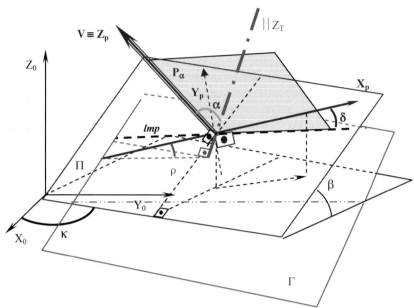

X_0, Y_0, Z_0: Absolute coordinate system
X_p, Y_p, Z_p: Contact point coordinate system; Z_p is similar to the normal vector V to the tangent plane Π at the contact point; X_p is defined just after the feed sense is selected.
$\| Z_T$: Parallel to the tool axis
Π: Tangent plane to the surface on the contact point
Γ: Plane perpendicular to the tool axis Z_H
P_α: Plane defined by the normal to Π in the contact point and parallel to the tool axis
lmp: Maximum slope line of Π with respect to Γ. It is the intersection of P_α with Π.
α: Angle on P_α defined by $Z_p \equiv V$ and $\| Z_H$
ρ: Angle on the perpendicular plane to Γ containing X_p, between X_p and the plane Γ. It is introduced in the basic forces model for calculations.
δ: Angle on the tangent plane Π between *lmp* and the tentative feed sense X_p

Figure 8.11. Angles of the five-axis milling tilt (α) and lead (δ) angles are defined after testing the cutting force component perpendicular both to the surface and tool axis

8.7   Examples

8.7.1   Three-axis Mould

A mould for the injection of the front of an off-road vehicle was made on a three-axis high-speed machine, starting with a blank of heat-treated steel 40 CrMnNiMo8 with 32 HRC. Roughing tools were the round-insert type, given the large volume to be removed. Cutting conditions are shown in Table 8.3, with a time for roughing of 643 min, and 166 min for finishing. Results for roughness (0.6 μm Ra) and precision (mean error 12 μm) were very good, and no polishing by hand was necessary.

Figure 8.12. Mould of the front of an off-road vehicle; CAD image and workpiece after machining

Table 8.3. Operations and cutting parameters for a plastic injection mould

Operation	N (rpm)	F (mm/min)	a_p (mm)	a_e (mm)	Tools
Roughing	1,900	3,200	1	12	Bull-nose Ø40 mm, r 6 mm
Roughing	5,000	1,800	1	3	Ball-end Ø20 mm
Roughing	5,000	1,800	1	3	Ball-end Ø20 mm
Semi-finishing	5,000	1,800	1.5	1	Ball-end Ø20 mm
Semi-finishing	5,000	1,800	1.5	1	Ball-end Ø16 mm
Bitangencies for semi-finishing	7,000	1,900	Variable	1	Ball-end Ø12 mm
Semi-finishing	11,000	3,000	1	0.75	Ball-end Ø12 mm
Bitangencies for finishing	18,000	4,000	0.25	–	Ball-end Ø12 mm
Finishing	18,000	1,100	0.3	0.17	Ball-end Ø12 mm

However, roughing took more than 10 hours, so it might be more cost-effective to use a conventional centre than a high-speed one, applying method A shown in Figure 8.2 instead of the working scheme C.

The CAM (using Unigraphics™) methodology explained in Section 8.3 eliminated all risk of error from excessive cutting in the final mould. Cutting forces were less than 220 N in areas with slope 15°, a_p 0.3 mm, and a_e 0.17 mm. Such forces impair precision only slightly.

8.7.2 Five-axis Mould

This complex shape collects several features of the tempered insert blocks inserted into special punches and dies for AHSS bending. Inserts are tempered to 64 HRC, near the hardness limit of machining. Here, success in milling was achieved by estimation of deflection forces and selecting those tilt and feed sense angles that lead to dimensional errors lower than 50 μm. Finishing takes more than 20 hours in a five-axis machine. CAM programming work in this case required 40 hours.

Results for some zones are presented in Figure 8.13, which details tool, cutting conditions, strategy and tilt and feed angles. The measured error of the real surfaces with respect to the theoretical values is also indicated. This error was measured using a contact probe (Renishaw™) directly in the milling centre, calculated at each zone as the difference between one finishing pass and a contiguous one machined twice (in a small square). In this way, the second pass eliminates the stock allowance left by tool deflection in the first pass. This method is better than using a coordinate measurement machine. With this relative measurement, the effect of workpiece misalignment when the part is placed on the CMM is not decreased; even if a careful alignment procedure based on three reference workpiece planes is carried out, error derived from this fact is always present and of the same magnitude as the tool deflection error,

In this example, the results of workpiece inaccuracy caused by tool deflection are within the requirements of the part (50 μm); in the worst case, it is 42 μm in a zone where the tilt angle must be below 20° because of its depth. A higher precision is achieved in other zones.

Selected zone	Detail of the tool path	Cutting conditions	Strategy and dimensional error
	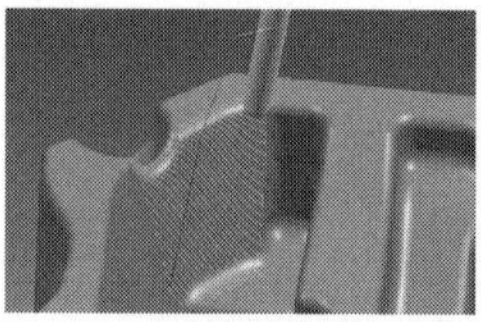	$F = 700$ m/min $N = 4100$ rpm $a_p = 0.3$ mm $a_e = 0.27$ mm Ball-end mill ⌀ 12 Overhang: 50 mm	Zig-zag cutting Feed sense: 30° Tool tilt angle: 15° *Dimensional error 12 μm*
	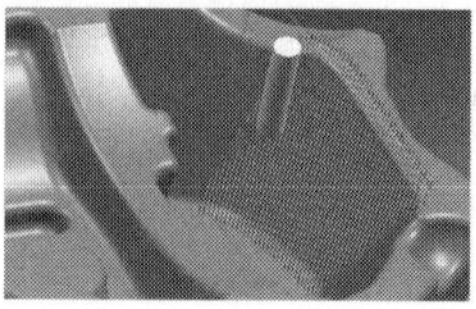	$F = 700$ m/min $N = 4100$ rpm $a_p = 0.3$ mm $a_e = 0.27$ mm Ball-end mill ⌀ 12 Overhang: 50 mm	Zig cutting, Downmilling Feed sense: 15° Tool tilt angle: 15° *Dimensional error 5 μm*
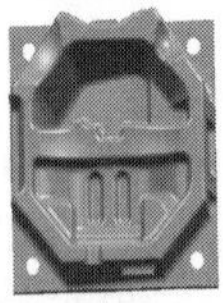	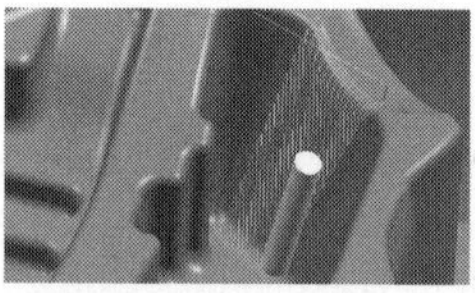	$F = 700$ m/min $N = 4100$ rpm $a_p = 0.3$ mm $ae = 0.27$ mm Ball-end mill ⌀ 12 Overhang: 50 mm	Zig-zag cutting Feed sense: 45° Tool tilt angle: 15° *Dimensional error 42 μm*
	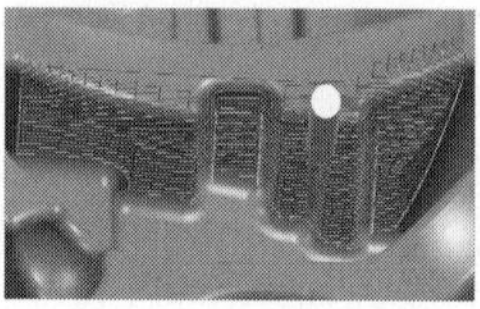	$F = 700$ m/min $N = 4100$ rpm $a_p = 0.3$ mm $ae = 0.27$ mm Ball-end mill ⌀ 12 Overhang: 55 mm	Zig-zag cutting Feed sense: 90° Tool tilt angle: 23° *Dimensional error 30 μm*

Figure 8.13. Test part of the 64 HRC workpiece, made in a five-axis machine

8.7.3   Three-axis Deep Mould

This example is an aluminium injection mould (Figure 8.14) for a kitchen handle. A special tapered ball-end tool, ∅2 mm with 30 mm overhang was needed. Final milling time was near 7 hours, and a carefully CAM programming of all operations were made. In this way, electro-discharge machining was avoided. All the operations are described in Table 8.4, along with their times.

This part can be considered itself an extreme case of the conventional scale, but tools ∅0.2 mm are being used in micromilling. In [9] some interesting examples about the limits of micromilling, and precision of micromilled parts, are collected.

Table 8.4. Operations and cutting parameters for a deep aluminium mould

Operation and tool	a_p	a_e	F (mm/min)	S (rpm)	Time
Roughing, bull-nose ∅8 radius 2	0.4	3	1,600	4,000	11 min
Finishing wall, bull-nose ∅8 radius 0.5	0.2	/	1,000	4,000	17 min
Roughing, ball-end ∅6	0.2	0.1	3,000	13,000	15 min
Bitangency, bull-nose ∅3 radius 0.5	0.15	1.5	360	6,000	11 min
Semifinishing, ball-end ∅6	/	0.1	2,000	13,000	10 min
Finishing, bull-nose ∅8 radius 0.5	/	1	450	3,500	5 min
Finishing, ball-end ∅4	/	0.1	2,000	18,000	10 min
Bitangencies finishing, bull-nose ∅3 radius 0.5	/	0.5	360	6,000	3 min
Corners, ball-end ∅2×16	0.1	0.1	900	15,900	1h
Finishing, ball-end ∅3	0.1	/	900	15,900	40 min
Roughing, ball-end ∅2×16	0.05	1	600	15,900	1h 40 min
Roughing, Tapered ball-end ∅2×30 tapering 0.5°	0.025	1	500	10,000	12 min
Roughing, ball-end ∅6	0.2	0.3	2,000	15,000	1h

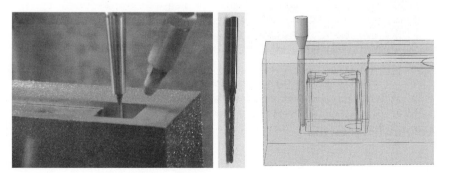

Figure 8.14. Mould for aluminium injection, with very deep cavities. Tapered ball-end mill

8.8   Present and Future

High-speed milling of a complex surface is a well-known technology for mould, die and blade manufacturers. In the last 10 years, it has replaced the slow electro-discharge machining for free-form milling.

Sub-micrograin-grade carbide tools are used to 300–400 m/min for finishing, with new developments in the quality and performance of tool coatings. Ball-end mill is the common type for sculpturing, but bull-nose milling can also be used in deep cavities. Tools with inserts are cost-effective for roughing, and solid tools (more precise and with lower run-out) for finishing.

CAM is important for defining correct CNC programs, especially for five-axis milling. Virtual verification of programs is highly recommended. At the same time, research has focussed on providing CAM users with utilities for the calculation of cutting forces, making possible the definition of tool paths and leading to minimal dimensional errors arising from tool deflection under the action of the cutting forces.

Three examples applying the new approach have been presented, one in three-axis, other in five-axis and finally one mould with very narrow and deep cavities.

In the next future no major changes are foreseen, but new emerging technologies such as laser sintering (see the example in Figure 8.15) and others commonly known as *rapid tooling* techniques will compete with machining in mould and blades manufacturing. Problems of this technology with surface roughness are now under study, achieving practical results with laser polishing. In [24] a reduction of Ra of 8–1.4 µm is reported, showing the possibilities of this technique.

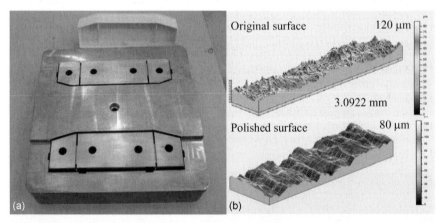

Figure 8.15. (a) Mould made by laser sintering, with very narrow slots. Upside, injected part (courtesy of AIJU centre). **(b)** Surface before and after laser polishing [24]

Acknowledgements

Special thanks go to E. Sasia for his technical suggestions over the years, and to Dr. M.A. Salgado for his work in sculptured machining.

References

[1]   Menzel C, Bedi S, Mann S (2004) Triple tangent flank milling of ruled surfaces, Comput Aided Des 36, 289–296

[2]   López de Lacalle LN, Lamikiz A, Sánchez JA, Arana JL (2002) Improving the surface finish in high speed milling of stamping dies. J Mater Process Technol 123(2): 292–302

[3]   Siller H, Rodriguez CA, Ahuett H (2006) Cycle time prediction in high-speed milling operations for sculptured surface finishing. J Mater Proces Technol 174 (1–3): 355–362

[4]   Starrag (2007) 4/5 Axis Machining for complex surfaces, available at http://www.starragheckert.com

[5]   López de Lacalle LN, Lamikiz A, Sánchez JA, Salgado MA (2004) Effects of tool deflection in the high-speed milling of inclined surfaces. Int J Adv Manuf Technol 24(9–10):621–631

[6]   Choi BK, Jerard RB (1999) Sculptured Surface Machining: Theory and Applications. Springer, Berlin Heidelberg New York

[7]   López de Lacalle LN, Lamikiz A, Sánchez JA, Salgado MA (2007) Toolpath selection based on the minimum deflection cutting forces in the programming of complex surfaces milling. Int J Mach Tools Manuf 47(2): 388–400

[8]   Salgado M, López de Lacalle LN, Lamikiz A, Muñoa M, Sánchez JA (2005) Evaluation of the stiffness chain on the deflection of end-mills under cutting forces. Int J Mach Tool Manuf 45:727–739

[9]   Uriarte L, Herrero A, Zatarain M, Santiso G, López de Lacalle LN, Lamikiz A, Albizuri A (2007) Error budget and stiffness chain assessment in a micromilling machine equipped with tools less than 0.3 mm in diameter. Prec Eng 31: 1–12

[10]   López de Lacalle LN, Lamikiz A, Muñoa J, Salgado MA, Sánchez JA (2006) Improving the high-speed finishing of forming tools for advanced high-strength steels (AHSS). Int J Adv Manuf Technol 29:1–2

[11]   Feng HY, Menq CH (1994) The prediction of cutting forces in the ball-end milling process–I. Model formulation and model building procedure. Int J Mach Tools Manuf 34: 697–710

[12]   Engin S, Altintas Y (2001) Mechanics and dynamics of general milling cutters. Part I and II,. Int J Mach Tool Manuf 41: 2195–2212

[13]   Lazoglu I (2003) Sculpture surface machining: a generalized model of ball-end milling force system. Int J Mach Tool Manuf 43, 453–462

[14]   Roth D, Gray PJ, Ismail F, Bedi S (2007) Mechanistic modelling of 5-axis milling using an adaptive and local depth buffer. Comput-Aided Des 39(4): 302–312

[15]   Erdim H, Lazoglu I, Ozturk B (2006) Feedrate scheduling strategies for free-form surfaces. Int J Mach Tool Manuf 46:747–757

[16]   Lamikiz A, López de Lacalle LN, Sánchez JA, Bravo U (2005) Calculation of the specific cutting coefficient and geometrical aspects in sculptured surface machining. Mach Sci Technol 9(3):411–436

[17]   Lamikiz A, López de Lacalle LN, Sánchez JA, Salgado MA (2004) Cutting force estimation in sculptured surface milling. Int J Mach Tools Manuf 44,1519–1526

[18]   Lee P, Altintas Y (1996) Prediction of ball-end milling forces from orthogonal cutting data. Int J Mach Tool Manuf 36(9):1059–1072

[19]   Bouzakis KD, Aichouh P, Efstathiou K (2003) Determination of the chip geometry, cutting force and roughness in free form surfaces finishing milling, with ball end tools. Int J Mach Tools Manuf 43 (5):499–514

[20] Tae-Sung Jung, Min-Yang Yang and Kang-Jae Lee (2005) A new approach to ana-
lysing machined surfaces by ball-end milling, part II: Roughness prediction and experimental verification. Int J Adv Manuf Technol 25(9–10): 841–849
[21] Jenq-Shyong Chen, Yung-Kuo Huang, Mao-Son Chen (2005) A study of the surface
scallop generating mechanism in the ball-end milling process. Int J Mach Tools
Manuf 45(9):1077–1084
[22] López de Lacalle LN, Lamikiz A, Muñoa J, Sánchez JA (2005) Quality improvement
of ball-end milled sculptured surfaces by ball burnishing. Int J Mach Tools Manuf
45(15):1659–1668
[23] López de Lacalle LN, Lamikiz A, Sanchez JA, Salgado MA (2007) Toolpath selection based on the minimum deflection cutting forces in the programming of complex
surfaces milling. Int J Mach Tool Manuf 47(2):388–400
[24] Lamikiz A, Sánchez JA, López de LacalleLN, Del Pozo D, Etayo JM, López JM
(2007) Laser polishing techniques for roughness improvement on metallic surfaces.
Int J Nanomanuf 1(4): 490–498

9

Grinding Technology and New Grinding Wheels

M.J. Jackson

Center for Advanced Manufacturing, College of Technology, Purdue University, 401 North Grant Street, West Lafayette, Indiana, IN 47907, USA.
E-mail: jacksomj@purdue.edu

High-speed grinding offers great potential for significant advances in component quality combined with high productivity. A factor behind the process has been the need to increase productivity for conventional finishing processes. In the course of process development it has become evident that high-speed grinding enables the configuration of new process sequences with high performance capabilities. By using appropriate grinding machines and grinding tools, it is possible to expand the scope of grinding to high-efficiency machining of soft, ductile materials. In this chapter, a basic examination of process mechanisms that relates the configuration of grinding tools and the requirements for grinding soft materials is discussed.

9.1   Introduction

There are three processes that have become established for high-speed grinding:

- High-speed grinding with aluminium oxide (Al_2O_3) grinding wheels
- High-speed grinding with cubic boron nitride (CBN) grinding wheels
- Grinding with aluminium oxide grinding wheels in conjunction with continuous dressing techniques

Material removal rates resulting in a proportional increase in productivity for component machining have been achieved for each of these fields of technology in industrial applications. High equivalent chip thickness (h_{eq}) values between 0.5 and 10 μm are a characteristic feature of high-speed grinding. CBN high-speed grinding is employed for a large proportion of these applications. An essential characteristic of the technology is that the enhanced performance of CBN is utilized when high cutting speeds are employed. CBN grinding tools for high-speed machining are subject to certain requirements regarding resistance to wear. Good damping characteristics, high rigidity and high thermal conductivity are also desirable. Such tools normally consist of a body of high mechanical strength and a comparably thin coating of abrasive attached to the body using a high-strength adhesive. The suitability of cubic boron nitride as an abrasive material for high-speed machining of ferrous materials is attributed to its extreme hardness and its thermal and chemical durability when bonded with the correct composition of glass.

High cutting speeds are attainable with metal bonding systems. One method that uses such bonding systems is electro-plating, where grinding wheels are produced with a single-layer coating of abrasive CBN grain material. The electro-deposited nickel bond displays outstanding grain retention properties. This provides a high level of grain projection and very large chip spaces. Cutting speeds exceeding 300 ms^{-1} are possible, and the service life ends when the abrasive layer wears out. The high roughness of the cutting surfaces of electroplated CBN grinding wheels has disadvantageous effects. The high roughness is accountable to exposed grain tips that result from different grain shapes and grain diameters. Although electroplated CBN grinding wheels are not considered to be dressed in the conventional sense, the resultant workpiece surface roughness can nevertheless be influenced within narrow limits by means of a so-called touch-dressing process. This involves removing the peripheral grain tips from the abrasive coating by means of very small dressing in-feed steps in the range of dressing depths of cut between 2 to 4 µm, thereby reducing the effective roughness of the grinding wheel.

Multi-layer bonding systems for CBN grinding wheels include sintered metal bonds, resin bonds and vitrified bonds. Multi-layer metal bonds possess high bond hardness and wear resistance. Profiling and sharpening these tools is a complex process, however, on account of their high mechanical strength. Synthetic resin bonds permit a broad scope of adaptation for bonding characteristics. However, these tools also require a sharpening process after dressing. The potential for practical application of vitrified bonds has yet to be fully exploited. In conjunction with suitably designed bodies, new bond developments permit grinding wheel speeds of up to 200 ms^{-1}. In comparison with other types of bonds, vitrified bonds permit easy dressing while at the same time possessing high levels of resistance to wear. In contrast to impermeable resin and metal bonds, the porosity of the vitrified grinding wheel can be adjusted over a broad range by varying the formulation and the manufacturing process. As the structure of vitrified bonded CBN grinding wheels results in an increased chip space after dressing, the sharpening process is simplified, or can be eliminated in numerous applications.

9.2   High-efficiency Grinding Using Conventional Abrasive Wheels

9.2.1   Introduction

High-efficiency grinding practices using conventional aluminium oxide grinding wheels has been successfully applied to grinding external profiles between centres and in the centreless mode, grinding internal profiles, threaded profiles, flat profiles, guide tracks, spline shaft profiles and gear tooth profiles. Operations that are carried out using high-performance conventional grinding wheels as a matter of routine are the grinding of:

- **Auto engine:** crankshaft main and connecting-rod bearings, camshaft bearings, piston ring grooves, valve rocker guides, valve head and stems, head profile, grooves, expansion bolts
- **Auto gearbox:** gear wheel seats on shafts, pinion gears, splined shafts, clutch bearings, grooves in gear shafts, synchromesh rings, oil-pump worm wheels
- **Auto chassis:** steering knuckle, universal shafts and pivots, ball tracks, ball cages, screw threads, universal joints, bearing races, cross pins
- **Auto steering:** ball joint pivots, steering columns, steering worms, servo steering pistons and valves
- **Aerospace industry:** turbine blades, root and tip profiles, fir-tree root profiles [1, 2]

9.2.2 Grinding Wheel Selection

The selection of grinding wheels for high-performance grinding applications is focussed on three basic grinding regimes: rough, finish and fine grinding. The grain size of the grinding wheel is critical in achieving a specified workpiece surface roughness. The grain size is specified by the mesh size of a screen through which the grain can just pass while being retained by the next smaller size. The general guidelines for high-performance grinding require a 40–60 mesh for rough grinding; a 60–100 mesh for finish grinding; and a 100–320 mesh grain for fine grinding. When selecting a particular grain size one must consider that large grains allow the user to remove material economically by making the material easier to machine by producing longer chips. However, finer grain sizes allow the user to achieve lower surface roughness, and achieve greater accuracy by producing shorter chip sizes with a greater number of sharp cutting points. Table 9.1 shows the relationship between abrasive grain size and workpiece surface roughness for aluminium oxide grains. Grinding wheel specifications are

Table 9.1. Relationship between abrasive grain size and workpiece surface roughness

Surface roughness, R_a (μm)	Abrasive grain size (US mesh size)
0.7–1.1	46
0.35–0.7	60
0.2–0.4	80
0.17–0.25	100
0.14–0.2	120
0.12–0.17	150
0.–0.14	180
0.08–0.12	220

specific to a particular operation, and grinding wheels are formulated to account for the differences in grinding operations. Figure 9.1 shows a micrograph of a standard structure grinding wheel that is used for applications used in the automotive industry. Figure 9.2 shows a grinding wheel that is formulated for use in creep feed grinding operations that are specific to the grinding of aerospace components.

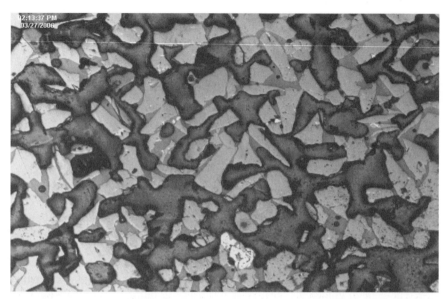

Figure 9.1. Micrograph of conventional grinding wheel with standard porosity

Figure 9.2. Micrograph of conventional grinding wheel with distributed porosity

The wheel has a specially developed structure that is used for increased metal removal rates. The suggested operating conditions are also supplied by applications' engineers who spend their time optimizing grinding specifications for a variety of tasks that are supported by case studies of similar operations [2]. A list of web sites for those companies who supply grinding wheels and expertise is shown at the end of this chapter.

9.2.3 Grinding Machine Requirements for High-efficiency Dressing

Diamond dressing wheels require relative motion between the dressing wheel and the grinding wheel. Dressing form wheels require relative motion generated by the path of the profile tool for the generation of the grinding wheel form, and the relative movement in the peripheral direction. Therefore, there must be a separate drive for the dressing wheel. The specification of the drive depends on the following factors: grinding wheel specification and type, dressing roller type and specification, dressing feed, dressing speed, dressing direction and the dressing speed ratio. A general guide for drive power is that 20W per mm of grinding wheel contact is required for medium to hard vitrified aluminium oxide grinding wheels.

The grinding machine must accommodate the dressing drive unit so that dressing wheels rotate at a constant speed between itself and the grinding wheel. This means that grinding machine manufacturers must coordinate the motion of the grinding wheel motor and dressing wheel motor. For form profiling, the dressing wheel must also have the ability to control longitudinal feed motion in at least two axes. The static and dynamic rigidity of the dressing system has a major effect on the dressing system. Profile rollers are supported by roller bearings in order to absorb rather high normal forces. Slides and guides on grinding machines are classed as weak points and should not be used to mount dressing drive units. Therefore, dressing units should be firmly connected to the bed of the machine tool. Particular importance must be attached to geometrical run-out accuracy of the roller dresser and its accurate balancing. Tolerances of 2 μm are maintained with high-accuracy profiles with radial and axial run-out tolerances not exceeding 2 μm. The diameter of the mandrel should be as large as possible to increase its stiffness; typically, roller dresser bores are in the range 52–80 mm diameter. The class of fit between the bore and the mandrel must be H3/h2 with a 3–5 μm clearance. The characteristic vibrations inherent to roller dresser units are bending vibrations in the radial direction and torsional vibrations around the base plate. Bending vibrations generate waves in the peripheral direction, while torsional vibrations generate axial waves and distortions in profile. The vibrations are caused by rotary imbalance and the dressing unit should be characterized in terms of resonance conditions. A separate cooling system should also be designed in order to prevent the dressing wheel from losing its profile accuracy due to thermal drift [1].

9.2.4 Diamond Dressing Wheels

The full range of diamond dressing wheels is shown in Table 9.2. There are five basic types of dressing wheels for conventional form dressing of conventional

Table 9.2. Types of dressing rolls and wheels

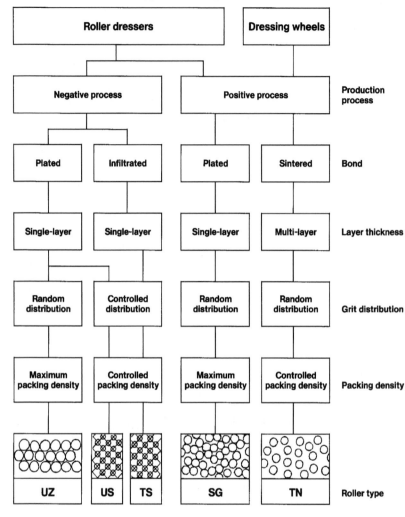

grinding wheels. The types described here are those supplied by Winter Diamond Tools (www.winter-diamantwerkz-saint-gobain.de). Table 9.2 shows the five types of roller dressing wheels:

- **UZ type (reverse plated – random diamond distribution):** The diamond grains are randomly distributed at the diamond roller dresser surface. The diamond spacing is determined by the grain size used and the close-packed diamond layers give a larger diamond content than a hand-set diamond dressing roll. The manufacturing process is independent of profile shape. The process permits concave radii to be greater than 0.03 mm and convex radii to be greater than 0.1 mm. The geometrical and dimensional accuracy of these dressers is achieved by re-working the diamond layer

- **US type (reverse plated – hand-set diamond distribution):** Unlike the UZ design, the diamonds are hand set, which means that certain profiles cannot be produced. However, the diamond spacing can be changed and profile accuracy can be changed by re-working the diamond layer. Convex and concave radii greater than 0.3 mm can be achieved;
- **TS type (reverse sintered – hand-set diamond distribution):** Diamonds are hand set, which means that certain profiles cannot be produced. However, the diamond spacing can be changed and profile accuracy can be changed by re-working the diamond layer. Concave radii greater than 0.3 mm can be produced;
- **SG type (direct plated – random diamond distribution, single layer):** The diamonds grains are randomly distributed. Convex and concave radii greater than 0.5 mm are possible.
- **TN type (sintered – random diamond distribution, multi-layer):** The diamond layer is built up in several layers providing a long-life dressing wheel. The profile accuracy can be changed by re-working the diamond layer.

The minimum tolerances that are attainable using diamond dressing rolls and wheels are shown in Table 9.3. The charts show tolerances of engineering interest for each type of dressing wheel.

Table 9.3. Minimum tolerances attainable with dressing wheels

Table 9.3. *(continued)*

Symmetry tolerance ⟂ α 1 to α 2 referred to leg length L	(diagram)	L: ≤1, ≤5, >5 / Tw: 7', 4', 2' A	—	L: ≤1, ≤5, >5 / Tw: 10', 7', 3' A	—	
Rectangularity tolerance ⊥ of diamond studded plane faces	(diagram)	—	see TS	—	face equal to workpiece / face not equal to workpiece, tolerance per 1 mm face: 0,001 | 0,005	—
Ts Step-Tolerance in difference between two associated diameters on different dressing rolls	(diagram)	± 0,002	± 0,05	± 0,002	± 0,01	
Ts Step-Tolerance in difference between two associated diameters on one truing roll / Linear dimensional tolerance Tᴸ of two associated faces	(diagram)	± 0,002 / ± 0,005	± 0,05	± 0,002 / ± 0,005	—	
Linear dimensional tolerance Tᴸ of two opposite faces	(diagram)	± 0,02 A	—	± 0,003 B	—	
Dimensional tolerance on pitch Tᴸ / Cylindrical shape tolerance for profile	(diagram)	single pitch P:±0,002, P total for ≤16:±0,002, >16: pro 10 mm = 0,00125, profile 0,002 per 10 mm thread length	—	—	—	
Straightness tolerance —	(diagram)	L: ≤50, ≤80, ≤130 / Tz: 0,002, 0,003, 0,004	—	L: ≤50, ≤80, ≤130 / Tz: 0,002, 0,003, 0,004	—	

9.2.5 Application of Diamond Dressing Wheels

The general guide and limits to the use of diamond dressing wheels are shown in Table 9.4, which shows the relationship between application and the general specification of dressing wheels.

Table 9.4. General specification and application of diamond dressing wheels

Application	Wheel spec., grain size, hardness	Surface roughness, R_a (µm)	Output	TS type, diamond size (mesh)	UZ type, diamond size (mesh)
Rough grinding	40–60, K–L	0.8–3.2 (hand set) 0.4–1.0 (random)	High	80–100 (hand set) 100–150 (random)	Not applicable
Transmission, gear box	60–80, J–M	0.2–16	High	200–250	200–250
Bearings, CV joints	80–120, J–M	0.2–1.6	High	200–300	200–300
Creep feed, continuous dressing	Porous structure	0.8–1.6	Low	Limited	250–300

9.2.6 Modifications to the Grinding Process

When the grinding wheel and dressing wheel have been specified for a particular grinding operation, adjustments can be made during the dressing operation, which affects the surface roughness condition of the grinding wheel. The key factors that affect the grinding process during dressing are: dressing speed ratio, V_r/V_s, between the dressing wheel and grinding wheel; dressing feed rate, a_r, per grinding wheel revolution; and the number of running-out, or dwell, revolutions of the dressing wheel, n_a. By changing the dressing conditions it is possible to rough and finish grind using the same diamond dressing wheel and the same grinding wheel. By controlling the speed of the dressing wheel, or by reversing its rotation, the effective roughness of the grinding wheel can be varied in the ratio of 1:2.

9.2.7 Selection of Grinding Process Parameters

The aim of every grinding process is to remove the grinding allowance in the shortest time possible while achieving the required accuracy and surface roughness on the workpiece. During the grinding operation the following phases are usually present:

- Roughing phase, during which the grinding allowance removed is characterized by large grinding forces, causing the workpiece to become deformed
- Finishing phase for improving surface roughness
- Spark-out phase, during which the workpiece distortion is reduced to a point where form errors are insignificant.

Continuous dressing coupled with one or more of the following phases have contributed to high-efficiency grinding of precision components:

- Single-stage process with a sparking-out regime
- Two-stage process with a defined sparking out regime coupled with a slow finishing feed rate
- Three- and multi-stage processes with several speeds and reversing points
- Continuous matching of feed rate to the speed of deformation reduction of the workpiece

For many grinding processes using conventional vitrified grinding wheels, starting parameters are required that are then optimized. A typical selection of starting parameters is shown below:

- Metal removal rates for plunge grinding operations: when the diameter of the workpiece is greater than 20 mm then the following metal removal rates (Q'_w in mm³/mm.s) are recommended; roughing – 1 to 4, finishing – 0.08 to 0.33. When the workpiece is less than 20 mm, roughing – 0.6 to 2, finishing – 0.05 to 0.17;
- Speed ratio, q, is the relationship between grinding wheel speed and workpiece speed. For thin-walled, heat-sensitive parts q should be in the range

105–140, for soft and hard steels, q should be 90–135, for high metal removal rates q should be 120–180, and for internal grinding q should be in the range 65–75;

- Overlap factors should be in the range 3–5 for wide grinding wheels and 1.5–3 for narrow grinding wheels;
- Feed rates should be between 0.002 and 0.006 mm per mm traverse during finish grinding;
- Number of sparking-out revolutions should be 3–10, depending on the rigidity of the workpiece.

However, it is important that the user ensures that the specification of grinding wheel and dressing wheel are correct for the grinding task prior to modifying the grinding performance [1–3].

9.2.8   Selection of Cooling Lubricant Type and Application

The most important aspect of improving the quality of workpieces is the use of high-quality cooling lubricant. In order to achieve a good surface roughness of less than 2 μm, a paper-type filtration unit must be used. Air cushion deflection plates improve the cooling effect. The types of cooling lubricant in use for grinding include emulsions, synthetic cooling lubricants and neat oils:

- Emulsions: oils emulsified in water are generally mineral based and are concentrated in the range 1.5–5%. In general, the 'fattier' the emulsion the better the surface finish, but this leads to high normal forces and roundness is impaired. Also, they may become susceptible to bacteria;
- Synthetic cooling emulsions: chemical substances dissolved in water at concentrations of 1.5–3%. These are resistant to bacteria and are good wetting agents. They allow grinding wheels to act more aggressively but tend to foam and destroy seals;
- Neat oil: enabled the highest metal removal rates with a low tendency to burn the workpiece. Neat oils are difficult to dispose of and present a fire hazard.

There are general rules concerning the application of cooling lubricants but the reader is advised to contact experienced grinding applications engineers from grinding wheel suppliers and lubricant suppliers. A list of suppliers is shown in the internet resource section of this chapter.

9.3   High-efficiency Grinding Using CBN Grinding Wheels

9.3.1   Introduction

High-efficiency grinding practice using super-abrasive CBN grinding wheels has been successfully applied to grinding external profiles between centres and in the centreless mode, grinding internal profiles, threaded profiles, flat profiles, guide tracks, spline shaft profiles and gear tooth profiles. These operations require dressing of the grinding wheel with a highly accurate and precise rotating wheel that is studded with diamond. Operations that are carried out using high-performance conventional grinding wheels as a matter of routine are the grinding of:

- **Auto engine:** crankshaft main and connecting-rod bearings, camshaft bearings, piston ring grooves, valve rocker guides, valve head and stems, head profile, grooves, expansion bolts;
- **Auto gearbox:** gear wheel seats on shafts, pinion gears, splined shafts, clutch bearings, grooves in gear shafts, synchromesh rings, oil-pump worm wheels;
- **Auto chassis:** steering knuckle, universal shafts and pivots, ball tracks, ball cages, screw threads, universal joints, bearing races, cross pins;
- **Auto steering:** ball joint pivots, steering columns, steering worms, servo steering pistons and valves;
- **Aerospace industry:** turbine blades, root and tip profiles, fir-tree root profiles [3].

9.3.2 Grinding Wheel Selection

The selection of the appropriate grade of vitrified CBN grinding wheel for high-speed grinding is more complicated than for aluminium oxide grinding wheels. Here, the CBN abrasive grain size is dependent on the specific metal removal rate, surface roughness requirement and equivalent grinding wheel diameter. As a starting point when specifying vitrified CBN wheels, Figure 9.3 shows the relationship between CBN abrasive grain size, equivalent diameter and specific metal removal rate for cylindrical grinding operations.

However, the choice of abrasive grain is also dependent on the surface roughness requirement and is restricted by the specific metal removal rate. Table 9.5 shows the relationship between CBN grain size and their maximum surface roughness and specific metal removal rates. The workpiece material has a significant influence on the type and volume of vitrified bond used in the grinding wheel. Table 9.6 shows the wheel grade required for a variety of workpiece materials that are based on cylindrical (crankshaft and camshaft) grinding operations [1–3].

Considering the materials shown in Table 9.6, chilled cast iron is not burn sensitive and has a high specific grinding energy owing to its high carbide content. Its hardness is approximately 50 HRc and the maximum surface roughness achieved on machined camshafts is 0.5 μm Ra, therefore a standard structure bonding system is used that is usually between 23 and 27 vol.% of the wheel. The CBN grain content is usually 50% by volume, and wheel speeds are usually up to 120 m/s. Nodular cast iron is softer than chilled cast iron and is not burn sensitive. However, it does tend to load the grinding wheel. Camshaft lobes can have hardness values as low as 30 HRc and this tends to control wheel specification. High stiffness crankshafts and camshafts can tolerate a 50 vol.% abrasive structure containing 25 vol.% bond. High loading conditions and high contact re-entry cam forms require a slightly softer wheel where the bonding system occupies 20 vol.% of the entire wheel structure. Low-stiffness camshafts and crankshafts require lower CBN grain concentrations (37.5 vol.%) and a slightly higher bond volume (21 vol.%). Very low-stiffness nodular iron components may even resort to grinding wheels containing higher strength bonding systems containing sharper CBN abrasive grains operating at 80 m/s.

The stiffness of the component being ground has a significant effect on the workpiece/wheel speed ratio. Figure 9.4 demonstrates the relationship between this ratio and the stiffness of the component. Steels such as AISI 1050 can be ground in the hardened and the soft state. Hardened 1050 steels are in the range 68–62 HRc. They are burn sensitive and as such wheels speeds are limited to 60 m/s. The standard structure contains the standard bonding systems up to 23 vol.%. The abrasive grain volume contained is 37.5 vol.%. Lower-power machine tools usually have grinding wheels where a part of the standard bonding system contains hollow glass spheres (up to 12 vol.%) exhibiting comparable grinding ratios to the standard structure system. These specifications also cover most powdered metal components based on AISI 1050 and AISI 52100 ball bearing steels.

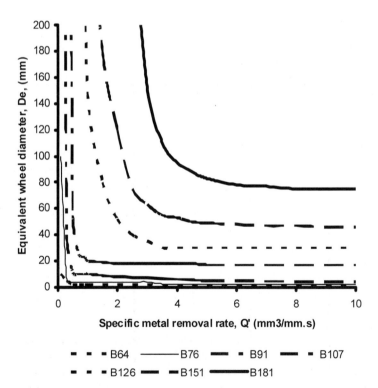

Figure 9.3. Chart for selecting CBN abrasive grit size as function of the equivalent grinding wheel diameter, D_e, and the specific metal removal rate, Q'_w

Table 9.5. CBN abrasive grain selection chart based on camshaft and cranks applications

CBN grain size	Surface roughness, R_a (μm)	Maximum specific metal removal rate, $Q'_{w\,max}$ (mm³/mm.s)
B46	0.15–0.3	1
B54	0.25–0.4	3
B64	0.3–0.5	5
B76	0.35–0.55	10
B91	0.4–0.6	20
B107	0.5–0.7	30
B126	0.6–0.8	40
B151	0.7–0.9	50
B181	0.8–1	70

Table 9.6. Vitrified CBN grinding wheel specification chart and associated grinding wheel speeds based on camshaft and crankshaft grinding applications

Workpiece material	Grinding wheel speed, v_s (m/s)	Vitrified CBN wheel specification	Application details
Chilled cast iron	120	B181R200VSS	High Q'_w
		B126P200VSS	Medium Q'_w
		B107N200VSS	Low Q'_w
Nodular cast iron	80	B181P200VSS	Little or no wheel loading
		B181K200VSS	Wheel loading significant
		B181L150VSS	Low stiffness workpiece
		B181L150VDB	Very low-stiffness workpiece
AISI 1050 steel (hardened)	80	B126N150VSS	Standard specification wheel
		B126N150VTR	For use on low-power machine tools
AISI 1050 steel (soft condition)	120	B181K200VSS	Standard specification wheel
High-speed tool steel	60	B107N150VSS	Standard specification wheel
Inconel (poor grindability)	50	B181T100VTR B181T125VTR B181B200VSS	Form dressing is usually required with all wheel specifications.

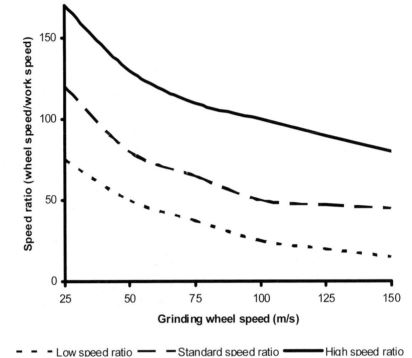

Figure 9.4. Work speed selection chart for camshaft and crankshaft grinding operations

Softer steels are typically not burn sensitive, but do tend to burr when ground. Maximum wheel and work speeds are required in order to reduce equivalent chip thickness. High-pressure wheel scrubbers are required in order to prevent the grinding wheel from loading. The grinding wheel specification is based on an abrasive content in the region of 50 vol.% and a bonding content of 20 vol.% using the standard bonding system operating at 120 m/s. Tool steels are very hard and grinding wheels should contain 23 vol.% standard bonding system and 37.5 vol.% CBN abrasive working at speeds of 60 m/s. Inconel materials are extremely burn sensitive, are limited to wheel speeds of 50 m/s and have large surface roughness requirements, typically 1 µm Ra. These grinding wheels contain porous glass sphere bonding systems with 29 vol.% bond, or 11 vol.% bond content using the standard bonding system. Figure 9.5 shows a micrograph with standard porosity that is formed by pressing to a known volume. Figure 9.6 shows a recent development known as induced porosity in which CBN bond formulations are supplemented with hollow alumina pores that create porosity at the contact zone between grinding wheel and workpiece. The level of porosity created at the contact zone is dependent upon the material removal rate.

Figure 9.7 shows a micrograph detailing open and distributed porosity that is similar to that created in porous conventional grinding wheels to grind aerospace materials. Developments in bonding and bonding systems have recently been described by Jackson and Mills [4] and Jackson [5, 6].

Figure 9.5. Micrograph of vitrified CBN grinding wheel with standard porosity

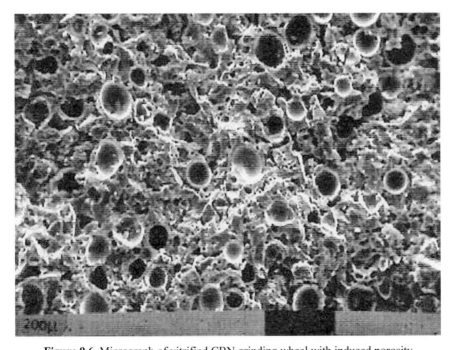

Figure 9.6. Micrograph of vitrified CBN grinding wheel with induced porosity

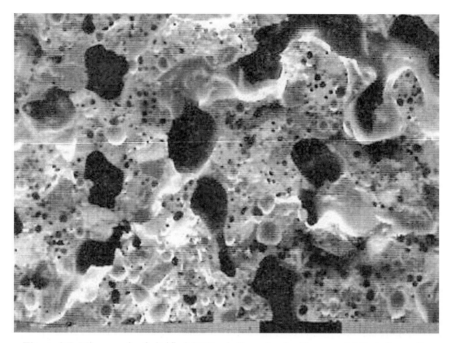

Figure 9.7. Micrograph of vitrified CBN grinding wheel with open distributed porosity

High-efficiency grinding with internal grinding wheels is limited by the bursting speed of the grinding wheel and the effects of coolant. The quill is an inherently weak part of the system and must be carefully controlled when using CBN in order to avoid the problems associated with changes in normal grinding forces. Quills should have a large diameter and a short length, and should be made from a stiff material such as Ferro-TiC, which has a relatively low density.

9.3.3   Grinding Machine Requirements for High-efficiency CBN Grinding

The advantages of high-speed CBN grinding can only be realised in an effective manner if the machine tool is adapted to operate at high cutting speeds. In order to attain very high cutting speeds, grinding wheel spindles and bearings are required to operate at speeds on the order of 20,000 rev min^{-1}.The grinding wheel/spindle/ motor system must run with extreme accuracy and minimum vibration in order to minimise the level of dynamic process forces. Therefore, a high level of rigidity is required for the entire machine tool. Balancing of high-speed grinding wheels is also necessary at high operating speeds using dynamic balancing techniques. These techniques are required so that workpiece quality and increased tool life is preserved.

Another important consideration is the level of drive power required when increases in rotational speed become considerable. The required total output is composed of the cutting power, P_c, and the power loss, P_l,

$$P_{total} = P_c + P_l \qquad (9.1)$$

The cutting power is the product of the tangential grinding force and the cutting speed,

$$P_c = F'_t . v_c \qquad (9.2)$$

The power loss of the drive is comprised of the idle power of the spindle, P_L, and power losses caused by the coolant, P_{KSS}, and by spray cleaning of the grinding wheel, P_{SSP}, thus,

$$P_l = P_L + P_{KSS} + P_{SSP} \qquad (9.3)$$

The grinding power, P_c, increases by a relatively small amount when the cutting speed increases and all other grinding parameters remain constant. However, this means that the substantial power requirement that applies at maximum cutting speeds results from a strong increase in power due to rotation of the grinding wheel, the supply of coolant and the cleaning of the wheel. The quantities and pressures of coolant supplied to the grinding wheel and the wheel cleaning process are the focus of attention by machine tool designers. The power losses associated with the rotation of the grinding wheel are supplemented by losses associated with coolant supply and wheel cleaning. The losses are dependent on machining parameters, implying that machine settings and coolant supply need to be optimised for high-speed grinding. In addition to the advantage of effectively reducing the power required for grinding, optimisation of the coolant supply also offers ecological benefits as a result of reducing the quantities of coolant required. Various methods of coolant supply are available such as the free-flow nozzle that is conventionally used, the shoe nozzle that ensures *reduced quantity lubrication* and the mixture nozzle that ensures *minimum quantity lubrication*. The common task is to ensure that an adequate supply of coolant is presented at the grinding wheel–workpiece interface. The systems differ substantially regarding their operation and the amount of energy required supplying the coolant.

A shoe nozzle, or supply through the grinding wheel, enables coolant to be directed into the workpiece–wheel contact zone. A substantial reduction in volumetric flow can be achieved in this way. In comparison to the shoe nozzle, supply through the grinding wheel requires more complex design and production processes for the grinding wheel and fixtures. An advantage of this supply system is that it is independent of a particular grinding process. Both systems involve a drastic reduction in supply pressures as the grinding wheel effects acceleration of the coolant. A more effective reduction in the quantity of the coolant results in *minimal quantity coolant* supply, amounting to several millilitres of coolant per hour. As the cooling effect is reduced, dosing nozzles are used exclusively to lubricate the contact zone.

9.3.4 Dressing High-efficiency CBN Grinding Wheels

Dressing is the most important factor when achieving success with CBN grinding wheels. CBN abrasive grains have distinct cleavage characteristics that directly affect the performance of dressing tools. A dress depth of cut of 1 µm produces a microfractured grain, whilst a depth of cut of 2–3 µm produces a macrofractured grain. The latter effect produces a rough workpiece surface and shorter CBN tool life. The active surface roughness of the grinding wheel generally increases as grinding proceeds, which means that the vitrified CBN grinding wheel must be dressed in such a way that microfracture of the CBN grains is achieved. This means that touch sensors are needed in order to find the relative position of grinding wheel and dressing wheel because thermal movements in the machine tool far exceed the in-feed movements of the dressing wheel. The design of the dressing wheel for vitrified CBN is based on traversing a single row of diamonds so that the overlap factor can be accurately controlled. The spindles used for transmitting power to such wheels are usually electric spindles because they provide high torque and do not generate heat during operation. Therefore, they assist in the accurate determination of the relative position of dressing wheel and grinding wheel.

9.3.5 Selection of Dressing Parameters for High-efficiency CBN Grinding

Selecting the optimum dressing condition may appear to be complex owing to the combination of abrasive grain sizes, grinding wheel–dressing wheel speed ratio, dressing depth of cut, diamond dressing wheel design, dresser motor power, dressing wheel stiffness and other factors. It is not surprising to learn that a compromise may be required. For external grinding, the relative stiffness of external grinding machines absorbs normal forces without significant changes in the quality of the ground component. This means that the following recommendations can be made: dressing wheel/grinding wheel velocity ratio between +0.2 to +0.5; dressing depth of cut per pass 0.5–3 μm; total dressing depth 3–10 μm; traverse rate calculated by multiplying the dressing wheel r.p.m. by the grain size of CBN and multiplying by 0.3–1, depending on the initial condition of the dressing wheel. It is also right to specify the dressing wheel to be a single row, or a double row, of diamonds. For internal grinding, the grinding system is not so stiff, which means that any change in grinding power will result in significant changes in normal grinding forces and quill deflection. In order to minimize the changes in power, the following dressing parameters are recommended: dressing wheel/grinding wheel velocity ratio +0.8; dressing depth of cut per pass 1–3 μm; total dressing depth 1–3 μm; traverse rate calculated by multiplying the dressing wheel r.p.m. by the grain size of CBN (which means that each CBN grain is dressed once). It is also right to specify the dressing wheel to be a single row of diamonds set on a disc.

9.3.6 Selection of Cooling Lubrication for High-efficiency CBN Grinding Wheels

The selection of the appropriate cooling lubrication for vitrified CBN are considered to be environmentally unfriendly. Although neat oil is the best solution, most applications use soluble oil with sulphur- and chlorine-based extreme pressure additives. Synthetic cooling lubricants have been used but lead to loading of the wheel and excessive wear of the vitrified bond. Using air scrapers and jets of cooling lubricants normal to the grinding wheel surface enhances grinding wheel life.

Pressurized shoe nozzles have also been used to break the flow of air around the grinding wheel. In order to grind soft steel materials, or materials that load the grinding wheel, high-pressure low-volume scrubbers are used to clean the grinding wheel. It is clear that the experience of application engineers is of vital importance when developing a vitrified CBN grinding solution to manufacturing process problems.

9.4 Internet Resources

Grinding wheel suppliers:

> www.citcodiamond.com – suppliers of CBN and diamond grinding and dressing tools
> www.tyrolit.com – suppliers of CBN and diamond grinding and dressing tools
> www.noritake.com – suppliers of CBN and diamond grinding and dressing tools
> www.sgabrasives.com – suppliers of CBN and diamond grinding and dressing tools
> www.nortonabrasive.com – suppliers of CBN and diamond grinding and dressing tools
> www.wendtgroup.com – suppliers of CBN and diamond grinding and dressing tools
> www.rappold-winterthur.com/usa – suppliers of CBN and diamond grinding and dressing tools

Dressing wheel suppliers:

> www.citcodiamond.com – suppliers of CBN and diamond grinding and dressing tools
> www.tyrolit.com – suppliers of CBN and diamond grinding and dressing tools
> www.noritake.com – suppliers of CBN and diamond grinding and dressing tools
> www.sgabrasives.com – suppliers of CBN and diamond grinding and dressing tools
> www.nortonabrasive.com – suppliers of CBN and diamond grinding and dressing tools
> www.wendtgroup.com – suppliers of CBN and diamond grinding and dressing tools
> www.rappold-winterthur.com/usa – suppliers of CBN and diamond grinding and dressing tools

Grinding machine tool suppliers:

> www.landis-lund.co.uk – suppliers of camshaft and crankshaft grinding machine tools

www.toyoda-kouki.co.jp – suppliers of camshaft and crankshaft grinding machine tools

www.toyodausa.com – suppliers of camshaft and crankshaft grinding machine tools

www.weldonmachinetool.com – suppliers of cylindrical grinders and other special purpose grinding machine tools

www.landisgardner.com – suppliers of cylindrical grinders and other special purpose grinding machine tools

www.unova.com – suppliers of cylindrical grinders and other special purpose grinding machine tools

www.voumard.ch – suppliers of internal grinding machine tools

www.bryantgrinder.com – suppliers of internal grinding machine tools

For a complete list of grinding machine suppliers contact:

www.techspec.com/emdtt/grinding/manufacturers – index of grinding machine tool suppliers and machine tool specifications.

Case studies:

www.abrasives.net/en/solutions/casestudies.html – case studies containing a variety of grinding applications using conventional and superabrasive materials

www.winter-diamantwerkz-saint-gobain.de – case studies of CBN grinding wheel and diamond dressing wheel applications

www.weldonmachinetool.com/appFR.htm – grinding case studies

Cooling lubricant suppliers:

www.castrol.com – suppliers of grinding oils/fluids and lubrication application specialists

www.hays-siferd.com – suppliers of grinding oils/fluids and lubrication application specialists (formerly Master Chemicals)

www.quakerchem.com – suppliers of grinding oils/fluids and lubrication application specialists

www.mobil.com – suppliers of grinding oils/fluids and lubrication application specialists

www.exxonmobil.com – suppliers of grinding oils/fluids and lubrication application specialists

www.nocco.com – suppliers of grinding oils/fluids and lubrication application specialists

www.marandproducts.com – suppliers of grinding oils/fluids and lubrication application specialists

www.metalworkinglubricants.com – suppliers of grinding oils/fluids and lubrication application specialists

References

[1]   Jackson MJ, Mills B (2000) Materials selection applied to vitrified alumina and c.B.N. grinding wheels. J Mater Process Technol 108: 114–124.
[2]   Grinding Data Book (2000) Unicorn International, Stafford, United Kingdom.
[3]   Jackson MJ, Davis CJ, Hitchiner MP, Mills B (2001) High-speed grinding with c.B.N. grinding wheels – applications and future developments. J Mater Process Technol 110: 78–88.
[4]   Jackson MJ, Mills B (2004) Microscale wear of vitrified abrasive materials. J Mater Sci 39: 2131–2143.
[5]   Jackson MJ (2004) Fracture dominated wear of sharp abrasive grains and grinding wheels. Proc Inst Mech Eng (London): Part J – J Eng Tribol 218: 225–235.
[6]   Jackson MJ (2006) Tribological design of grinding wheels using X-ray diffraction techniques. Proc Inst Mech Eng (London): Part J – J Eng Tribol 220 (1): 1–17.

10

Micro and Nanomachining

M.J. Jackson

Center for Advanced Manufacturing, College of Technology, Purdue University,
401 North Grant Street, West Lafayette, Indiana, IN 47907, USA.
E-mail: jacksomj@purdue.edu

Recent advances in miniaturization have led to the development of microscale components, usually in silicon. However, components made from engineering materials require shaping processes other than those established for processing silicon. Therefore, traditional machining processes require further development in order to machine components that are fit for purpose at the micro and nanoscales. This chapter provides a timely review of the current developments and recent advances in the area of micro and nanomachining.

10.1 Introduction

There is a substantial increase in the specific energy required with a decrease in chip size during machining. It is believed that this is due to the fact that all metals contain defects such as grain boundaries, missing and impurity atoms *etc.*, and when the size of the material removed decreases the probability of encountering a stress-reducing defect decreases. Since the shear stress and strain in metal cutting is unusually high, discontinuous microcracks usually form on the primary shear plane. If the material is very brittle, or the compressive stress on the shear plane is relatively low, microcracks will grow into larger cracks, giving rise to discontinuous chip formation. When discontinuous microcracks form on the shear plane they will weld and reform as strain proceeds, thus joining the transport of dislocations in accounting for the total slip of the shear plane. In the presence of a contaminant, such as carbon tetrachloride vapour at a low cutting speed, the re-welding of microcracks will decrease, resulting in a decrease in the cutting force required for chip formation. A number of special experiments that support the transport of microcracks across the shear plane, and the important role compressive stress plays on the shear plane are explained. An alternative explanation for the size effect in cutting is based on the belief that shear stresses increase with increasing strain rate. When an attempt is made to apply this to metal cutting, it is assumed in the analysis that the von Mises criterion applies to the shear plane. This is inconsistent with the experimental findings by Merchant [9]. Until this difficulty is resolved with the experimental verification of the strain rate approach, it should be assumed that the strain rate effect may be responsible for some portion of the size effect in metal cutting.

10.2 Machining Effects at the Microscale

It has been known for a long time that a size effect exists in metal cutting, where the specific energy increases with decrease in deformation size. Backer *et al.* [1] performed a series of experiments in which the shear energy per unit volume deformed (u_S) was determined as a function of specimen size for a ductile metal (SAE 1112 steel). The deformation processes involved were as follows, listed with increasing size of specimen deformed:

- surface grinding
- micromilling
- turning
- tensile test

The surface grinding experiments were performed under relatively mild conditions involving plunge-type experiments in which a wheel with an 8 inch (20.3 cm) diameter wheel was directed radially downward against a square specimen of length and width 0.5 in (1.27 cm). The width of the wheel was sufficient to grind the entire surface of the work at different down feed rates (t). The vertical and horizontal forces were measured by a dynamometer supporting the workpiece. This enabled the specific energy (u_S) and the shear stress on the shear plane (τ) to be obtained for different values of undeformed chip thickness (t). The points corresponding to a constant specific energy below a value of down feed of about 28 μinch (0.7 μm) are on a horizontal line due to a constant theoretical strength of the material being reached when the value of t goes below approximately 28 μinch (0.7 μm). The reasoning in support of this conclusion is presented in Backer *et al.* [1].

In the micromilling experiments, a carefully balanced 6-inch (152 cm) carbide-tipped milling cutter was used with all but one of the teeth relieved so that it operated as a fly milling cutter. Horizontal and vertical forces were measured for a number of depths of cut (t) when machining the same sized surface as in grinding. The shear stress on the shear plane (τ) was estimated by a rather detailed method presented in Backer *et al.* [1]. Turning experiments were performed on a 2.25-inch (5.72-cm) diameter SAE 1112 steel bar pre-machined in the form of a thin-walled tube having a wall thickness of 0.2 inch (5 mm). A carbide tool with 0° rake angle was operated in a steady-state two-dimensional orthogonal cutting mode as it machined the end of the tube. Values of shear stress on the shear plane (τ) versus undeformed chip thickness were determined for experiments at a constant cutting speed and different values of axial infeed rate and for variable cutting

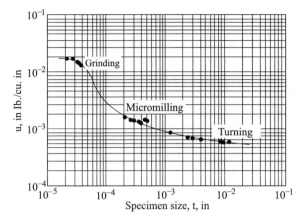

Fig. 10.1. Variation of the shear stress on shear plane when cutting SAE 1112 steel [1]

speeds and a constant axial in-feed rate. The grinding, micromilling and turning
results are shown in Figure 10.1.

A true stress–strain tensile test was performed on a 0.505-inch (1.28 cm) diameter
by 2-inch (5.08 cm) gage length specimen of SAE 1112 steel. The mean shear stress
at fracture was 22,000 psi (151.7 MPa). This value is not shown in Figure 10.1 since
it falls too far to the right. Taniguchi discussed the size effect in cutting and forming.
His version of the figure is presented in Figure 10.2 [2]. Shaw [3] discusses the origin
of the size effect in metal cutting, which is believed to be primarily due to short-range
inhomogeneities present in all engineering metals.

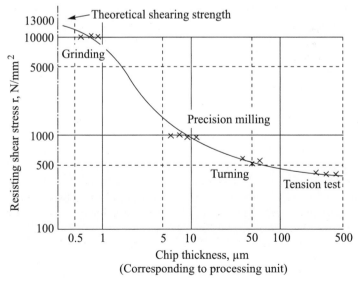

Fig. 10.2. Relationship between chip thickness and resisting shear stress for Figure 10.1 as
modified by Taniguchi [2]

When the back of a metal cutting chip is examined at very high magnification by means of an electron microscope individual slip lines are evident, as shown in Figure 10.3. In deformation studies, Heidenreich and Shockley [4] found that slip does not occur on all atomic planes but only on certain discrete planes. In experiments on deformed aluminium single crystals the minimum spacing of adjacent slip planes was found to be approximately 50 atomic spaces while the mean slip distance along the active slip planes was found to be about 500 atomic spaces. These experiments further support the observation that metals are not homogeneous and suggest that the planes along which slip occurs are associated with inhomogeneities in the metal. Strain is not uniformly distributed in many cases. For example, the size effect in a tensile test is usually observed only for specimens less than 0.1 inch (2.5 mm) in diameter. On the other hand, a size effect in a torsion test occurs for considerably larger samples due to the greater stress gradient present in a torsion test than in a tensile test. This effect and several other related ones are discussed in detail by Shaw [3].

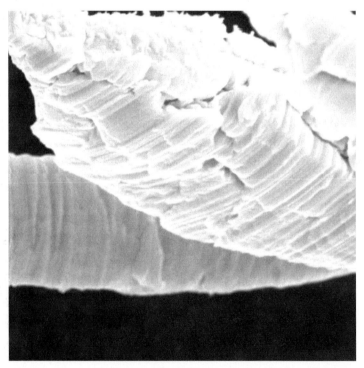

Figure 10.3. Back free surface of chip showing regions of discontinuous strain or microfracture

10.2.1 Shear Angle Prediction

There have been many notable attempts to derive an equation for the shear angle (ϕ) shown in Figure 10.4 for steady-state orthogonal cutting. Ernst and Merchant [5] presented the first quantitative analysis. Figure 10.5 shows the forces acting on a chip at the tool point, where R is the resultant force on the tool face, R' is the resultant force in the shear plane, N_C and F_C are the components of R normal to and parallel to the tool face, N_S and F_S are the components of R' normal to and parallel to the cutting direction, F_Q and F_P are the components of R normal to and parallel to the cutting direction and β (called the friction angle) is $\tan^{-1} F_C/N_C$.

Assuming the shear stress on the shear plane (τ) to be uniformly distributed it is evident that:

$$\tau = \frac{F_S}{A_S} = \frac{R' \cos(\phi + \beta - \alpha) \sin\phi}{A} \tag{10.1}$$

where A_S and A are the areas of the shear plane and that corresponding to the width of cut (b) times the depth of cut (t). Ernst and Merchant [5] reasoned that τ should be an angle such that τ would be a maximum and a relationship for ϕ was obtained by differentiating Equation (10.1) with respect to ϕ and equating the resulting expression to zero, producing

$$\phi = 45 - \frac{\beta}{2} + \frac{\alpha}{2} \tag{10.2}$$

However, it is to be noted that in differentiating, both R' and β were considered independent of ϕ.

Merchant [6] presented a different derivation that also led to Equation (10.2). This time an expression for the total power consumed in the cutting process was first written as

$$P = F_P V = (\tau A V)\frac{\cos(\beta - \alpha)}{\sin\phi\cos(\phi + \beta - \alpha)} \tag{10.3}$$

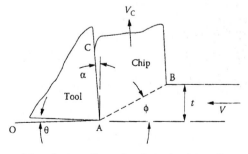

Figure 10.4. Nomenclature for two-dimensional steady-state orthogonal cutting process

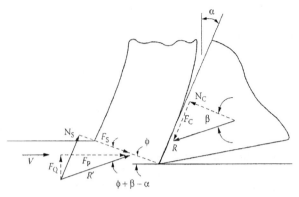

Figure 10.5. Cutting forces at the tool tip for the cutting operation shown in Figure 10.4

It was then reasoned that ϕ would be such that the total power would be a minimum. An expression identical to Equation (10.2) was obtained when P was differentiated with respect to ϕ, this time considering τ and β to be independent of ϕ. Piispanen [7] had done this previously in a graphical way. However, he immediately carried his line of reasoning one step further and assumed that the shear stress τ would be influenced directly by normal stress on the shear plane as follows:

$$\tau = \tau_0 + K_\sigma \tag{10.4}$$

where K is a material constant. Piispanen then incorporated this into his graphical solution for the shear angle. Upon finding Equation (10.2) to be in poor agreement with experimental data Merchant also independently (without knowledge of Piispanen's work at the time) assumed the relationship given in Equation (10.4), and proceeded to work this into his second analysis as follows. From Figure 10.5 it may be seen that

$$\sigma = \tau \tan(\phi + \beta - \alpha) \tag{10.5}$$

or, from Equation (10.4)

$$\tau_0 = \tau + K\tau \tan(\phi + \beta - \alpha) \tag{10.6}$$

Hence,

$$\tau = \frac{\tau_0}{1 - K \tan(\phi + \beta - \alpha)} \tag{10.7}$$

When this is substituted into Equation (10.3) we have,

$$P = \frac{\tau_0 A V \cos(\beta - \alpha)}{\left[1 - K \tan(\phi + \beta - \alpha)\right] \sin\phi \cos(\phi + \beta - \alpha)} \tag{10.8}$$

Now, when P is differentiated with respect to ϕ and equated to zero (with τ_0 and p considered independent of ϕ we obtain,

$$\phi = \frac{cot^{-1}(K)}{2} - \frac{\beta}{2} + \frac{\alpha}{2} = \frac{C-\beta+\alpha}{2}$$

(10.9)

Merchant called the quantity $cot^{-1} K$, the machining constant C. The quantity C is seen to be the angle the assumed line relating τ and ϕ makes with the τ axis [6–7, 9]. Merchant [8] determined the values of C given in Table 10.1 for materials of different chemistry and structure being turned under finishing conditions with different tool materials. From this table it is evident that C is not a constant. Merchant's empirical machining C that gives rise to Equation (10.9) with values of ϕ is in reasonably good agreement with experimentally measured values.

While it is well established that the rupture stress of both brittle and ductile materials is increased significantly by the presence of compressive stress (known as the Mohr effect), it is generally believed that a similar relationship for flow stress does not hold. However, an explanation for this paradox with considerable supporting experimental data is presented below. The fact that this discussion is limited to steady-state chip formation rules out the possibility of periodic gross cracks being involved. However, the role of microcracks is a possibility consistent with steady-state chip formation and the influence of compressive stress on the flow stress in shear. A discussion of the role microcracks can play in steady-state chip formation is presented in the next section. Hydrostatic stress plays no role in the

Table 10.1. Values of C in Equation (10.9) for a variety of work and tool materials in finish turning without a cutting fluid

Work material	Tool material	C (degrees)
SAE 1035 Steel	HSS*	70
SAE 1035 Steel	Carbide	73
SAE 1035 Steel	Diamond	86
AISI 1022 (leaded)	HSS*	77
AISI 1022 (leaded)	Carbide	75
AISI 1113 (sul.)	HSS*	76
AISI 1113 (sul.)	Carbide	75
AISI 1019 (plain)	HSS*	75
AISI 1019 (plain)	Carbide	79
Aluminium	HSS*	83
Aluminium	Carbide	84
Aluminium	Diamond	90
Copper	HSS*	49
Copper	Carbide	47
Copper	Diamond	64
Brass	Diamond	74

*HSS = high-speed steel

plastic flow of metals if they have no porosity. Yielding then occurs when the von Mises criterion reaches a critical value. Merchant [9] indicated that Barrett [10] found that for single-crystal metals τ_S is independent of $\varnothing$ and was also observed when plastics such as celluloid are cut. In general, if a small amount of compressibility is involved, yielding will occur when the von Mises criterion reaches a certain value.

However, based on the results of Table 10.1 the role of compressive stress on shear stress on the shear plane in steady-state metal cutting is substantial. The fact that there is no outward sign of voids or porosity in steady-state chip formation of a ductile metal during cutting and yet there is a substantial influence of normal stress on shear stress on the shear plane represents an interesting paradox. It is interesting to note that Piispanen [7] had assumed that shear stress on the shear plane would increase with normal stress and had incorporated this into his graphical treatment.

10.2.2 Plastic Behaviour at Large Strains

There has been little work done in the region of large plastic strains. Bridgman [11] used hollow tubular notched specimens to perform experiments under combined axial compression and torsion. The specimen was loaded axially in compression as the centre section was rotated relative to the ends. Strain was concentrated in the reduced sections and it was possible to crudely estimate and plot shear stress versus shear strain with different amounts of compressive stress on the shear plane. From these experiments Bridgman concluded that the flow curve for a given material was the same for all values of compressive stress on the shear plane, a result consistent with other materials experiments involving much lower plastic strains. However, the strain at gross fracture was found to be influenced by compressive stress. A number of related results are considered in the following subsections.

10.2.3 Langford and Cohen's Model

Langford and Cohen [12] were interested in the behaviour of dislocations at very large plastic strains and whether there was saturation relative to the strain hardening effect with strain, or whether strain hardening continued to occur with strain to the point of fracture. Their experimental approach was an interesting and fortunate one. They performed wire drawing on iron specimens using a large number of progressively smaller dies with remarkably low semi die angle (1.5°) and a relatively low (10%) reduction in area per die pass. After each die pass, a specimen was tested in uniaxial tension and a true stress–strain curve obtained. The drawing and tensile experiments were performed at room temperature and low speeds to avoid heating and specimens were stored in liquid nitrogen between experiments to avoid strain aging effects. All tensile results were then plotted in a single diagram, the strain used being that introduced in drawing (0.13 per die pass) plus the plastic strain in the tensile test. The general overlap of the tensile stress–strain curves gives an overall strain-hardening envelope, which indicates that the wire drawing and tensile deformations are approximately equivalent relative to strain hardening [13].

Blazynski and Cole [14] were interested in strain hardening in tube drawing and tube sinking. Drawn tubes were sectioned and tested in plane-strain compression. Up to a strain of about unity the usual strain-hardening curve was obtained, which is in good agreement with the generally accepted equation,

$$\sigma = \sigma_1 \varepsilon^n$$

(10.10)

However, beyond a strain of unity, the curve was linear, corresponding to the equation,

$$\sigma = A + B\varepsilon, (\varepsilon < 1)$$

(10.11)

where A and B are constants. It may be shown that,

$$A = (1 - n)\,\sigma_1$$

(10.12)

$$B = n\sigma_1$$

(10.13)

From transmission electron micrographs of deformed specimens, Langford and Cohen found that cell walls representing concentrations of dislocations began to form at strains below 0.2 and became ribbon shaped with decreasing mean linear intercept cell size as the strain progressed. Dynamic recovery and cell wall migration resulted in only about 7% of the original cells remaining after a strain of 6. The flow stress of the cold-worked wires was found to vary linearly with the reciprocal of the mean transverse cell size [15].

10.2.4 Walker and Shaw's Model

Acoustic studies were performed on specimens of the Bridgman type but fortunately, lower levels of axial compressive stress than Bridgman had used were employed in order to more closely simulate the concentrated shear process of metal cutting. The apparatus used, which was capable of measuring stresses and strains as well as acoustic signals arising from plastic flow, is described in the dissertation of T.J. Walker [16]. Two important results were obtained:

1. A region of rather intense acoustical activity occurred at the yield point followed by a quieter region until a shear strain of about 1.5 was reached. At this point there was a rather abrupt increase in acoustical activity that continued to the strain at fracture, which was appreciably greater than 1.5
2. The shear stress appeared to reach a maximum at strain corresponding to the beginning of the second acoustic activity ($\gamma \approx 1.5$).

The presence of the notches in the Bridgman specimen made interpretation of the stress–strain results somewhat uncertain. Therefore, a new specimen was designed which substituted simple shear for torsion with normal stress on the shear plane. By empirically adjusting the distance Δx to a value of 0.25 mm it was possible to confine all the plastic shear strain to the reduced area, thus making it possible to readily determine the shear strain ($\gamma \approx \Delta y / \Delta x$). When the width of the minimum section was greater or less than 0.25 mm, the extent of plastic strain observed in a transverse micrograph at the minimum section either did not extend completely across the 0.25 mm dimension or was beyond this width.

Similar results were obtained for non-resulphurised steels and other ductile metals. There is little difference in the curves for different values of normal stress on the shear plane (σ) to a shear strain of about 1.5 [17]. This is in agreement with Bridgman. However, beyond this strain the curves differ substantially with compressive stress on the shear plane. At large strains, τ was found to decrease with increase in γ, a result that does not agree with Bridgman [11].

It is seen that, for a low value of normal stress on the shear plane of 40 MPa, strain hardening appears to be negative at a shear strain of about 1.5; that is, when the normal stress on the shear plane is about 10% of the maximum shear stress reached, negative strain hardening sets in at a shear strain of about 1.5. On the other hand, strain hardening remains positive up to a normal strain of about 8 when the normal stress on the shear plane is about equal to the maximum shear stress.

10.2.5 Usui's Model

In Usui *et al.* [18] an experiment is described designed to determine why CCl_4 is such an effective cutting fluid at low cutting speeds. Since this also has a bearing on the role of microcracks in large-strain deformation, it is considered here.

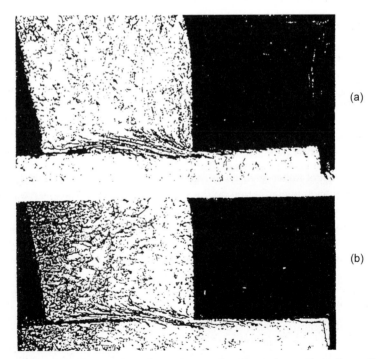

(a)

(b)

Figure 10.6. Photomicrographs of specimens that have been sheared a distance approximately equal to the shear plane length: (a) in air and (b) with a drop of CCl_4 applied

A piece of copper was prepared [18]. The piece that extends upward and appears to be a chip is not a chip but a piece of undeformed material left there when the specimen was prepared. A vertical flat tool was then placed precisely opposite the free surface and fed horizontally. Horizontal (F_P) and vertical (F_Q) forces were recorded as the shear test proceeded. It was expected that the vertical piece would fall free from the lower material after the vertical region had been displaced a small percentage of its length. However, it went well beyond the original extent of the shear plane and was still firmly attached to the base. This represents a huge shear strain since the shear deformation was confined to a narrow band. When a single drop of CCl_4 was placed before the shear test was conducted the protrusion could be moved only a fraction of the displacement in air before gross fracture occurred on the shear plane. Figure 10.6 shows photomicrographs of experiments without and with CCl_4. It is apparent that CCl_4 is much more effective than air in preventing microcracks from re-welding.

10.2.6 Saw-tooth Chip Formation in Hard Turning

Saw-tooth chip formation for hard steel discussed by Vyas and Shaw [19] is another example of the role microcracks play. In this case gross cracks periodically form at the free surface and run down along the shear plane until sufficient compressive stress is encountered to cause the gross crack to change to a collection of isolated microcracks.

10.2.7 Fluid-like Flow in Chip Formation

An interesting paper was presented by Eugene [20]. Using an unusual apparatus, water was pumped into a baffled chamber that removed eddy currents and then caused flow under gravity past a simulated tool. Powdered bakelite was introduced to make the streamlines visible as the fluid flowed past the tool. The photographs taken by the camera were remarkably similar to quick stop photomicrographs of actual chips. It was thought by this author at the time that any similarity between fluid flow and plastic flow of a solid was not to be expected. That was long before it was clear that the only logical explanation for the results of Bridgman and Merchant involved microfracture [21]. A more recent paper was presented that again suggests that metal cutting might be modelled by a fluid [22]. However, this paper was concerned with ultra-precision machining (depths of cut <4 μm) and potential flow analysis was employed instead of the experimental approach taken by Eugene.

It is interesting to note that chemists relate the flow of liquids to the migration of vacancies (voids) just as physicists relate ordinary plastic flow of solid metals to the migration of dislocations. Henry Eyring and co-workers [24–26] studied the marked changes in volume, entropy and fluidity that occur when a solid melts. For example, a 12% increase in volume accompanies melting of argon, suggesting the removal of every eighth molecule as a vacancy upon melting. This is consistent with X-ray diffraction of liquid argon that showed good short-range order but poor long-range order. The relative ease of diffusion of these vacancies accounts for the increased fluidity that accompanies melting. A random distribution of vacancies is also consistent with the increase in entropy observed on melting. Eyring's theory of fluid flow was initially termed the *hole theory of fluid flow* but later became known as *the significant structure theory*, which is the title of Eyring–Jhon book [25]. According to this theory the vacancies in a liquid move through a sea of molecules. Eyring's theory of liquid flow is mentioned here since it explains why the flow of a liquid approximates the flow of metal passed a tool in chip formation. In this case microcracks (voids) move through a sea of crystalline solid.

10.3 Size Effects in Micromachining

It is appropriate at this point to mention that an alternative explanation for the increase in hardness that occurs when the indentation size is reduced in metals has recently been introduced [27–38]. It is based on the fact that there is an increase in the strain gradient with a reduction in indentation size. This has been extended by Dinesh *et al.* [39] to explain the size effect in machining. In the Dinesh *et al.* [39] analysis the size effect in hardness is related to that in cutting by assuming the von Mises criterion is applicable. Based on the experiments of Merchant, it is evident that this is not applicable in steady-state chip formation.

In this strain gradient theory two types of dislocations are proposed: geometrically necessary dislocations (ρ_g) that are responsible for work hardening and statistically stored dislocations (ρ_S) that are affected by a strain gradient. When $\rho_g >> \rho_S$ conventional plasticity pertains (strain rate unimportant) but when $\rho_g << \rho_S$ a constitutive equation including strain rate should be included. The impression one obtains in reading Dinesh *et al.* [39] is that the strain gradient approach is uniquely responsible for the size effect in cutting. In their concluding remarks it is suggested that it should be possible to verify the validity of the strain rate formulation by experiments designed to test predictions of this approach. This has not yet been done and until it is it will not be possible to determine whether the influence of strain rate is significant in chip formation applications. In any case, it is believed that the explanation presented here based on the influence of defects and normal stress on the shear plane is sufficiently well supported by the experiments described that it should not be considered insignificant.

10.4 Nanomachining

Nanomachining can be classified into four categories:

- Deterministic mechanical nanometric machining. This method utilizes fixed and controlled tools, which can specify the profiles of three-dimensional components by a well-defined tool surface and path. The method can remove materials in amounts as small as tens of nanometres. It includes typically diamond turning, micro milling and nano/micro grinding, *etc.*;

- Loose abrasive nanometric machining. This method uses loose abrasive grits to removal a small amount materials. It consists of polishing, lapping and honing, *etc.*
- Non-mechanical nanometric machining: focussed ion beam machining, micro-EDM, and excimer laser machining; and
- Lithographic methods, which employ masks to specify the shape of the product. Two-dimensional shapes are the main outcome; severe limitations occur when three-dimensional products are attempted. This category mainly includes X-ray lithography, LIGA and electron beam lithography.

Mechanical nanometric machining has more advantages than other methods since it is capable of machining complex 3D components in a controllable way. The machining of complex surface geometry is just one of the future trends in nanometric machining, which is driven by the integration of multiple functions in one product. For instance, the method can be used to machine micromoulds and dies with complex geometric features and high dimensional and form accuracy, and even nanometric surface features. The method is indispensable to manufacturing complex micro and miniature structures, components and products in a variety of engineering materials. This section focusses on nanometric cutting theory, methods and its implementation and application perspectives [40–44].

10.4.1 Nanometric Machining

Single-point diamond turning and ultra-precision grinding are two major nanometric machining approaches. They are both capable of producing extremely fine cuts. Single-point diamond turning has been widely used to machine non-ferrous metals such as aluminium and copper. An undeformed chip thickness about 1 nm is observed in diamond turning of electroplated copper [45]. Diamond grinding is an important process for the machining of brittle materials such as glasses and ceramics to achieve nanometre levels of tolerances and surface finish. A repeatable optical quality surface roughness (surface finish < 10 nm Ra) has been obtained in nanogrinding of hard steel by Stephenson *et al.* using 76 μm grit CBN wheel on the ultra-precision grinding machine tool [46]. Recently, diamond fly-cutting and diamond milling have been developed for machining non-rotational non-symmetric geometry, which has enlarged the product spectrum of nanometric machining [47]. In addition the utilization of ultra-fine grain hard metal tools and diamond coated microtools represents a promising alternative for microcutting of even hardened steel [48].

Early applications of nanometric machining are in the mass production of some high-precision parts for microproducts or microsystems. In fact, microproducts, or microsystems, will be the first path that enables nanoproducts to enter the marketplace since microproducts, or microsystems, have been dominating nanotechnology application markets worldwide [49]. It is also anticipated that requirements for microproducts will increase worldwide. It is very interesting to see that the information technology (IT) peripheral market is still the biggest market for microproducts. In 2005, the total turnover of microproducts reached US $38 billion, which is twice the total turnover in 2000. Nanometric machining can be applied in bulk machining of silicon, aluminium substrates for computer memory disks, *etc.* In other areas such as biomedical, automotive, household and telecommunications the total turnovers of microproducts are still steadily growing. Nanometric machining is also very promising in the production of sensors, accelerometer, actuators, micro-mirror, fibre-optic connectors and microdisplays. In fact applications of nanoproducts will enhance the performance of microproducts in terms of sensitivity, selectivity and stability [50]. However, by 2006, IT is expected to lose this predominant position owing to new micro-electro-mechanical systems (MEMS)-based applications in sectors such as biotechnology and communication (optical and radiofrequency switching, for example, will become a major growth area) [51]. Nanometric machining still has priority in this application area. The microproducts are normally integrated products of some electronics, mechanical parts and optical parts while in miniature or microscale dimensions. In fact only a small number of microproducts rely solely on electronics. The mechanical and optical parts are of significant importance for microproducts. The indispensable advantage of nanometric machining is its applicability to manufacture complex three-dimensional (3D) components/devices including micromoulds, dies and embossing tooling for cheap mass production of optical and mechanical parts. Therefore, it is undoubtedly one of the major enabling technologies for commercialization of nanotechnology in the future.

10.4.2 Theoretical Basis of Nanomachining

Scientific study of nanometric machining has been undertaken since the late 1990s. Much attention has been paid to its study, especially with the advance of nanotechnology [52]. This scientific study will result in the formation of the theoretical basis of nanometric machining, which will enable better understanding of nanometric machining physics and the development of controllable techniques to meet the advanced requirements of nanotechnology and nanoscience.

10.4.2.1 Cutting Force and Energy

In nanomanufacturing, the cutting force and cutting energy are important issues. They are important physical parameters for understanding cutting phenomena as they clearly reflect the chip removal process. From the aspect of atomic structures cutting forces are the superposition of the interactions forces between workpiece atoms and cutting tool atoms. Specific energy is an intensive quantity that characterizes the cutting resistance offered by a material [53]. Ikawa *et al.* and Luo *et al.* have acquired the cutting forces and cutting energy by molecular dynamics simulations [52–55]. Moriwaki and Lucca have carried out experiments to measure the cutting forces in nanometric machining [52]. Figure 10.7 shows the simulation and experimental results in nanometric cutting. Figure 10.7(a) illustrates the linear relation that exists between the cutting forces per unit width and the depth of cut in both simulations and experiments. The cutting forces per width increase with increasing depth of cut.

The difference in the cutting force between the simulations and the experiments is caused by the different cutting edge radii applied in the simulations. In

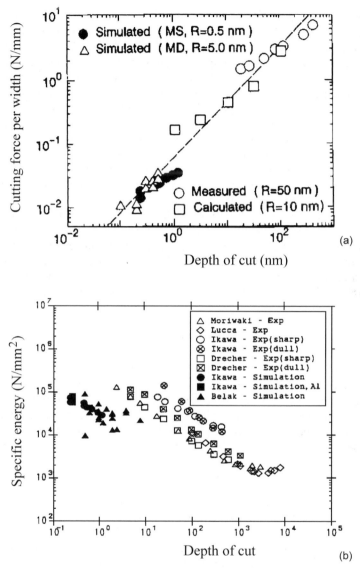

Figure 10.7. Comparison of results between simulations and experiments: (a) cutting force per width against depth of cut; (b) specific energy against depth of cut [52]

nanometric machining the cutting edge radius plays an important role since the depth of cut is similar in scale. For the same depth of cut higher cutting forces are needed for a tool with a large cutting edge radius compared with a tool with a small cutting edge radius. The low cutting force per width is obviously the result of fine cutting conditions, which will decrease the vibration of the cutting system and thus improve the machining stability and also result in better surface roughness. A linear relationship between the specific energy and the depth of cut can

Table 10.2. Material properties under different machining units [56]

	1 nm – 0.1 μm	0.1 μm – 10 μm	10 μm – 1 mm
Defects/impurities	Point defect	Dislocation/crack	Crack/grain boundary
Chip removal unit	Atomic cluster	Sub-crystal	Multi-crystals
Brittle fracture limit	10^4 J/m^3 – 10^3 J/m^3	10^3 J/m^3 – 10^2 J/m^3	10^2 J/m^3 – 10^1 J/m^3
	Atomic crack	Microcrack	Brittle crack
Shear failure limit	10^4 J/m^3 – 10^3 J/m^3	10^3 J/m^3 – 10^2 J/m^3	10^2 J/m^3 – 10^1 J/m^3
	Atomic-dislocation	Dislocation slip	Shear deformation

also be observed in Figure 10.7(b). The figure shows that the specific energy increases with decreasing depth of cut, because the effective rake angle is different for different depths of cut. For small depths of cut the effective rake angle will increase with decreasing depth of cut. Large rake angle results in an increase in the specific energy. This phenomenon is often called the *size effect*, which can be clearly explained by the material data listed in Table 10.2, according to which, in nanometric machining only point defects exist in the machining zone in a crystal, so more energy is required to initiate an atomic crack or atomic dislocation. For decreasing depth of cut the chance that the cutting tool will meet point defects decreases, resulting in the increasing specific cutting energy.

If the machining unit is reduced to 1 nm, the workpiece material structure at the machining zone may approach atomic perfection, so more energy will be required to break the atomic bonds. On the other hand when the machining unit is higher than 0.1 μm, the machining points will fall into the distribution distances of some defects such as dislocations, cracks and grain boundaries. These pre-existing defects will ease the deformation of workpiece material and result in a comparatively low specific cutting energy. Nanometric cutting is also characterized by a high ratio of the normal to the tangential component in the cutting force [53, 55], as the depth of cut is very small in nanometric cutting, and the workpiece is mainly processed by the cutting edge. The compressive interactions will thus become dominant in the deformation of the workpiece material, which will therefore result in an increase of the friction force at the tool–chip interface and a relative high cutting ratio. Usually the cutting force in nanometric machining is very difficult to measure due to its small amplitude compared with the noise (mechanical or electronic) [52]. A piezoelectric dynamometer or load cell is used to measure the cutting forces because of their high sensitivity and natural frequency [57].

10.4.2.2 Cutting Temperatures

In molecular dynamics simulation, the cutting temperature can be calculated under the assumption that the cutting energy is totally transferred into cutting heat and results in an increase of the cutting temperature and kinetic energy of system. The lattice vibration is the major form of thermal motion of atoms. Each atom has three degrees of freedom. According to the theorem of equipartition of energy, the average kinetic energy of the system can be expressed as:

$$\bar{E}_k = \frac{3}{2}Nk_BT = \sum_i \frac{1}{2}m(V_i^2)$$

(10.14)

where $\bar{E}_k$ is average kinetic energy in equilibrium state, K_B is Boltzmann's constant, T is temperature, m_i and V_i are the mass and velocity of an atom, respectively, and N is the number of atoms. The cutting temperature can be deduced as:

$$T = \frac{2\bar{E}_k}{3Nk_B}$$

(10.15)

Figure 10.8 shows the variation of the cutting temperature on the cutting tool in a molecular dynamics simulation of nanometric cutting of single-crystal aluminium. The highest temperature is observed at the cutting edge, although the temperature at the flank face is also higher than that at the rake face. The temperature distribution suggests that a major heat source exists at the interface between the cutting edge and the workpiece and that the heat is conducted from there to the rest of the cutting zone in the workpiece and cutting tool. The reason is that, because most cutting action take place at the cutting edge of the tool, the dislocation deformations of workpiece materials will transfer potential energy into the kinetic energy and result in a rise in temperature. The comparatively high temperature at

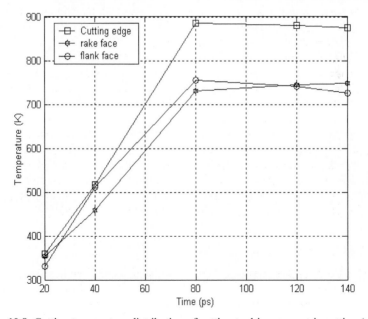

Figure 10.8. Cutting temperature distribution of cutting tool in nanometric cutting (cutting speed = 20 m/s, depth of cut = 1.5 nm, cutting edge radius = 1.57 nm) [58]

the tool flank face is obviously caused by friction between the tool flank face and the workpiece. The released energy due to the elastic recovery of the machined surface also contributes to the temperature increase at the tool flank face. Although there is also friction between the tool rake face and the chip, the heat will be taken away from the tool rake face by the removal of the chip.

Therefore, the temperature at the tool rake face is lower than that at the tool cutting edge and tool flank face. The results shows that the cutting temperature in diamond machining is quite low in comparison with that in conventional cutting, due to the low cutting energy as well as the high thermal conductivity of diamond and the workpiece material. The cutting temperature is considered to govern the wear of a diamond tool in a molecular dynamics simulation study by Cheng *et al.* [58]. More in-depth experimental and theoretical studies are needed to find the quantitative relationship between the cutting temperature and tool wear, although there is considerable evidence of chemical damage to diamond, in which temperature plays a significant role [52].

10.4.2.3 Chip Formation

Chip formation and surface generation can be simulated by molecular dynamics (MD) simulation. Figure 10.9 shows an MD simulation of a nanometric cutting

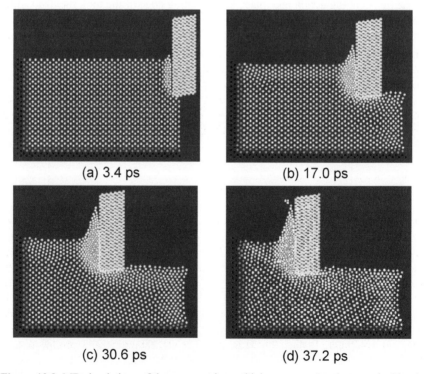

(a) 3.4 ps   (b) 17.0 ps

(c) 30.6 ps   (d) 37.2 ps

Figure 10.9. MD simulations of the nanometric machining process (cutting speed = 20 m/s, depth of cut = 1.4 nm, cutting edge radius = 0.35 nm) [58]

process on single crystal aluminium. From Figure 11.6(a) it is observed that after the initial plough of the cutting edge the workpiece atoms are compressed in the cutting zone near to the rake face and the cutting edge. The disturbed crystal lattices of the workpiece and even the initiation of dislocations can be observed in Figure 10.9(b). Figure 10.9(c) shows that the dislocations have piled up to form a chip. The chip is removed with the unit of an atomic cluster as shown in Figure 10.9(d). The lattice-disturbed workpiece material is observed on the machined surface.

Based on this visualisation of the nanometric machining process, the mechanism of chip formation and surface generation in nanometric cutting can be explained. Owing to the ploughing of the cutting edge, the attractive force between the workpiece atoms and the diamond tool atoms becomes repulsive. Because the cohesion energy of diamond atoms is much larger than that of Al atoms, the lattice of the workpiece is compressed. When the strain energy stored in the compressed lattice exceeds a specific level, the atoms begin to rearrange so as to release the strain energy. When the energy is not sufficient to perform the rearrangement, some dislocation activity is generated. Repulsive forces between compressed atoms in the upper layer and the atoms in the lower layer are increasing, so the upper atoms move along the cutting edge, and at the same time the repulsive forces from the tool atoms cause the resistance for the upward chip flow to press the atoms under the cutting line. With the movement of the cutting edge, some dislocations move upward and disappear from the free surface as they approach the surface.

This phenomenon corresponds to the process of the chip formation. As a result of the successive generation and disappearance of dislocations, the chip seems to be removed steadily. After the passing of the tool, the pressure at the flank face is released. The layers of atoms move upwards and result in elastic recovery, so the machined surface is generated. The conclusion can therefore be drawn that the chip removal and machined surface generation have the nature of dislocation slip movement inside the workpiece material crystal grains. In conventional cutting the dislocations are initiated from existing defects between the crystal grains, which will ease the movement of dislocation and result in smaller specific cutting forces compared with that in nanometric cutting.

The height of the atoms on the surface layer of the machined surface creates the surface roughness. For this, two-dimensional (2D) MD simulation R_a can be used to assess the machined surface roughness. The surface integrity parameters can also be calculated based on the simulation results. For example, the residual stress of the machined surface can be estimated by averaging the forces acting on the atoms in a unit area on the upper layer of the machined surface. Molecular dynamics (MD) simulation has been proved to be a useful tool for the theoretical study of nanometric machining [59]. At present the MD simulation studies on nanometric machining are limited by the computing memory size and speed of the computer. It is therefore difficult to enlarge the dimensions of the current MD model on a personal computer. In fact, the machined surface topography is produced as a result of the copy of the tool profile on a workpiece surface that has a specific motion relative to the tool. The degree of surface roughness is governed by both the controllability of the machine tool motion (or the relative motion between tool and workpiece) and the transfer characteristics (or the fidelity) of the tool profile to the workpiece [52]. A multi-scale analysis model, which can fully model the machine tool and cutting tool motion, environmental effects and the tool–workpiece interactions, is greatly needed to predict and control the nanometric machining process in a determinative manner.

10.4.2.4 Minimum Undeformed Chip Thickness

The minimum undeformed chip thickness is an important issue in nanometric machining because it is related to the ultimate machining accuracy. In principle the minimum undeformed chip thickness will be determined by the minimum atomic distance within the workpiece. However, in ultra-precision machining practice, it strongly depends on the sharpness of the diamond cutting tool, the capability of the ultra-precision machine tool and the machining environment. The diamond turning experiments of non-ferrous work materials carried out at Lawrence Livermoore National Laboratory (LLNL) show that a minimum undeformed chip thickness down to 1 nm is attainable with a specially prepared fine diamond cutting tool on a highly reliable ultra-precision machine tool [45]. Based on the tool wear simulation, the minimum undeformed chip thickness is further studied in this chapter. Figure 10.10 illustrates chip formation of single-crystal aluminium with a tool cutting edge radius of 1.57 nm. No chip formation is observed when the undeformed chip thickness is 0.25 nm, but the initial stage of

(a)   (b)

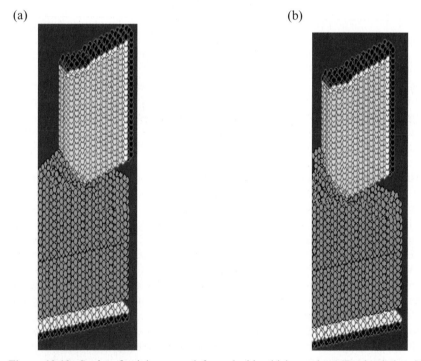

Figure 10.10. Study of minimum undeformed chip thickness by MD simulation [58]: (a) undeformed chip thickness 0.25 nm, (b) undeformed chip thickness 0.26 nm

Table 10.3. Minimum undeformed chip thickness against the tool cutting edge radius and cutting forces [58]

Cutting edge radius (nm)	1.57	1.89	2.31	2.51	2.83	3.14
Minimum undeformed chip thickness (nm)	0.26	0.33	0.42	0.52	0.73	0.97
Ratio of minimum undeformed chip thickness to tool cutting edge radius	0.17	0.175	0.191	0.207	0.258	0.309
Ratio of tangential cutting force to normal cutting force	0.92	0.93	0.92	0.92	0.94	0.93

chip formation is apparent when the undeformed chip thickness is 0.26 nm. In nanometric cutting, as the depth of cut is very small, chip formation is related to the force conditions at the cutting edge. Generally, chip formation is mainly a function of tangential cutting force.

The normal cutting force makes little contribution to the chip formation since it has the tendency to cause the atoms of the surface to penetrate the bulk of the workpiece. In theory, the chip is formed under the condition that the tangential cutting force is larger than the normal cutting force. The relationships between the minimum undeformed chip thickness, the cutting edge radius, and the cutting forces are studied via MD simulations. The results are highlighted in Table 10.3. These data show that the minimum undeformed chip thickness is about one-third to one-sixth of the tool cutting edge radius. The chip formation will be initiated when the ratio of tangential cutting force to normal cutting force is larger than 0.92.

10.4.2.5 Critical Cutting Radius

It is widely accepted that the sharpness of the cutting edge of a diamond cutting tool directly affects the machined surface quality. Previous MD simulations show that, the sharper the cutting edge is, the smoother the machined surface becomes. However, this conclusion is based on no tool wear. To study the real effects of cutting edge radius, the MD simulations on nanometric cutting of single-crystal aluminium are carried out using a tool wear model [59].

In these simulations the cutting edge radius of the diamond cutting tool varies from 1.57 to 3.14 nm with depth of cut of 1.5, 2.2 and 3.1 nm respectively. The cutting distance is fixed at 6 nm. The root-mean-square deviation of the machined surface and mean stress on the cutting edge are listed in Table 10.4. Figure 10.11 shows a visualization of the simulated data, which clearly indicates that the surface roughness increases with the decreasing cutting edge radius when the cutting edge radius is smaller than 2.31 nm. The tendency is obviously caused by the rapid tool wear when a cutting tool with a small cutting edge radius is used. However, when the cutting edge is larger than 2.31 nm, the cutting edge is under compressive stress and no tool wear occurs. Therefore, it shows the same tendency that the surface roughness increases with decreasing tool cutting edge radius, as in the previous MD simulations.

Table 10.4. The relationship between cutting edge radius and machined surface quality [58]

	Cutting edge radius (nm)	1.57	1.89	2.31	2.51	2.83	3.14
Depth of cut: 1.5 nm	S_q (nm)	0.89	0.92	0.78	0.86	0.98	1.06
Depth of cut: 2.2 nm	S_q (nm)	0.95	0.91	0.77	0.88	0.96	1.07
Depth of cut: 3.1 nm	S_q (nm)	0.97	0.93	0.79	0.87	0.99	1.08
Mean stress at cutting edge (GPa)		0.91	0.92	−0.24	−0.31	−0.38	−0.44

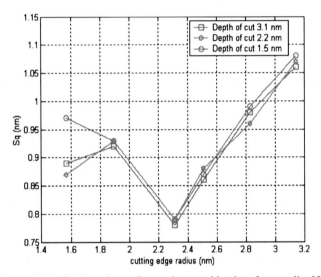

Figure 10.11. Cutting edge radius against machined surface quality [58]

The MD simulation results also illustrate that it is not true that, the sharper the cutting edge, the better the machined surface quality. The cutting edge is destined to wear and results in degradation of the machined surface quality if its radius is smaller than a critical value. However, when the cutting edge radius is larger than a critical value, compressive stress will take place at the tool edge and the tool condition is more stable. As a result a high-quality machined surface can be achieved. Therefore, there is a critical cutting edge radius for stably achieving high-quality machined surfaces. For cutting single-crystal aluminium this critical cutting edge radius is at 2.31 nm. The MD simulation approach is applicable for acquiring the critical cutting edge radius for nanometric cutting of other materials.

10.4.2.6 Workpiece Materials

In nanometric machining the microstructure of the workpiece material will play an important role in affecting the machining accuracy and machined surface quality.

For example, when machining polycrystalline materials, the difference in the elastic coefficients at the grain boundary and interior of the grain causes small steps formed on the cut surface since the respective elastic *rebound* varies [60]. The study by Lee and Chueng shows that the shear angle varies with the crystallographic orientation of the materials being cut. This will produce a self-excited vibration between the cutting tool and the workpiece and result in a local variation of surface roughness of a diamond-turned surface [61].

A material's destructive behaviour can also be affected by nanometric machining. In nanometric machining of brittle materials it is possible to produce plastically deformed chips, if the depth of cut is sufficiently small [62]. It has been shown that a brittle-to-ductile transition exists when cutting brittle materials at low load and penetration levels [63]. The transition from ductile to brittle fracture has been widely reported and is usually described as the *critical depth of cut* [62], which is generally small, up to 0.1–0.3 μm. This will result in relatively slow material removal rates [62].

However, ductile-mode machining is a cost-effective technique for producing high-quality spherical and non-spherical optical surfaces, with or without the need for lapping and polishing [62]. The workpiece materials should also have a low affinity with the cutting tool material. If bits of the workpiece material are deposited onto the tool, this will cause tool wear and adversely affect the finished surface (surface finish and surface integrity). Therefore, the workpiece materials chosen must possess an acceptable machinability on which a nanometric surface finish can thus be achieved. Diamond tools are widely used in nanometric machining because of their excellent characteristics. The materials currently turned with diamond tools are listed in Table 10.5. The materials that can be processed via ductile-mode grinding with diamond wheels are listed in Table 10.6.

Table 10.5. Current diamond-turned materials [62]

Semiconductors	Metals	Plastics
Cadmium telluride	Aluminium and alloys	Acrylic
Gallium arsenide	Copper and alloys	Fluoroplastics
Germanium	Electroless nickel	Nylon
Lithium niobate	Gold	Polycarbonate
Silicon	Magnesium	Polymethylmethacrylate
Zinc selenide	Silver	Propylene
Zinc sulphide	Zinc	Styrene

Table 10.6. Materials that can be processed via ductile-mode diamond grinding [62]

Ceramics/intermetallics		Glasses
Aluminium oxide	Tatanium aluminide	BK7 or equivalent
Nickel aluminide	Tatanium carbide	SF10 or equivalent
Silicon carbide	Tungsten carbide	ULE or equivalent
Silicon nitride	Zirconia	Zerodur or equivalent

10.4.3 Comparison of Nanometric Machining and Conventional Machining

Table 10.7 summarizes a comparison of nanometric machining and conventional machining in all major aspects of cutting mechanics and physics.

Table 10.7. Ccomparison of nanometric machining with conventional machining [55]

		Nanometric machining	Conventional machining
Fundamental cutting principles		Discrete molecular mechanics/ micromechanics	Continuum elastic/ plastic /fracture mechanics
Workpiece material		Heterogeneous (presence of microstructure)	Homogeneous (ideal element)
Cutting physics		Atomic cluster or microelement model $$q_i = \frac{\partial H}{\partial p_i} \quad i = 1, 2, \dots N$$ $$p_i = -\frac{\partial H}{\partial q_i}$$	Shear plane model (continuous points in material)
		First principal stress $$\sigma = \frac{1}{S}\sum_{i=1}^{N_A}\sum_{j=1}^{N_B} f_{ij} - \frac{1}{S}\sum_{i=1}^{N_A}\sum_{j=1}^{N_B} f_{0ij}$$ (crystal deformation included)	Cauchy stress principle $$\tau_s = \frac{F_s}{A}$$ (constant)
Cutting force and energy	Energy consideration	Interatomic potential functional $$U(r^N) = \sum_i \sum_{<i} u(r_{ij})$$	Shear/friction power $$P_s = F_s \cdot V_s$$ $$P_u = F_u \cdot V_c$$
	Specific energy	High	Low
	Cutting force	Interatomic forces $$F_I = \sum_{j \neq i}^{N} F_{ij} = \sum_{j \neq i}^{N} -\frac{du(r_{ij})}{dr_{ij}}$$	Plastic deformation/ friction $$F_c = F(b, d_c, \tau_s, \beta_a, \phi_c, \alpha_r)$$
Chip formation	Chip initiation	Inner crystal deformation (point defects or dislocation)	Inter crystal deformation (grain boundary void)
	Deformation and stress	Discontinuous	Continuous
Cutting tool	Cutting edge radius	Significant	Ignored
	Tool wear	Clearance face and Cutting edge	Rake face

This comparison highlighted is by no means comprehensive, but rather provides a starting point for further study on the physics of nanometric machining.

Acknowledgements

The author thanks Professor Kai Cheng for assistance in preparing this chapter and for the use of nanomachining research material provided by him and Dr. Luo, and the late Professor Milton Shaw for his encouragement and help in preparing this and other chapters on machining before his untimely death.

References

[1]   Backer WR, Marshall ER, Shaw MC (1952) Trans ASME 74: 61.
[2]   Taniguchi N (1994) Precis Eng 16: 5–24.
[3]   Shaw MC (1952) J Franklin Inst 254(2): 109.
[4]   Heidenreich RO, Shockley W (1948) Report on Strength of Solids, Phys. Soc. of London, 57.
[5]   Ernst HJ, Merchant ME (1941) Trans Am Soc Metals 29: 299.
[6]   Merchant ME (1945) J Appl Phys 16: 267–275.
[7]   Piispanen V (1937) Teknillinen Aikakaushehti (Finland) 27: 315.
[8]   Merchant ME (1950) Machining Theory and Practice. Am Soc Metals 5–44.
[9]   Merchant ME (1945) J Appl Phys 16: 318–324.
[10]   Barrett CS (1943) Structure of Metals. McGraw Hill, NY, p. 295.
[11]   Bridgman PW (1952) Studies in Large Plastic Flow and Fracture. McGraw Hill, NY.
[12]   Langford G, Cohen M (1969) Trans ASM 62: 623.
[13]   V. Piispanen (1948) J Appl Phys 19: 876.
[14]   Blazynski TZ, Cole JM (1960) Proc Inst Mech Engrs 1(74): 757.
[15]   Shaw MC (1950) J Appl Phys 21: 599.
[16]   Walker TJ (1967) PhD Dissertation, Carnegie-Mellon University, PA.
[17]   Walker TJ, Shaw MC (1969) Advances in Machine Tool Design and Research. Pergamon pp. 241–252.
[18]   Usui E, Gujral A, Shaw MC (1960) Int J Mach Tools Res 1: 187–197.
[19]   Vyas A, Shaw MC (1999) Trans ASME-J Mech Sci 21(1): 63–72.
[20]   Eugene F (1952) Ann CIRP 52(11): 13–17.
[21]   Shaw MC (1980) Int J Mech Sci 22: 673–686.
[22]   Kwon KB, Cho DW, Lee SJ, Chu CN (1999) Ann CIRP 47(1): 43–46.
[23]   Eyring H, Ree T, Harai N (1958) Proc Nat Acad Sci 44: 683.
[24]   Eyring H, Ree T (1961) Proc Nat Acad Sci 47: 526–537.
[25]   Eyring H, Jhon MS (1969) Significant Theory of Liquids. Wiley, NY.
[26]   Kececioglu D (1958) Trans ASME 80: 149–168.
[27]   Kececioglu D (1958) Trans ASME 80: 541–546.
[28]   Kececioglu D (1960) Trans ASME J Eng Ind 82: 79–86.
[29]   Anderson TL (1991) Fracture Mechanics. CRC, Florida.
[30]   Zhang B, Bagchi A (1994) Trans ASME J Eng Ind 116: 289.
[31]   Argon AS, Im J, Safoglu R (1975) Metall Trans 6A: 825.
[32]   Komanduri R, Brown RH (1967) Metals Mater 95: 308.
[33]   Drucker DC (1949) J Appl Phys 20: 1.

[34] Fleck NA, Muller GM, Ashby MF, Hutchinson JM (1994) Acta Metall Materialia 41(10): 2855.
[35] Stelmashenko NA, Walls MG, Brown LM, Milman YV (1993) Acta Metall Materialia 41 (10): 2855.
[36] Ma Q, Clarke DR (1995) J Mater Res 46(3): 477.
[37] Nix WD, Gao H (1998) J Mech Phys Solids 1(4): 853.
[38] Gao H, Huang Y, Nix WD, Hutchinson JW (1999) J Mech Phys Solids 47: 1239.
[39] Dinesh D, Swaminathan S, Chandrasekar S, Farris TN (2001) Proc ASME-IMECE, NY: 1–8.
[40] Committee on Technology National Science and Technology Council (2000) "National Nanotechnology initiative: Leading to the next industrial revolution, Washington D.C.
[41] Snowdon K, McNeil C, Lakey J (2001) Nanotechnology for MEMS components. mstNews 3, 9–10.
[42] El-Fatatry A, Correial A (2003) Nanotechnology in microsystems: potential influence for transmission systems and related applications. mstNews 3: 25–26.
[43] Werner M, Köhler T, Grünwald W (2001) Nanotechnology for applications in microsystems. mstNews 3: 4–7.
[44] El-Hofy H, Khairy A, Masuzawa T, McGeough J (2002) Introduction. In: McGeough J eds. Micromachining of Engineering Materials. Marcel Dekker, New York, NY.
[45] Donaldson R, Syn C, Taylor J, Ikawa N, Shimada S (1987) Minimum thickness of cut in diamond turning of electroplated copper. UCRL-97606.
[46] Stephenson DJ, Veselovac D, Manley S, Corbett J (2001) Ultra-precision grinding of hard steels. Precis Eng 15: 336–345.
[47] Rübenach O, Micro technology – applications and trends. Euspen online traininglecture.http://www.euspen.org/training/lectures/course2free2view/02MicroTechApps/de molecture.asp (accessed July 2007).
[48] http://www.lfm.uni-bremen.de/html/res/res001/res108.html (accessed July 2007).
[49] Weck M (2000) Ultraprecision machining of microcomponents. Machine Tools 1: 113–122.
[50] Schütze A, Lutz-Günter J (2003) Nano sensors and micro integration. mstNews 3: 43–45.
[51] El-Fatatry A, Correial A (2003) Nanotechnology in Microsystems: potential influence for transmission systems and related applications. mstNews 3: 25.
[52] Ikawa N, Donaldson R, Komanduri R, König W, Mckeown PA, Moriwaki T, Stowers I (1991) Ultraprecision metal cutting – the past, the present and the future. Ann CIRP 40(2): 587–594.
[53] Shaw MC (1996) Principles of Abrasive Processing. Oxford University Press, New York, NY.
[54] Komanduri R, Chandrasekaran, Raff L (1998) Effects of tool geometry in nanometric cutting: a molecular dynamics simulation approach. Wear 219: 84–97.
[55] Luo X, Cheng K, Guo X, Holt R (2003) An investigation on the mechanics of nanometric cutting and the development of its test-bed. Int J Prod Res 41 (7): 1449–1465.
[56] Taniguchi N (1996) Nanotechnology. Oxford University Press, New York, NY.
[57] Dow T, Miller E, Garrard K (2004) Tool force and deflection compensation for small milling tools. Precis Eng 28 (1): 31–45.
[58] Cheng K, Luo X, Ward R, Holt R (2003) Modelling and simulation of the tool wear in nanometric cutting. Wear 255: 1427–1432.
[59] Shimada S (2002) Molecular dynamics simulation of the atomic processes in microcutting. In McGeough J (ed.) Micromachining of Engineering Materials. Marcel Dekker, New York, NY, pp. 63–84.

[60] Nakazawa H (1994) Principles of Precision Engineering. Oxford University Press, New York, NY.
[61] Lee W, Cheung CA (2001) Dynamic surface topography model for the prediction of nano-surface generation in ultra-precision machining. Int J Mech Sci 43: 961–991.
[62] Corbett J (2002) Diamond Micromachining. In McGeough J (ed.) Micromachining of Engineering Materials. Marcel Dekker, New York, NY, pp. 125–146.

11

Advanced (Non-traditional) Machining Processes

V.K. Jain

Department of Mechanical Engineering, Indian Institute of Technology Kanpur, Kanpur-208016, India.
E-mail: vkjain@iitk.ac.in

While making a part from raw material, one may require bulk removal of material, forming cavities/holes and finally finishing as per the parts requirements. Many advanced finishing processes have been employed to make circular and/or non-circular cavities and holes in difficult-to-machine materials. Some of the processes employed for hole making are electro-discharge machining, laser beam machining, electron beam machining, shaped tube electro-chemical machining and electro-chemical spark machining. With the demand for stringent technological and functional requirements of the parts from the micro- to nanometre range, ultra-precision finishing processes have evolved to meet the needs of the manufacturing scientists and engineers. The traditional finishing processes of this category have various limitations, for example, complex shapes, miniature sizes, and three-dimensional (3D) parts cannot be processed/finished economically and rapidly by traditional machining/finishing processes. This led to the development of advanced finishing techniques, namely abrasive flow machining, magnetic abrasive finishing, magnetic float polishing, magneto-rheological abrasive finishing and ion beam machining. In all these processes, except ion beam machining, abrasion of the workpiece takes place in a controlled fashion such that the depth of penetration in the workpiece is a small fraction of a micrometre so that the final finish approaches the nano range. The working principles and the applications of some of these processes are discussed in this chapter.

11.1 Introduction

The expectations from present-day manufacturing industries are very high, *viz.* high economic manufacturing of high-performance precision and complex parts made of very hard high-strength materials. Every customer demands products to their own taste/choice, hence there is a need for high-quality low-cost parts made in small batches and large variety. Furthermore, there is a trend in the market for miniaturization of parts with high degree of reliability. The traditional machining methods, even with added CNC features, are unable to meet such stringent demands of various industries such as aerospace, electronics, automobiles, *etc*. As a result, a new class of machining processes has evolved over a period of time to meet such demands, named non-traditional, unconventional, modern or advanced machining processes [1–3]. These advanced machining processes (AMP) become still more important when one considers precision and ultra-precision machining. In some AMPs, material is removed even in the form of atoms or molecules individually or in groups. These advanced machining processes are based on the direct application of energy for material removal by mechanical erosion, thermal erosion or electro-chemical/chemical dissolution.

Developments in materials science have led to the evolution of difficult-to-machine, high-strength temperature-resistant materials with many extraordinary qualities. Nanomaterials and smart materials are the demands of the day. To make different products in various shapes and sizes, traditional manufacturing techniques are often found not fit for purpose. One needs to use non-traditional or advanced manufacturing techniques in general and advanced machining processes in particular [1]. The latter includes both bulk material removal advanced machining processes as well as advanced fine finishing processes. Bulk material removal activities can be divided mainly in two categories, hole or cavity making, and shaping. Furthermore, the need for high precision in manufacturing was felt by manufacturers the world over, to improve interchangeability of components, enhance quality control and increase wear/fatigue life [4, 5].

The first three sections of this chapter deal with the working principles and parametric analysis of some of the important hole making and shaping processes, and some of the applications of each of them. The last section deals with some of the advanced fine finishing processes and their special applications. Figure 11.1 shows the classification of various AMP. In mechanical-type AMP, a mechanical force is employed to remove/erode material from the workpiece. In thermo-electric/thermal processes, it is the heat that is responsible for the thermal erosion of material from the workpiece surface. In electro-chemical and chemical machining processes, it is an electro-chemical or chemical reaction that removes material from the workpiece. Only a few of these processes are discussed, in brief, in this chapter. Furthermore, none of these processes is unique such that it can be satisfactorily employed in all machining situations.

Shaping and sizing are not the only requirements of a part. Surface integrity in general, and surface finish in particular, is equally important. Traditionally, abrasives either in loose or bonded form whose geometry varies continuously in an unpredictable manner during the process are used for final finishing purposes. Nowadays, new advances in materials syntheses have enabled the production of ultra-fine abrasives in the nanometre range. With such abrasives, it has become possible to achieve nanometre surface finishes and dimensional tolerances. There are processes (ion beam machining and elastic emission machining) that can give ultra-precision finish of the order of size of an atom or molecule of a substance. In some cases, the surface finish (center line average (CLA) value) obtained has been reported to be even smaller than the size of an atom. Various processes have been

Mechanical advanced machining processes

Bulk material removal processes
- Abrasive jet machining (AJM)
- Ultrasonic machining (USM)
- Water jet machining (WJM)
- Abrasive water jet machining (AWJM)

Micro/nanofinishing processes
- Abrasive flow machining (AFM)
- Magnetic abrasive finishing (MAF)
- Magneto-rheological finishing (MRF)
- Magneto-rheological abrasive flow finishing (MRAFF)
- Magnetic float polishing (MFP)
- Elastic emission machining (EMM)
- Ion beam machining (IBM)

Thermal advanced machining processes

- Plasma arc machining (PAM)
- Laser beam machining (LBM)
- Electron beam machining (EBM)
- Electro-discharge machining (EDM)

Electro-chemical and chemical advanced machining processes

- Electrochemical machining (ECM)
- Chemical machining (ChM)
- Biochemical machining (BM)

Classification of advanced machining Processes

Figure 11.1. Classification of advanced machining processes

employed for finishing purposes, like abrasive flow machining (AFM), magnetic abrasive flow machining (MAFM), magnetic abrasive finishing (MAF), magnetic float polishing (MFP), magneto-rheological abrasive flow finishing (MRAFF), elastic emission machining (EEM) and ion beam machining (IBM).

11.2 Mechanical Advanced Machining Processes (MAMP)

Mechanical-type advanced machining processes (MAMP) are of various types, as shown in Figure 11.1. This section deals with two commonly used MAMP, *viz.* ultrasonic machining (USM) and abrasive water jet cutting (AWJC).

11.2.1 Ultrasonic Machining (USM)

The ultrasonic machining (USM) process is normally employed for hard and/or brittle materials (irrespective of their electrical conductivity) usually having hard-

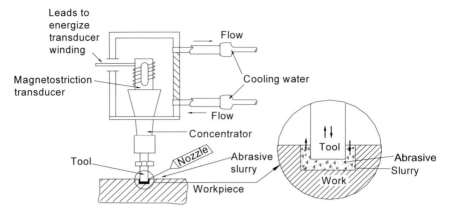

Figure 11.2. A schematic diagram of ultrasonic machining

ness > 40 RC. As shown in Figure 11.2, a slurry (a mixture of fine abrasive particles and water) is supplied in the gap between tool and workpiece [1]. The tool
vibrates at a very high frequency (≥ 16 kHz) created by ultrasonic transducer which
converts high-frequency electrical signal into high-frequency linear mechanical
motion (or vibration). These vibrations are transmitted to the tool via mechanical
amplifier. The tool and tool holder are designed to vibrate at their resonance frequency so that the maximum material removal rate (MRR) can be achieved.

The individual abrasive grains that come into contact with the vibrating tool
acquire high velocity and are propelled towards the work surface. High-velocity
bombardment of the work surface by the abrasive particles gives rise to the
formation of a multitude of tiny highly stressed regions, leading to cracking and
fracture of the work surface, resulting into material removal. The magnitude of
the induced stress into the work surface is proportional to the kinetic energy
($1/2\, mv^2$; m =particle mass, v =particle velocity) of the particles hitting the
work surface. Thus, a brittle material can be more easily machined than a ductile
material. As the material is removed from the work surface, the gap between the
tool bottom face and the work surface being machined increases, hence the machining efficiency goes down. To maintain the high efficiency of USM, the tool
is constantly fed towards the workpiece such that the surface recession rate (or
linear material removal rate – MRR_l) is equal to the tool feed rate (f).

The size of the cavity produced during USM is slightly larger than the tool dimensions (or tapered, Figure 11.3). A cylindrical solid tool of diameter D produces a circular hole of diameter $D + \Delta D$, where ΔD depends upon various process parameters and the location where it is being measured. The drilled hole also
has a small taper (angle θ). The value of taper angle can be reduced by giving
a reverse taper on the tool.

Ultrasonic machining machines are available in the power output range of 40 W to
2.4 kW. These machines usually have five sub-systems, namely, power supply, transducer, tool holder, tool and tool feed system, and slurry and slurry supply system.

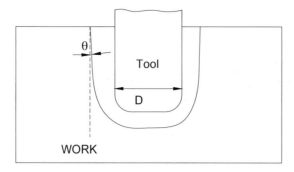

Figure 11.3. Tapered hole produced by USM (θ is the taper angle)

A high-power sine wave generator converts low-frequency (60 Hz) electrical power to high-frequency (≈ 20 kHz) electrical power. This high-frequency electrical signal is transmitted to the transducer, which converts it into high-frequency low-amplitude vibration. In USM, either of two types of transducers are used, *i.e.*, piezoelectric (for low powers up to 900 W) or magneto-strictive type (for high powers up to 2.4 kW). Magneto-strictive transducers are made of nickel or nickel alloy sheets and their efficiency (20–35%) is much lower than the piezoelectric transducers' efficiency (up to 95%), hence cooling is essential in the case of magneto-strictive transducer to remove waste heat. The maximum change in length (or amplitude of vibration) that can be usually achieved is 25 μm.

The tool holder holds and connects the tool to the transducer (Figure 11.2), transmits the energy and, in some cases, amplifies the amplitude of vibration. Amplifying tool holders give as much as six times increased tool motion, and yield an MRR up to ten times higher than non-amplifying tool holders. The material of the tool should therefore have good acoustic properties, and high resistance to fatigue cracking. Commonly used materials for the tool holder are Monel (for low-amplitude applications), titanium and stainless steel. Tools are usually made of relatively ductile materials (brass, stainless steel, mild steel, *etc.*) to minimize tool wear rate (TWR). The ratio of TWR and MRR depends upon the type of abrasive, workpiece material and the tool material. The surface finish of the tool affects the surface finish obtained on the workpiece.

Hardness, particle size, usable life time and cost are used as criteria for selecting abrasive grains for USM. Commonly used abrasives (in order of increasing hardness are Al_2O_3, SiC and boron carbide (B_4C). Abrasive hardness should be greater than the hardness of the workpiece material. The MRR and surface finish obtained during USM also depend on the size of the abrasive particles. Fine grains result in a low MRR and good surface finish while the reverse is true with coarse grains. The mesh size of commonly used grits range from 240 to 800. Abrasive slurry consists of water and abrasives usually in the ratio 1:1 (by weight). However, this can vary depending upon the type of operation. Thinner (or lower-concentration) mixtures are used while drilling deep holes, or machining complex cavities so that the slurry flow is more efficient. The slurry, which is stored in a reservoir, is pumped to the gap formed between the tool and the work.

11.2.1.1 Process Parameters, Capabilities and Applications

USM process performance depends on the abrasive (material, size, shape and concentration), the tool and tool holder (tool material, frequency of vibration and amplitude of vibration) the and workpiece material (hardness). An increase in the amplitude of vibration increases the linear material removal rate (MRR_l) for different pressures (Figure 11.4). An increase in grit size also increases MRR_l but exhibits an optimum value (Figure 11.5). However, increase in MRR_l also results in higher value of surface roughness (R_a) (or poorer surface finish). As the cutting depth increases, the flow of slurry through the cutting zone becomes inefficient, hence MRR_l decreases further. Materials that can be easily machined by this process include ceramics, glass, carbides, *etc.*, which cannot be efficiently machined by traditional methods. It is also quite useful for electrically non-conductive ceramics and fragile components. It can drill multiple holes at a time. Following are some of the capabilities of this process:

- Aspect ratio (ratio of hole length to diameter): 40:1
- Hole depth: 51–152 mm (with special flushing arrangement)
- MRR_l: 0.025–25 mm/min
- Surface finish: 0.25–0.75 μm
- Surface texture: non-directional
- Accuracy or radial overcut: 1.5–4.0 times the mean abrasive grain size

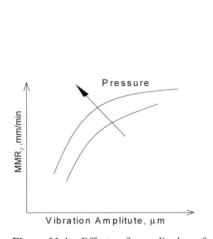

Figure 11.4. Effect of amplitude of vibration on penetration rate (MRR_l) for different pressures

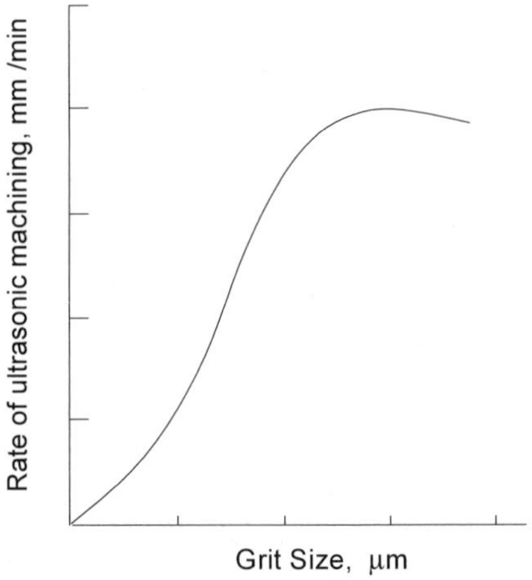

Figure 11.5. Effect of amplitude of vibration on penetration rate (MRR_l) for different pressures

11.2.2 Abrasive Water Jet Cutting (AWJC)

The abrasive water jet cutting (AWJC) process is a high-potential process applicable to both metals as well as non-metals. In this process, a high-velocity

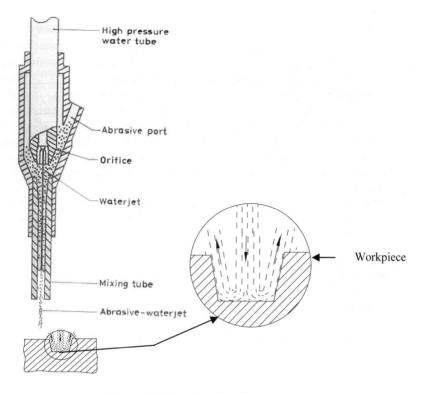

Figure 11.6. Details of abrasive water jet nozzle

water jet mixed with fine abrasive particles hits the workpiece surface (Figure 11.6). The velocity of the abrasive mixed water jet is very high, hence the kinetic energy with which the abrasive particles and the water jet hit the workpiece surface is very high (as high as 900 m/s in special cases) and hence it leads to the erosion of the work surface. Here, a part of the momentum of water jet is transferred to the abrasives, hence the velocity of abrasives rises rapidly.

Depending upon the type of the workpiece material being cut and the depth at which cutting is taking place, material removal occurs due to erosion, shear or failure under a rapidly changing localized stress field. The pressure at which a water jet operates is about 400 MPa, which is sufficient to produce a jet velocity of 900 m/s. Such a high-velocity jet is able to cut materials such as ceramics, composites, rocks, metals *etc*. [6]. Material removal by erosion takes place in the upper part of the workpiece while it occurs by deformation wear at the lower part of the workpiece being cut. The AWJC process can easily cut both electrically non-conductive and conductive, and difficult-to-machine materials. This process does not produce dust, thermal defects, and fire hazards. Recycling of water and abrasives is possible to some extent. It is a good process for shaping and cutting of composite materials, and creates almost no delamination.

11.2.2.1 AWJC Machine

Abrasive water jet cutting machines have four basic elements: a pumping system, abrasive feed system, abrasive water jet nozzle and catcher. The pumping system produces a high-velocity water jet by pressurizing water up to as high as 400 MPa using a high-power motor. The water flow rate can be as high as 3 gallons per minute. To mix the abrasives into this high-velocity water jet, the abrasive feed system supplies a controlled quantity of abrasives through a port. The abrasive water jet nozzle mixes abrasives and water (in mixing tube) and forms a high-velocity water abrasive jet. Sapphire, tungsten carbide, or boron carbide can be used as the nozzle material. There are various kinds of water abrasive jet nozzles. Another element of the system is a catcher, for which two configurations are commonly known: a long narrow tube placed under the cutting point to capture the used jet with the help of obstructions placed alternately in the opposite direction and a deep water-filled settling tank placed directly underneath the workpiece in which the abrasive water jet dies out.

11.2.2.2 Process Parameters, Capabilities and Applications

Independent process parameters include water (pressure, flow rate), abrasive (type, size, flow rate), nozzle, traverse rate, the stand-off distance and workpiece material. For a specified workpiece material and process parameters, there is a minimum pressure (i.e., critical pressure or threshold pressure) below which no cutting will take place [6]. The machined depth increases as the pressure increases, and this relationship becomes steeper as the abrasive flow rate increases. An increase in abrasive flow rate increases the machined depth, but beyond the critical value of abrasive flow rate the machined depth starts to decrease. Various types of abrasive particles can be used during AWJC, viz. Al_2O_3, SiC, silica sand, garnet sand, etc. It is also found that, with an increase in traverse rate (relative motion between the water abrasive jet and workpiece) and stand-off distance (the distance between the nozzle tip and the workpiece surface being cut) the machined depth decreases. Under certain circumstances, more than one pass of cutting may be required, in which case less power is consumed compared with single-pass cutting. A theoretical model has also been proposed [7] to predict the penetration of an abrasive jet in a piercing operation.

The capabilities and specification of AWJC process are:

- Pressure   : Up to 415 MPa
- Jet velocity   : Up to 900 m/s
- Abrasive   : Al_2O_3. SiC, silica sand
- Abrasive mesh size : 60 – 300
- Nozzle material   : Sapphire, WC, B_4C
- Water requirement : Up to 3 gallons/min

This process has been applied to cut both metals as well as non-metals. AWJC process is good to cut honeycomb material, corrugated structures, etc. This process is gaining acceptability as a standard cutting tool in industries such as aerospace, nuclear, oil, foundry, automotive, construction etc.

11.3 Thermoelectric Advanced Machining Processes

Application of AMP is quite common in making holes in difficult-to-machine materials as well as shaping and sizing a part. Sometimes holes with a high aspect ratio or a large number of holes in a workpiece without burrs and without residual stresses are needed. Some processes are good only for electrically conductive materials while others are excellent for making thousands of holes in a square centimetre area of metallic as well as non-metallic materials. Processes such as EDM, travelling-wire EDM, LBM and EBM fall into the category of thermal AMP in which material removal takes place by melting or melting and vaporization. The energy source for material removal is in the form of heat.

In this section only two thermoelectric processes are discussed: electrical discharge machining (EDM), including travelling-wire EDM (TW-EDM), and laser beam machining (LBM).

11.3.1 Electric Discharge Machining (EDM) and Wire EDM

The working principle of EDM process can be understood from Figure 11.7(a). Dielectric flows through the gap between the electrodes (usually with the tool as the cathode and the workpiece as the anode), which are connected to a pulsed direct-current (DC) power supply. This produces sparks between the electrodes, which melt and sometimes vaporize material from both the tool and the workpiece. To improve the accuracy of axisymmetric profiles, orbital EDM is advocated. Figure 11.7(a) shows an inter-electrode gap between the tool and the workpiece in which dielectric is flushed at high pressure. As shown in Figure 11.7(b), once the power supply is on, the capacitor keeps charging until the breakdown voltage (V_b) is attained and then sparking takes place at a point of least electrical resistance. After each discharge, the capacitor recharges (Figure 11.7(b)) and the spark energy is shared mainly by workpiece, tool, dielectric and debris (removed material). Radiation losses are also present. The flowing dielectric in the IEG cools the tool and workpiece, cleans the IEG and localizes the spark energy into a small cross-sectional area.

The spark radius is usually very small (100–200 μm) [8] however the spark energy density is very high, hence the electrode's material melts and vaporizes in the localized area. The craters formed in this way spread over the entire surface of the workpiece under the tool. The cavity produced in the workpiece is approximately the replica of the tool. However, tool wear should be minimized by selecting optimum machining parameters and appropriate polarity. The material eroded from the electrodes is known as debris, which is a mixture of irregularly shaped particles and spherical particles. A very small gap (usually ≤ 100 μm) between the two electrodes is maintained with the help of a servo system. Figure 11.7(c) shows a photograph of an EDM m/c. It shows various important elements of an EDM m/c.

An allied EDM process has been developed, known as travelling-wire EDM (TW-EDM). In travelling-wire EDM, in place of a solid tool, a wire is used as the cathode (Figure 11.7(d)). The wire travels through the workpiece in its axial direction while the workpiece is fed in the X and Y directions to give it different shapes. The TW-EDM system is computer controlled and it can machine complicated 3D shapes.

(a)

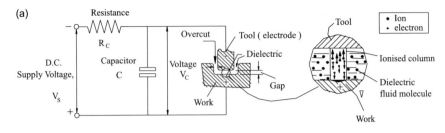

(b)

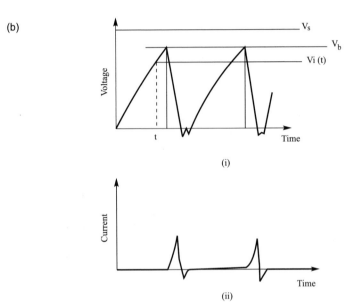

(i)

(ii)

(c)

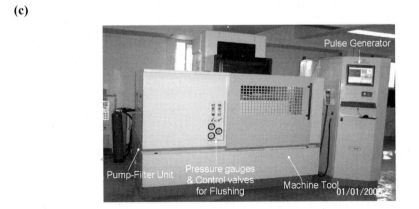

(d)

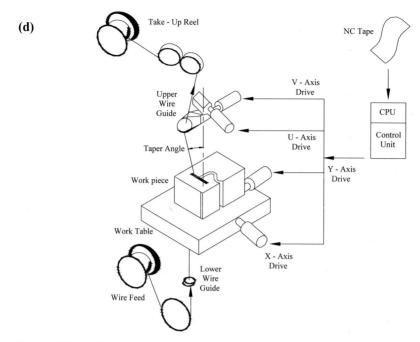

Figure 11.7. (a) Electric discharge machining using relaxation circuit **(b)** (i) Voltage versus time relationship, (ii) Current versus time relationship in EDM using relaxation circuit. V_s = supply voltage, V_b = breakdown voltage, V_i = instantaneous voltage **(c)** EDM machine (Courtesy: Electronica Machine Tools Limited, Pune, India) **(d)** Schematic illustration of travelling wire electric discharge machining process

Usually, 80–100 V DC pulses at approximately 5 kHz are passed through the electrodes. Negatively charged particles (electrons) break loose from the cathode surface under the influence of electric field forces. These electrons collide with the neutral molecules of the dielectric (kerosene, oil, or water) and produce electrons and ions, resulting in further ionization. In this ionized narrow channel, there is a continuous flow of a considerable number of electrons towards the anode and of ions towards the cathode. When these particles hit the electrodes their kinetic energy (KE) is converted into heat energy due to the bombardment of electrons onto the anode and ions on the cathode. Finally, localized high temperature at the electrodes results in the melting and/or vaporization of electrodes. Usually, a component made by EDM is machined in two stages, *viz.* rough machining (high MRR and high Ra value) and finish machining (low MRR and low Ra value).

The power supply of an EDM process should be able to control the parameters of voltage, current, duration and frequency of a pulse, and duty cycle (the ratio of on-time to pulse time). Figure 11.8 shows the on-time, off-time, pulse time and duty cycle for an ideal pulse. Also, the filtration of dielectric fluid before re-circulation is essential so that changes in its properties are minimal during the process. Effective flushing of dielectric is an important parameter and it removes by-products from the gap. Various methods of flushing are shown in Figure 11.9.

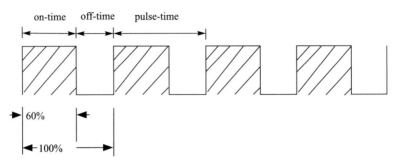

Figure 11.8. An illustration of on-time, off-time, pulse time and duty cycle

An increase in current results in an increase in MRR as well as an increased value of surface roughness, as shown in Figures 11.10(a,b). However, an increase in spark frequency results in an improved surface finish through a lower Ra value, as shown in Figure 11.10(c).

In TW-EDM, to minimize frequent breakage of wire and its rethreading at high load, stratified wire (copper core with zinc clad) is used (Figure 11.11); it also has the advantage of not being too expensive. One can easily obtain a surface finish of the order of 0.1 μm using TW-EDM. Nowadays, CNC-EDM and CNC-TW-EDM machines are equipped with an automatic tool changer, automatic wire feeder, automatic workpiece changer, automatic pallet changer and automatic position compensator. With the adoption of such functions in EDM, flexible manufacturing cells (FMC) are under design consideration [9].

To enhance the capabilities of EDM process, hybrid EDM processes have been developed. For example, electric discharge diamond grinding (EDDG → EDM + grinding) [10] and ultrasonic-assisted EDM [11] are two such hybrid EDM processes. EDDG of very hard materials such as WC is advantageous because electric discharges thermally soften work material, thus facilitating grinding. Continuous in-process dressing of grinding wheel gives stable grinding performance [1]. It is observed that specific energy in EDDG decreases with increasing pulse current.

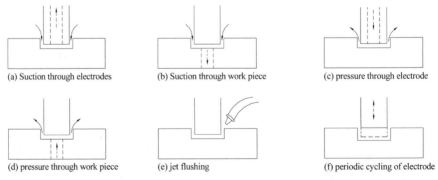

Figure 11.9. Various methods for dielectric flushing: **(a)** suction through electrode **(b)** suction through workpiece, **(c)** pressure through velectrode, **(d)** pressure through workpiece, **(e)** jet flushing, **(f)** periodic cycling of electrode [Courtesy: HMT, Bangalore, India]

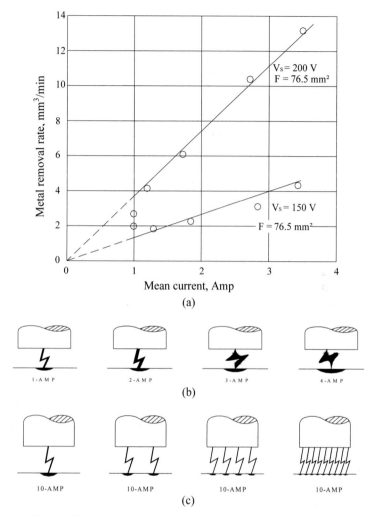

Figure 11.10. (a) Effect of mean current on MRR for different voltages. Dielectric = kerosene, tool = brass, work material = low carbon steel, F = Tool Electrode Area [12]. **(b)** Effect of current on surface finish (or crater size). **(c)** Effect of frequency of sparking on surface finish (or crater size)

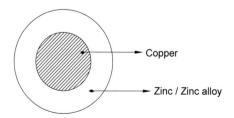

Figure 11.11. Stratified wire used in TW-EDM

11.3.1.1 Process Capabilities and Applications

EDM can be used only for electrically conductive materials, and its performance is not substantially affected by mechanical, physical and metallurgical properties of workpiece material. It can perform various kinds of operations such as drilling, cutting, 3D shaping and sizing (wire EDM) and spark-assisted grinding (EDDG). It gives good repeatability and accuracy of the order of 25–125 μm. The tolerances that can be achieved are ±2.5 μm. This has been used to produce as good as 100:1 aspect ratio holes. Under normal conditions, the volumetric material removal rate (MRR$_v$) is in the range of 0.1–10 mm^3/min. The surface finish produced during EDM is usually in the range 0.8–3 μm, depending upon the machining conditions used. The machined surface normally has a recast layer which should be removed before fitting the part into the assembly or sub-assembly. It is commonly used for making hardened steel dies and moulds and has numerous applications in various types of industries.

11.3.2 Laser Beam Machining (LBM)

The acronym laser means light amplification by stimulated emission of radiation. In laser beam machining, a laser beam is focussed onto the target/workpiece surface (Figure 11.12(a)), resulting in an energy density of the order of 10^3 W/mm^2 (or more in some cases), which is enough to melt and vaporize materials such as diamond. Holes drilled using a laser beam system are normally not straight sided, as shown in Figure 11.12(b). Laser light is monochromatic, coherent and gives very low divergence. An LBM machine has high capital and operating costs, and very low machining efficiency (<1%). Industrial lasers operate either in continuous wave (CW) or in pulse mode [13]. CW lasers are used for processes such as

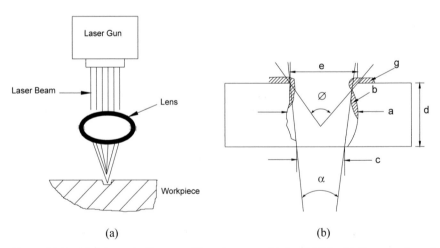

(a)   (b)

Figure 11.12. (a) Schematic diagram of laser beam machining (LBM). **(b)** Cross-section of a hole drilled by LBM: a – diameter of the mid span, b – thickness of the recast layer, c – exit diameter; d – hole depth, e – inlet diameter; g – thickness of the surface debris, φ - inlet cone angle, α - taper angle

welding, laser chemical vapour deposition (LCVD) and surface hardening. These applications require uninterrupted supply of energy for melting and phase transformation. Controlled pulse energy is desirable for processes such as cutting, drilling, marking *etc.*, so that the heat-affected zone (HAZ) is minimal.

LBM is capable of machining refectory, brittle, hard, metallic and non-metallic materials. As long as the laser beam path is not obstructed, it can be used to machine in otherwise inaccessible areas. LBM is suitable for drilling very small-diameter holes with a reasonable large aspect ratio. LBM is used for drilling, trepanning, trimming, marking, welding and similar other operations. It is used for both micromachining as well as macromachining. LBM has also been used for 3D machining, namely, threading, turning, grooving *etc.* [13].

11.4 Electrochemical Advanced Machining Processes

11.4.1 Electrochemical Machining (ECM)

ECM works on the principle of Faraday's laws of electrolysis. The noteworthy feature of electrolysis is that electrical energy is used to produce a chemical reaction, therefore the machining process based on this principle is known as electrochemical machining (ECM). In ECM, a small DC potential (5–30 V) is applied across two electrodes (the workpiece being the anode and the tool the cathode). Electrolyte flows in the gap between these two electrodes (Figure 11.13). Transfer of electrons between ions and electrodes completes the electrical circuit. During the flow of current in the circuit, metal is detached atom by atom from the anode surface and appears in the electrolyte as ions (Fe^{2+}). These ions form the precipitate of metal hydroxides ($Fe(OH)_2$) (Figure 11.13). During the electrolysis of water, its molecules gain electrons from the cathode so that they separate into free hydrogen gas and hydroxyl ion as

$$H_2O + 2e^- \rightarrow H_2 \uparrow + 2OH^-.$$

As the anode dissolves, positively charged metal ions appear in the electrolyte, which combine with negatively charged hydroxyl ions to form metal hydroxides as

$$Fe - 2e^- \rightarrow Fe^{2+}$$

$$Fe^{2+} + 2OH^- \rightarrow Fe(OH)_2.$$

These hydroxides are insoluble in water hence they appear as a precipitate and do not affect the chemical reaction. Ferrous hydroxides may further react with water and oxygen to form ferric hydroxides.

$$Fe(OH)_2 + 2H_2O + O_2 \rightarrow 4Fe(OH)_3.$$

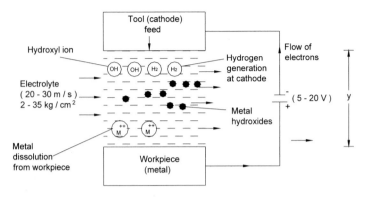

Figure 11.13. Schematic diagram of electrochemical machining

During ECM, metal from the anode is removed atom by atom by removing negative electrical charge that binds the surface atoms to their neighbours. The ionized atoms (say, Fe^{2+}) are attracted away from the workpiece by an electric field. The rate of removal of Fe^{2+} from the anode is governed by Faraday's laws of electrolysis. The amount of chemical change m (the substance deposited or dissolved) is proportional to the amount of charge passed through the electrolyte. The amount of change produced (substances deposited or dissolved) by the same quantity of charge is proportional to the chemical equivalent weights of the material. These laws can be expressed as follows:

$$m \propto I\,t$$

and

$$m \propto E$$

Therefore,

$$m = \frac{ItE}{F} = \frac{AIt}{Z.F} \tag{11.1}$$

where I is the current (A), t is the time (s), E is the gram-equivalent weight of material (or A/Z, where A is gram atomic weight and Z is the valency of dissolution), F is Faraday's constant (96500 As), and m is mass of the metal dissolved/deposited (g). Dividing both sides of the above equation by time t will give MRR_g (the material removal rate in units of g/s) as

$$MRR_g = \frac{m}{t} = \dot{m} = \frac{IE}{F}. \tag{11.2}$$

To obtain the volumetric material removal rate MRR_v, divide both sides of Equation (11.2) by the density of the metal (or alloy) as

$$MRR_v = \frac{MRR_g}{\rho_a} = \frac{IE}{\rho_a F} \tag{11.3}$$

This equation can be used to calculate the linear material removal rate MRR_l by dividing it by cross sectional area (A_r) of the tool (or smaller electrode):

$$MRR_l = \frac{IE}{\rho_a A_r F} \tag{11.4}$$

The accurate evaluation of m from Equation (1) is difficult when the anode is an alloy because the value of E for the alloy is not known. It can be evaluated either by the percent-by-weight method or the superposition-of-charge method. Furthermore, many elements have more than one valency of dissolution. The exact valency of dissolution under the given machining conditions should be known because $E = A/Z$. The current I is the function of applied voltage (V), interelectrode gap (IEG) and electrolyte conductivity (k). The gap between the bottom surface of the tool and the top dissolving surface of the workpiece is known as the inter-electrode gap (IEG), abbreviated as y. The smaller the IEG, the greater the current flow (or current density) will be. Furthermore, MRR_g depends on the current efficiency (η), which usually varies between 75% and 100%. Theoretically, the MRR depends upon the current passing through the workpiece and its chemical composition. To maintain a constant gap (or equilibrium gap), the tool should be fed towards the workpiece at the rate (f) at which the workpiece surface is recessing downwards (MRR_l). Hence,

$$f = MRR_l \tag{11.5}$$

11.4.2 ECM Machine

An ECM machine tool consists of four main subsystems: the power source, the electrolyte cleaning and supply system, the tool and tool feed system and the work and work holding system. The power source supplies a low-voltage (5–30 V) high-current (as high as 40 kA) rectified DC power supply. Figure 11.14 shows a photograph of a simple low-capacity ECM machine.

The electrolyte supply and cleaning system consists of a pump, filters, piping, control valves, heating/cooling coils, pressure gauge and a tank/reservoir. These elements should be made of anti-corrosive materials because the electrolyte used is corrosive in nature. Tools are also required to operate in a corrosive environment for a long period of time. Hence, they should also be made of anti-corrosive material having high thermal and electrical conductivity, and easy-to-machine characteristics (high machinability). Copper, brass and stainless steel are commonly used materials for making ECM tools. It is important to know that only electrically conductive materials can be machined by this process. However, work holding devices are made of electrically non-conducting materials having good thermal stability, corrosive resistance and low moisture absorption. Glass-fibre-reinforced plastics (GFRP), Perspex and plastics are some such materials that can be used to fabricate work holding devices.

To exploit the full potential of the process, many allied processes/operations have been developed using the principle of anodic dissolution described above.

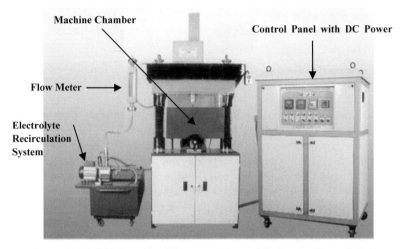

Figure 11.14. ECM machine (Courtesy: Metatech Industries, Pune, India)

Some of the allied ECM operations are electro-chemical boring, electro-chemical broaching, electro-chemical ballizing, electro-chemical drilling, electro-chemical deburring, electro-chemical die sinking, electro-chemical milling, electro-chemical sawing, electro-chemical micromachining, electro-chemical turning, electro-chemical trepanning, electro-chemical wire cutting, electro-stream drilling and shaped tube electromachining (STEM) [1]. Shaped tube electro-chemical drilling (STED) has been successfully used to drill small-diameter high-aspect-ratio holes in difficult to machine materials such as nimonic alloys (Figure 11.15) [14].

To enhance the capabilities of the base process (here ECM), two or more processes are combined together to make it a hybrid process. Some of these hybrid processes are electro-chemical grinding (ECG) [15], electro-chemical honing (ECH) [16], and electro-chemical spark machining [17–22]. Such hybrid processes give better performance than the constituent processes (ECM + grinding, or ECM + honing or ECM + EDM).

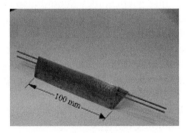

Figure 11.15. Photograph of a nimonic super alloy part in which a 100-mm-deep hole has been drilled using the STED process [14]

11.5 Fine Finishing Processes

11.5.1 Abrasive Flow Machining (AFM)

With today's focus on total automation in the flexible manufacturing systems, the abrasive flow machining (AFM) process offers both automation and flexibility. This process was developed basically to deburr, polish and radius difficult-to- reach surfaces and edges by flowing abrasive laden polymer, to and fro in two vertically opposed cylinders (Figure 11.16(a)) [23, 24]. The medium (a mixture of

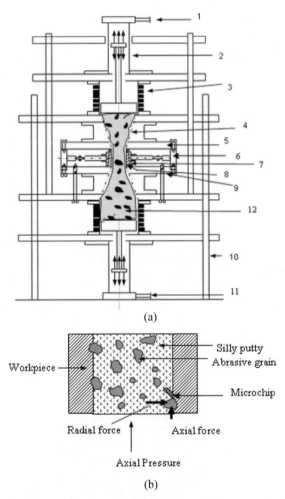

(a)

(b)

Figure 11.16. (a) Schematic diagram of abrasive flow machining setup: 1 – hydraulic oil inlet/outlet port, 2 – hydraulic cylinder, 3 - medium cylinder, 4 – smooth entry profile, 5 – Top cover plate, 6 – dynamometer for force measurement, 7 – central hub, 8 – split cylindrical fixture with workpiece, 9 – botSstom cover plate, 10 – support frame, 11 – hydraulic oil outlet/inlet port, 12 – medium with abrasive particles. **(b)** Types of forces acting on a grain [23]

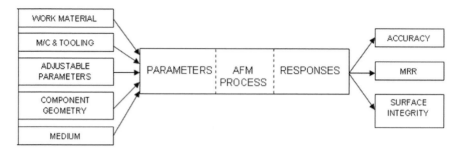

Figure 11.17. Variables and responses of AFM process

visco-elastic material, say, polyborosiloxane, additives and abrasive particles) enters the workpiece through the tooling. The abrasive particles (SiC, Al_2O_3, CBN, diamond) penetrate the workpiece surface depending upon the extent of the radial force

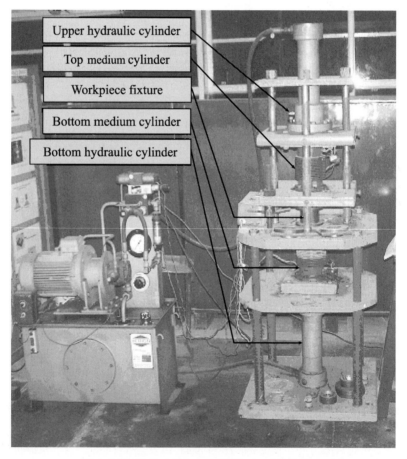

Figure 11.18. Experimental AFM setup [I.I.T. Kanpur]

acting on them. Due to the tangential/axial force, the material is removed in the form of microchips as shown in Figure 11.16(b). The medium is recirculated until the concentration of foreign particles increases to a level such that its finishing rate decreases substantially. At this time part of the used medium (say, 20%) is replaced by fresh medium. Figure 11.17 shows the variables and responses of the AFM process.

There are three major elements of the AFM system: the tooling, the machine, and the medium. Figure 11.18 shows a photograph of an AFM machine developed at IIT Kanpur for research and development activity. The tooling confines and directs the medium flow to the areas where abrasion is desired. The machine controls the extrusion pressure, which controls the medium flow rate. The abrasive-laden polymer is the medium, whose rheological properties determine the pattern and aggressiveness of the abrasive action. The rheological properties of the medium change during the AFM process and they substantially influence the process performance [25]. This process can be applied for finishing multiple pieces of the same shape and size (Fig-

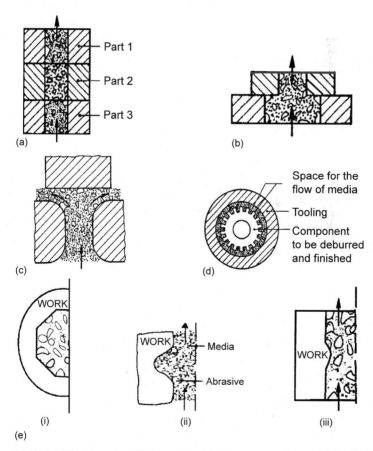

Figure 11.19. **(a)** Finishing of multiple parts having the same configuration. **(b)** Finishing of two parts but with different configurations. **(c)** Finishing and radiusing of an internal hole. **(d)** Tool for deburring and finishing of a gear using the AFM process. **(e)** medium acting as a self-deformable stone in the case of a complex, convex and concave component [1]

ure 11.19(a)), and also of different size (Figure 11.19(b)). It can radius the edges of a part (Figure 11.19(c)) and can finish a complete gear in one go (Figure 11.19(d)). The medium acts as a *self-deformable stone* and can finish concave, convex as well as complex-shaped components (Figure 11.19(e)). It also has some peculiar applications as well where no other traditional and advanced finishing processes can work. For example, a large number of small-diameter holes (say, diameter 3 mm and depth 30 mm) can be easily and simultaneously finished by this process, which is otherwise very difficult. This process can be used to control the surface finish of the cooling holes in a turbine blade, or the surface finish of stator and rotor blades of a turbine. The best surface finish achieved by this process is 50 nm.

11.5.2 Magnetic Abrasive Finishing (MAF)

In the magnetic abrasive finishing (MAF) process, granular magnetic abrasive particles (MAPs) [sintered ferromagnetic (iron) particles and abrasive grains (say, Al_2O_3, SiC, or diamond)] are used as cutting tools and the necessary finishing pressure is applied by an electromagnet-generated magnetic field. Figure 11.20(a) shows the working principle of the MAF process through a schematic diagram.

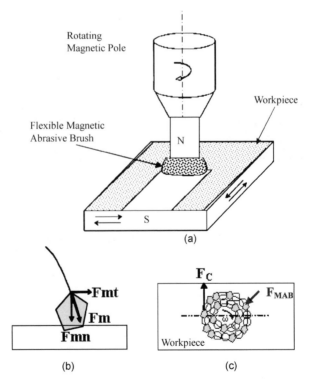

Figure 11.20. (a) Plane magnetic abrasive finishing of magnetic work material. **(b)** Schematic diagram showing the normal (F_{mn}) and tangential components (F_{mt}) of the magnetic force (F_m) acting on a magnetic abrasive particle. **(c)** FMAB and the cutting force (F_c) acting on an abrasive particle

The magnetic abrasive particles join each other magnetically between two magnetic poles (S and N) along the magnetic lines of force, forming a flexible magnetic abrasive brush (FMAB). When there is relative motion between the FMAB and the workpiece, abrasion takes place to give a polished finish. This finish can be as good as approaching to 50 nm, depending upon the machining conditions and workpiece material. The performance of the process depends on the parameters such as the MAP (type, size and mixing ratio), the working clearance (the gap between the workpiece surface being finished and the bottom face of magnetic pole), the rotational speed of the poles, the feed motion (x and y motion) of the workpiece, vibration (frequency and amplitude), the properties of the work material and the magnetic flux density [26–31].

Figure 11.20(b) shows the magnetic force (F_m) and its components: the magnetic normal force (F_{mn}) acting on a MAP due to the magnetic field, and the tangential force (F_c) acting mainly due to the rotation of the magnet (Figure 11.20(c)). There is a small contribution F_{mt} to this force F_c as a component of F_m (the magnetic force).

Figure 11.21(a) shows a schematic diagram of a magnetic abrasive finishing setup for plane surfaces. It shows the major elements of the setup, while Fig-

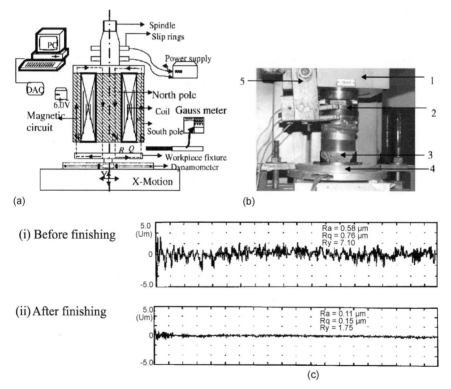

Figure 11.21. (a) Schematic diagram of plane magnetic abrasive finishing process setup. **(b)** Photograph of the plane MAF setup: 1 – column of a milling machine, 2 – slip ring, 3 – FMAB, 4 – ring dynamometer, 5 – slip ring attachment [28]. **(c)** Surface roughness plots (i) before, and (ii) after magnetic abrasive finishing of a stainless-steel workpiece

ure 11.21(b) shows a photograph of a part of a setup. Figure 11.21(c) shows the surface finish before (0.58 μm) and after (0.11 μm) MAF on a stainless-steel workpiece. This process can also be used to improve the surface properties by the process of diffusion of traces of MAP into the workpiece surface. Proper tooling can be developed to finish 3D intricate shapes because FMAB adapts to the shape of the workpiece, however it is not as flexible as the AFM medium.

Achieving uniform surface finish near the discontinuities or edges in the magnetic poles is difficult. At irregularities, the magnetic flux density is greater, hence the rate of change in surface roughness is also higher compared to other areas. There are various other areas in which research needs to be done, such as workpiece surface temperature and the forces acting on the workpiece during MAF processing. The effect of various parameters on responses such as MRR, surface roughness and out-of-roundness (in the case of cylindrical workpieces) have been reported [28, 29].

11.5.2.1 Internal Magnetic Abrasive Finishing

MAF can also be used for finishing internal surfaces. MAPs are blown with high pressure air jet. When the MAP jet comes into the domain of magnetic field, the particles slightly divert towards the source or pole of the magnetic field (the internal surface of the workpiece) and move forward due to an axial force (F_c), shearing the peaks off the internal surface. The path of the jet of the magnetic abrasive particles is slightly deviated due to the magnetic field force (F_m) (Figure 11.22) [32] so that the MAP flow abrades the internal surface of the tube. The velocity of

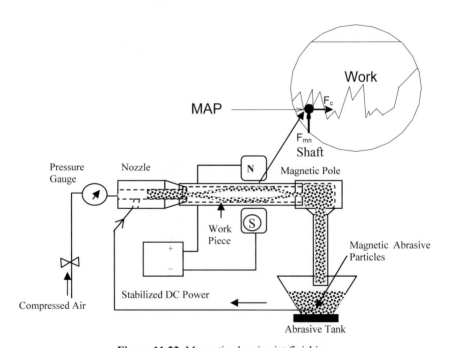

Figure 11.22. Magnetic abrasive jet finishing

the jet of MAPs is another important parameter of this process. However, this MAF technology needs to be brought out of the research laboratories so that it can be employed on the shop floor for industrial use. Results have been reported [26, 27] in the literature regarding finishing of stainless-steel rollers using MAF to obtain a final surface roughness R_a value of 7.6 nm from an initial surface roughness value of 220 nm.

11.5.3 Magnetic Float Polishing (MFP)

Various processes are available for finishing flat, complex-shaped and cylindrical surfaces but none of these processes qualifies for finishing spherically shaped workpieces, say, balls. A new process called magnetic float polishing (MFP) is applicable mainly for spherical balls. This process was developed for gentle finishing of very hard materials like ceramics, which develop defects during grinding, leading to fatigue failure. To achieve a low level of controlled forces, a magnetic field is used to support an abrasive slurry when finishing ceramic products, namely ceramic balls and rollers. This process is known as magnetic float polishing (MFP). This technique is based on the ferro-magnetic behaviour of the magnetic fluid that can levitate a non-magnetic float and suspended abrasive particles in it through the magnetic field. The levitation force applied by the abrasives is proportional to the magnetic field gradient and is extremely small and highly controllable. This is a good method for super-finishing of brittle materials with flat and spherical shapes [5, 27, 30, 31].

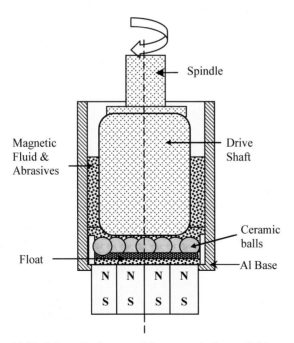

Figure 11.23. Schematic diagram of the magnetic float polishing apparatus

The setup consists of a magnetic fluid containing fine abrasive grains and extremely fine ferromagnetic particles in a carrier fluid (water or kerosene) (Figure 11.23). On the application of the magnetic field, the ferro-fluid is attracted downwards to the area of higher magnetic field. At the same time, a buoyancy force is exerted on the non-magnetic material, pushing it upward to the area of lower magnetic field. The abrasive grains, the ceramic balls, and the acrylic float inside the chamber, all being non-magnetic materials, are levitated by this magnetic buoyant force. The drive shaft is fed down to contact the balls and to press them down to reach the desired force level. The balls are polished by the relative motion between the balls and the abrasive particles.

11.6 Micromachining

Micromachining is sometimes misunderstood, as if it were performed only on microscale parts. Micromachining can be performed on both macro- and micro-components. Micromachining encompasses all those processes in which material is removed at the micron level. Micromachining processes are classified mainly into two classes: bulk micromachining, where the objective is to shape and size a component, usually a miniaturized component; This also includes drilling/machining of microholes and microcavities. Here, material removal takes place in the micron domain (say, microchips), hence leading to the term *micromachining*. The second class includes surface micromachining, where the objective is to improve the surface finish to the micron or sub-micron (nanoscale) range. Surface roughness values obtained by some of these processes have been reported to be as low as the size of or a fraction of an atom. Table 11.1 shows the capabilities of some of the micro/nanofinishing processes. To be more precise, sub-micromachining and sub-microfinishing are also known as nanomachining and nanofinishing, respectively. The majority of AMPs are capable of performing nanomachining/nanofinishing operations.

Normally, mechanical-type advanced machining processes (MAMP) are used for macromachining, so their process parameters should be scaled down so they can be used for micromachining of both macro- and microproducts. For example, ultrasonic micromachining and abrasive water jet micromachining are two such processes. However, almost all advanced finishing processes (AFM, MAF, MRF, MRAFF, *etc.*) are microfinishing and nanofinishing processes. In these processes, material is removed in the form of chips of micron/nanosize. These processes are capable of producing a surface finish in the sub-micron range (Table 11.1). There is another class of processes, known as mechanical-type advanced finishing processes, in which material is removed in the form of atoms or molecules (such as elastic emission machining and ion beam machining). Such processes are capable of producing surface finish Ra values smaller than the size of an atom (Table 11.1), that is, in the sub-nano range.

In the case of thermal AMP, material is removed by melting and/or vaporization. Some of these processes have been modified to machine at the micro- or nanoscale. If EDM is performed with lower machining parameters it can easily be used for microdrilling or microshaping of miniature components. The same is true

Table 11.1. Surface finish obtainable from some advanced finishing processes

No.	Finishing process	Workpiece	Ra (nm)
1.	Grinding	–	25 – 6250
2.	Honing11	–	25 – 1500
3.	Lapping	–	13 – 750
4.	Abrasive flow machining (AFM) with SiC abrasives	Hardened steel	50
5.	Magnetic abrasive finishing (MAF) with diamond abrasives	Stainless steel rods	7.6
6.	Magnetic float polishing (MFP) with CeO_2 abrasives	Si_3N_4	4.0
7.	Magneto-rheological finishing (MRF) with CeO_2	Flat BK7 glass	0.8
8.	Elastic emission machining (EEM) with ZrO_2 abrasives	Silicon	<0.5
9.	Ion beam machining (IBM)	Cemented carbide	0.1

with laser beam micromachining. However, electron beam machining performs operations at the micro/nanolevel depending upon the values of the parameters selected. However, the use of PAC does not seem feasible for micromachining.

ECM is mainly used for bulk material removal specially for macrocomponents, however it has been appropriately scaled down for performing electro-chemical micromachining (ECMM) operations on both macro- and microproducts. Photo-chemical machining and chemo-mechanical polishing are already popular micro-machining processes for macroproducts as well as for microproducts.

11.7 Finished Surface Characteristics

The performance of advanced fine abrasive finishing processes is evaluated by their achievable surface finish, form accuracy and resulting surface integrity. These fine abrasive finishing processes have been shown to yield accuracies in the nanometre range. The order of surface damage ranges from micrometres to nano-metres in these advanced finishing processes. This makes their measurement a formidable task. Table 11.1 summarises the attainable surface finish of advanced fine finishing processes.

References

[1] Jain VK (2002) Advanced Machining Processes. Allied, New Delhi.
[2] Benedict GF (1987) Nontraditional Manufacturing Processes. Marcel Dekker, New York, NY.
[3] McGeough JA (1988) Advanced Methods of Machining. Chapman & Hall, London.

[4]   McKeown PA (1987) The role of precision engineering in manufacturing of the future. Ann CIRP 36(2): 495–501.

[5]   Taniguchi N (1983) Current status in, and future trends of, ultraprecision machining. Ann CIRP 32(2): 573–582.

[6]   Hashish M (1989) A model for abrasive water jet (AWJ) machining. Trans ASME J Eng Mater Technol 111: 154–162.

[7]   Hemanth P, Naqash PV, Ramesh Babu NR (2005) A simplified model for predicting the penetration of an abrasive water jet in a piercing operation. Int J Manuf Technol Manage 7(2–4): 366–380.

[8]   Shankar P, Jain VK, Sundararajan T (1997) Analysis of spark profiles during EDM process. Mach Sci Technol 1(2): 195–217.

[9]   Kobayashi K (1995) The present and future developments of EDM and ECM. Proc. XI ISEM, 29–47.

[10]   Koshy P, Jain VK, Lal GK (1996) Mechanism of material removal in electric discharge diamond grinding. Int J Mach Tools Manuf 36(10): 1173–1185.

[11]   Murthy VSR, Philip PK (1987) Pulse train analysis in ultrasonic assisted EDM. Int J Mach Tools Manuf 27(4): 469–477.

[12]   Pandey PC and Shan HS (1980), Modern Machining Processes, published by Tata McGraw Hill, New Delhi.

[13]   Chryssolouris G (1990) Laser Machining – Theory and Practice. Springer-Verlag, New York, NY.

[14]   Bilgi DS, Jain VK, Shekhar R, Mehrotra S (2004) Electrochemical deep hole drilling in super alloy for turbine application. J Mater Proces Technol 149: 445–452.

[15]   Gedam A, Noble CF (1971) An assessment of the influence of some wheel variables in peripheral electrochemical grinding. Int J Mach Tool Des Res 11: 1–12.

[16]   Dubey AK, Jain NK, Shan HS (2006) Precision micro-finishing by electrochemical honing, Proceedings of ICOMAST 2006 held at Melaka, Malaysia.

[17]   Adhikari S, Jain VK (2007) Some new observations in electrochemical spark machining (ECSM) of quartz. J Mater Process Technol
       doi: 01016/jmatprotec.2007.08.071.

[18]   Gautam N, Jain VK (1998) Experimental investigations into ECSD process using various tool kinematics. Int J Mach Tools Manuf 38 (1–2): 15–27.

[19]   Bhattacharyya B, Doloi BN, Sorkhel SK (1999) Experimental investigations into electrochemical discharge machining (ECDM) of non-conductive ceramic materials. J Mater Process Technol 95: 145–154.

[20]   Jain VK, Chak SK (2000) Electrochemical spark trepanning of alumina and quartz. Mach Sci Technol 4:277–290.

[21]   Jain VK, Dixit PM, Pandey PM (1999) On the analysis of the electrochemical spark machining process. Int J Mach Tools Manuf 39: 165–186.

[22]   Jain VK, Rao PS, Choudhury SK, Rajurkar KP (1991) Experimental investigations into traveling wire electro-chemical spark machining (TW-ECSM) of composites. J Eng Ind 113: 75–84.

[23]   Gorana VK, Jain VK, Lal GK (2004) Experimental Investigation into cutting forces and active grain density during abrasive flow machining. Int J Mach Tools Manuf 44: 201–211.

[24]   Gorana VK, Jain VK, Lal GK (2006) Prediction of surface roughness during abrasive flow machining. Int J Adv Manuf Technol 31(1/2): 258–267.

[25]   Jain VK, Ranganatha C, Murlidhar K (2001) Evaluation of rheological properties of medium for AFM process. Mach Sci Technol 5(2): 151–170.

[26]   Fox M, Agarwal K, Shinmura T, Komanduri R (1994) Magnetic abrasive finishing of rollers.Ann CIRP 43: 181–184.

[27] Komanduri R (1996) On material removal mechanism in finishing of advanced ceramics and glasses. Ann CIRP 45(1): 509–514.

[28] Singh DK, Jain VK, Raghuram V (2004) Parametric study of magnetic abrasive finishing process. J Mater Process Technol 149: 22–29.

[29] Jain VK, Kumar P, Behra PK, Jayswal SC (2001a) Effect of working gap and circumferential speed on the performance of magnetic abrasive finishing process. Wear 250: 384–390.

[30] Tani Y, Kawata K (1984) Development of high efficient fine finishing process using magnetic fluid. Ann CIRP 33(1): 377–381.

[31] Umehara N (1994) Magnetic fluid grinding – A new technique for finish advanced ceramics. Ann CIRP 43(1): 85–188.

[32] Kim JD (1997) Development of magnetic abrasive jet machining system for internal polishing of circular tubes. J Mater Process Technol 71: 384–393.

12

Intelligent Machining: Computational Methods and Optimization

Sankha Deb and U.S. Dixit

Department of Mechanical Engineering, Indian Institute of Technology Guwahati, Guwahati-781 039, Assam, India.
E-mail: sankha.deb@iitg.ernet.in, uday@iitg.ernet.in

This chapter provides an introduction to intelligent machining. The various computational techniques to achieve the goal of intelligent machining are described. First, a description of neural networks and fuzzy set theory is presented. These are soft computing techniques. Afterwards the application of the finite element method to the machining processes is briefly mentioned. Finally, the optimization of machining processes is described.

12.1 Intelligent Machining

Machining processes are inherently complex, nonlinear, multivariate and often subjected to various unknown external disturbances. A machining process is usually performed by a skilled operator, who uses his decision-making capabilities based on the intuition and rules of thumb gained from experience. To develop a fully automated machining system with intelligent functions like those of expert operators is a very difficult problem. A fully automated system requires the capabilities for automatic control, monitoring and diagnostics of the machining processes. Such a system can be compared with a human operator. A number of sensors provide feedback to the system, in a way the sensory organs provide feedback to a human. Like the brain of a human, an automated system is equipped with a computer for processing the feedback information from sensors in real time and taking the appropriate decision to ensure optimal operating conditions. For the decision making, an automated system must have a model of the machining process. The model can be based on the physics of the process. The physics of the process is understood by a researcher and is converted in mathematical form, usually in the form of differential equations. The differential equations of the process are solved by a suitable technique such as the finite element method (FEM). In many instances, the physics of the process is not known properly. However, if there is a sufficient amount of data describing the behaviour of the process, a model can be developed based on the data, which is called modelling based on the data.

Modelling based on the physics is accomplished by the computational tools that may be called hard computing methods. On the other hand, modelling based on the data is accomplished by soft computing tools such as neural networks and fuzzy sets. Soft computing tools try to generate approximate solutions of the problem in the presence of uncertain or imprecise physics and/or the process variables. Soft computing differs from conventional (hard) computing in that, unlike hard computing, it is tolerant of imprecision, uncertainty, partial truth and approximation. As for modelling, the process optimization can also be carried out by using either conventional (hard) optimization algorithm or heuristic soft algorithm such as genetic algorithms (GAs). Similarly, the control of the process may be based on either conventional (hard) control techniques or soft control techniques such as fuzzy logic. A hybrid of hard and soft techniques may also be used.

Human beings themselves adapt to changes in the environment. In the same way, an adaptive control mechanism is an integral part of an intelligent machine. An adaptively controlled machine is able to adapt to the dynamic changes of the system caused by the variability of machining process due to changes in the cutting conditions such as the hardness of the work material, tool wear, deflection of the tool and the workpiece, and so on. The following are the main objectives of an adaptive control system:

- to adjust the machining parameters such as cutting speeds and feed rates and/or the motion of the cutter to optimize the machining process by maximizing some performance criteria based on the cost or the production;
- to satisfy various constraints against variations due to external factors and respond to such variations in the process in real time;
- to automatically improve the performance of the machining process through its learning capability.

An adaptive control system that can fulfil the third objective may be said to possess intelligence as one of the meanings of the word 'intelligence' is the capacity to acquire and apply knowledge.

For monitoring of machining processes and fault diagnostics, the intelligent machining system must be equipped with the knowledge of how to recognise failures, how to localize them and how to relate the faults and their effects to the operating state. Moreover, state classification and process intervention have to be completed in real time to avoid additional damage when a deviant state has been detected. There are various signals (force, torque, temperature, mechanical vibration, acoustic emission, *etc.*) which correlate with the condition of the machining process. The control and monitoring algorithms should be based on the simultaneous measurement and processing of different signals.

Figure 12.1 shows a scheme of an intelligent machining centre. The intelligent machining centre functions as follows. The design model of the component to be machined is provided to the computer. Using suitable software, the tool path is generated and the spindle speed and feed rate are decided. This information is fed to the machine controller, which provides signals to the drive motors. The adap-

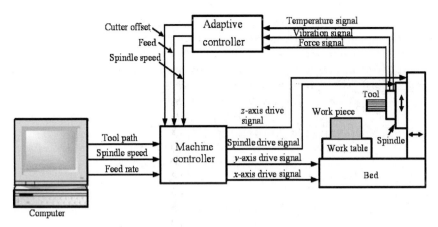

Figure 12.1. An intelligent horizontal machining centre

tive controller adjusts the feed rate, spindle speed and tool path, according to changes in the cutting conditions. The four different types of adaptive control system are: adaptive control optimization (ACO), adaptive control constraint (ACC), geometric adaptive control (GAC) and vibration adaptive control (VAC). The purpose of ACO is to search for the optimum values of feed rate and spindle speed that maximize some performance criteria such as minimum cost of machining or maximum production rate. ACC selects the feasible solution of the feed rate and spindle speed to satisfy the constraints in manufacturing by an algorithm that is rather simpler than ACO. GAC tries to obtain a highly accurate surface finish by adjusting the tool offset against the deflection of tools and work material caused by temperature rise and/or cutting force. The function of VAC is to avoid the vibration or chatter of tools, mainly due to regenerative oscillation, by adjusting the spindle speed or the resonance frequency of machine tools. Feedback about the process can be obtained by means of a number of different kinds of sensors, although the figure shows feedback only in the form of temperature, vibration and force signals.

An intelligent machining system should also possess a memory, like an expert operator. The knowledge available in machining data handbooks and that acquired from experience can be stored in the memory of the intelligent machining system. The system should have the capability to acquire data efficiently, store useful data by filtering out the noise and retrieve it efficiently. At the same time, it is desirable that the system is able to communicate with humans in some way, such that its decisions and actions become transparent to the users concerned. This implies that the acquired or available knowledge can be stored in the form of an expert system.

In this chapter, the computational techniques and optimization methods that can be used to develop intelligent machining systems are briefly described. Section 12.2 describes the application of neural networks in modelling the machining processes. Section 12.3 describes fuzzy-set-based modelling. Section 12.4 describes the hybrid system composed of neural networks and fuzzy set theory, either in the form of the mixture or the compound. Section 12.5 briefly discusses the application of FEM to the study of machining processes. Section 12.6 describes the optimization techniques useful for optimizing the machining processes. Both conventional (hard) and non-conventional (soft) techniques are touched upon. Finally, Section 12.7 discusses the challenges to be met for developing a truly intelligent machine tool.

12.2 Neural Network Modelling

It is well known that computers can perform a number of jobs at a much faster rate compared to human beings. However, there are many other tasks which a human being can perform in a faster and better way. One such task is recognising a face. A human being can recogonise a face quickly, because the information about vari- ous attributes of the face is processed parallely in the brain. The brain contains a neural network consisting of a number of interconnected information-processing units called neurons. A schematic drawing of a neural network showing only two neurons and their connections is shown in Figure 12.2. A neuron consists of a cell body called the soma, a number of fibres called dendrites and a single long fibre called the axon. The dendrites receive the electrical signals from the axons of other neurons. The axon transmits the electrical signal from one neuron to other neurons via the dendrites. The connection between the axon and the dendrite is called a synapse. At the synapses, the electrical signals are modulated by different amounts. The synapses release chemical substances that cause changes in the electrical potential of the soma. When the potential reaches its threshold, an elec- trical pulse called the action potential is sent down through the axon.

Artificial neural networks are a humble attempt to model biological neural net- works. Figure 12.3 shows a schematic diagram of an artificial neuron receiving signals (x_i, $i = 1$ to n) from a number of neurons and emitting signals (o_i, $I = 1$ to k) to be transmitted to a number of neurons. In an artificial neural network, the sig- nals are in the form of numerical values rather than electricity. The action of the synapses is simulated by multiplying each input value by a suitable weight w.

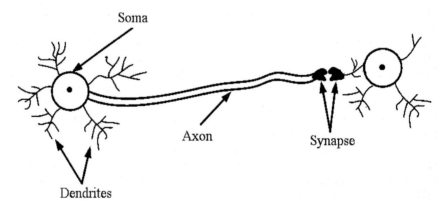

Figure 12.2. A schematic drawing of a biological neural network

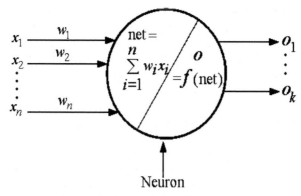

Figure 12.3. A typical artificial neuron

A biological neuron fires an output signal only when the total strength of the input signals exceeds a certain threshold. This phenomenon is modelled in an artificial neuron by calculating the weighted sum of the inputs to represent the total strength of the input signals, and applying a suitable activation (threshold) function to the sum to determine its output.

Neural networks can be classified based on their topology (architecture) and the method of training. The most common neural network architectures are: (1) feed-forward neural networks, (2) feedback neural networks, and (3) self-organizing neural networks. Feedforward neural networks are the most popular and widely used ones. There are two types of neural networks in this category: multi-layer perceptron (MLP) neural network and radial basis function (RBF) neural net-works. Figure 12.4 shows a feedforward architecture of a typical neural network consisting of three layers. Each layer contains a number of neurons, depicted by circles in the figure. Each layer has full interconnection to the next layer but no connection within a layer. The first layer of the network is known as the input layer whose neurons take on the values corresponding to different variables repre-senting the input pattern. The second layer is known as a hidden layer because its outputs are used internally and not used as the final output of the network. In MLP neural network, there may be more than two hidden layers. In RBF neural net-work, only one hidden layer is present. In Figure 12.4, only one hidden layer is shown. The final layer of network is known as the output layer. The values of the neurons of the output layer constitute the response of the neural network to an input pattern presented at the input layer. In MLP neural networks, the number of hidden layer in a network is an important design parameter. In both MLP and RBF neural networks the number of neurons in a hidden layer has to be chosen judi-ciously. The general rule is to design a neural network model which uses fewer parameters (weights and biases).

In recurrent or feedback neural networks, the outputs of some neurons are also fed back to some neurons in layers before them. Thus, signals can flow in both forward and backward directions, as shown in Figure 12.5. A recurrent network is said to have a dynamic memory. The output of such networks at a given instant reflects the current input as well as the previous inputs and outputs. An example of

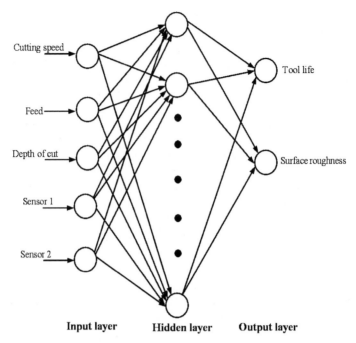

Figure 12.4. A typical feedforward neural network

the recurrent networks is the Hopfield network. Hopfield networks are typically used for classification problems with binary input pattern vectors. Another type of neural network is the self-organizing neural network that consists of neurons arranged in the form of a low-dimensional grid. Each input is connected to all the

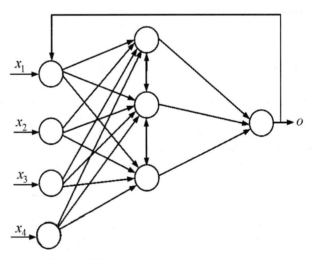

Figure 12.5. A typical recurrent neural netwok

output neurons. This type of network is useful in classifying high-dimensional data by constructing its own topology.

Neural networks need to be trained so that they produce proper response to a given input vector. The training process is an iterative process that adjusts the parameters (weights and biases) of the network until the network is able to produce the desired output from a set of inputs. The process of training the network can be broadly classified into supervised and unsupervised learning. A number of training algorithms based on the supervised learning are available of which the most common is the backpropagation algorithm. The backpropagation algorithm supplies the neural network with a sequence of input patterns and desired output (target) patterns, which together constitute the training exemplars. As an input pattern is presented to the neural network, the output response is calculated on a forward pass through the network. In Figure 12.4, the output of each neuron j in the hidden layer is computed according to the model of Figure 12.3. Given the input signal x_i at the input neurons, the output o_j is given by

$$o_j = f\left(\sum w_{ji} x_i\right), \tag{12.1}$$

where w_{ji} is the weight associated with the jth neuron of the hidden layer and the ith neuron of the input layer. The function f is called the activation function. Some commonly used activation functions are:

$$Log\ sigmoid:\ o = f(t) = \frac{1}{1+e^{-ct}}, \quad where\ c\ is\ a\ constant, \tag{12.2}$$

$$Tan\ hyperbolic:\ o = f(t) = tanh(ct/2),\ where\ c\ is\ a\ constant, \tag{12.3}$$

$$Identity:\ f(t) = t. \tag{12.4}$$

The output of each neuron in the output layer is computed in a similar manner. The final output is compared to the desired output, and error terms are calculated for each output neuron. A function of the errors of the neurons in the output layer is then propagated backwards through the network to each layer, and weights of each of the interconnected neurons are adjusted in such a way that the error between the desired output and the actual output is reduced. Any optimization method can be used to find out the weights that minimize the error. Early backpropagation algorithms were based on the steepest-descent algorithm, according to which the maximum decrease in the function is in a direction opposite to the direction of the gradient of the function. Another commonly used backpropagation algorithm is based on the Levenberg–Marquardt method, which is a combination of the steepest-descent method and the quasi-Newton method. At initial iterations, Cauchy's method is followed and the algorithm gradually moves towards the quasi-Newton method.

In back propagation algorithms using the steepest-descent or Levenberg–Marquardt method, the weights may correspond to a local minimum only. This problem can be solved by adding a momentum term to the training rule, which forces the weights to keep moving in the same direction in the error surface without becoming trapped in a local minimum. Momentum simply adds a fraction of the previous weight update to the current weights. This increases the size of the step taken towards minimum. It is therefore necessary to reduce the learning rate (a factor that determines how much change in the weights is carried out at each step) when using a lot of momentum. If a high learning rate is used, the algorithm may oscillate around the minimum and may become unstable. Too small a learning rate will take a long time to converge. The optimum value of learning rate is often found by trial and error.

A radial basis function neural network has weights only between the hidden layer and the output layer. There are no weights between the input layer and the hidden layer. With each node in the hidden layer, a centre x_k is associated, which is a vector with the same dimension as the input vector. Usually the centres are chosen in a random manner from the input dataset. The output o then becomes

$$o = \sum_{k=1}^{m} w_k g_k \left(\| \, x - x_k \, \|, s_k \right),$$ (12.5)

where x is the input vector, $\| \, x - x_k \, \|$ is the Euclidian norm, s_k is the spread parameter, w_k are weights, $g_k \left(\| \, x - c_k \, \|, s_k \right)$ is the radial basis function (RBF) and m is the number of centres, which is the same as the number of neurons in the hidden layer. Some of the most common RBFs are:

$$\text{Multiquadrics: } g_k(x) = \left(\| x - x_k \|^2 + s_k^2 \right)^{1/2};$$ (12.6)

$$\text{Inverse multiquadrics: } g_k(x) = \left(\| x - x_k \|^2 + s_k^2 \right)^{-1/2};$$ (12.7)

$$\text{Gaussians: } g_k(x) = e^{-s_k^2 \| x - x_k \|^2};$$ (12.8)

$$\text{Thin plate splines: } g_k(x) = \| x - x_k \|^2 \, log \| x - x_k \|.$$ (12.9)

Once the type of RBF function is decided and the centres are fixed, the output o given by Equation (12.5) becomes a linear combination of the weights. With the input and output dataset provided, the weights can be calculated using a multiple linear regression procedure so that the sum squared error between the predicted and the target output is minimized.

The process of training the neural network using supervised learning is applicable to problems where representative exemplars of both input pattern and output (target) patterns are known. In many problems where the target patterns are unknown, unsupervised learning is used. In the unsupervised learning process, the network is provided with a dataset containing input patterns but not with desired output patterns. The unsupervised learning algorithm then performs clustering of the data into similar groups based on the measured attributes or features of the given input patterns serving as inputs to the algorithms. After the training of the neural network by a suitable method, the trained network needs to be tested with unseen data (not included in the training dataset).

In the area of machining, neural networks have been used for the prediction of cutting forces [1,2], surface roughness [3–6], dimensional deviation [3,4], tool wear [6,7] and tool life [8,9]. Risbood *et al.* [4] fitted a neural network for the prediction of surface roughness and dimensional deviation in a turning process. For the wet machining of steel with a high-speed steel (HSS) tool with 8% cobalt, they used only 18 training data and 8 testing data for the prediction of surface roughness. The weight adjustment is carried out by training data and the trained neural network is tested by some data to ensure that there is no overfitting of the network. The multilayer perceptron neural network architecture used by the authors is shown in Figure 12.6. The input neurons correspond to feed f, cutting speed v, depth of cut d and acceleration of radial vibrations a of the turning tool, and the output neuron corresponds to the surface roughness R_a. The accelerations of the radial vibration were measured using an accelerometer. The training and testing data are shown in Table 12.1. While designing the network topology and training, it was ensured that the testing error was below 20% for each data. Later the fitted neural network was validated with 32 validation data. These data were not used in the training and testing. The results are shown in Table 12.2, where R_a is the experimentally obtained surface roughness and $\hat{R}_a$ is the neural network predicted surface roughness. It is seen that, in 24 out of 32 cases, the error in prediction of surface roughness is less than 20%. Only in four cases is the error more than 25%, with a maximum error of 31.9%. Figure 12.7 shows the performance of the neural network pictorially. When a graph is plotted between actual and predicted values of surface roughness, most of the points are found to fall in the neighbourhood of the line passing through the origin and equally inclined from the two axes. Sonar *et al.* [10] fitted a RBF neural network to the data and obtained almost same prediction accuracy in a shorter computational time.

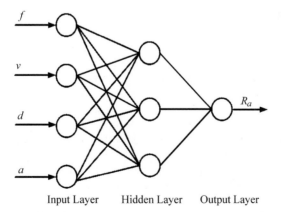

Figure 12.6. The neural network architecture for the prediction of surface roughness From Risbood *et al.* [4] copyright 2003, Elsevier

Table 12.1. Training and testing data for a wet turning operation, from Risbood *et al.* [4] copyright 2003, Elsevier

S. No.	v (m/min)	d (mm)	f (mm/rev)	a (m/s^2)	R_a (µm)
1	107.82	0.30	0.04	0.55	1.74
2	106.47	0.30	0.08	0.65	2.26
3	105.12	0.30	0.16	0.97	3.23
4	104.80	0.60	0.04	2.92	2.74
5	103.55	0.60	0.08	3.66	3.59
6	106.02	0.60	0.16	2.66	2.91
7	47.17	0.30	0.04	0.90	2.31
8	46.55	0.30	0.08	0.73	3.21
9	45.95	0.30	0.16	2.41	4.64
10	48.75	0.60	0.04	0.65	3.18
11	48.14	0.60	0.08	0.58	4.52
12	47.52	0.60	0.16	1.73	5.43
13	27.71	0.30	0.04	0.95	2.06
14	27.35	0.30	0.08	0.83	2.90
15	26.99	0.30	0.16	0.88	5.20
16	27.71	0.60	0.04	0.59	2.87
17	27.35	0.60	0.08	0.72	4.00
18	26.99	0.60	0.16	1.42	6.20
1[a]	74.13	0.40	0.10	2.48	4.80
2[a]	42.98	0.40	0.10	0.67	4.24
3[a]	36.87	0.50	0.12	1.07	4.55
4[a]	35.96	0.30	0.12	0.95	5.21
5[a]	78.10	0.60	0.12	2.10	4.57
6[a]	76.43	0.60	0.05	1.23	2.91
7[a]	73.95	0.30	0.05	0.76	2.52
8[a]	72.92	0.60	0.04	0.79	2.81

[a] test datasets

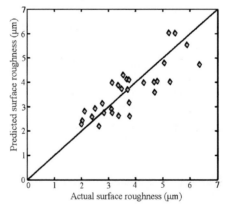

Figure 12.7. Predicted versus actual surface roughness in turning by HSS tool in the presence of coolant. From Risbood *et al.* [4] copyright 2003, Elsevier

Table 12.2. Results of experiments carried out to test the performance of fitted neural network model for the wet turning operation. from Risbood *et al.* [4] copyright 2003, Elsevier

S. no.	F (mm/rev)	v (m/min)	d (mm)	a (m/s^2)	R_a (μm)	$\hat{R}_a$ (μm)	% error
1	0.08	34.71	0.60	0.63	3.67	4.15	−13.10
2	0.04	23.45	0.30	0.42	1.99	2.28	−14.57
3	0.04	64.56	0.40	0.66	2.13	2.81	−31.90
4	0.12	32.87	0.30	0.48	4.29	3.98	7.23
5	0.04	54.28	0.60	0.55	2.78	3.12	−12.23
6	0.16	29.39	0.60	1.12	5.22	6.02	−15.32
7	0.12	20.88	0.40	0.67	4.79	4.04	15.65
8	0.08	54.91	0.40	2.80	3.53	4.33	−22.66
9	0.06	38.50	0.30	0.47	3.37	2.64	21.66
10	0.08	34.71	0.60	0.57	3.73	4.13	−10.72
11	0.04	23.45	0.30	0.52	1.99	2.26	−13.56
12	0.04	64.56	0.40	0.54	2.84	2.75	3.16
13	0.12	32.87	0.30	0.55	5.26	4.03	23.38
14	0.04	54.28	0.60	0.71	3.76	3.17	15.61
15	0.16	29.39	0.60	0.98	5.44	6.02	−10.66
16	0.12	20.88	0.40	0.73	3.76	4.09	−8.77
17	0.08	54.91	0.40	0.73	3.48	3.76	−8.05
18	0.06	38.50	0.30	0.56	3.77	2.63	30.23
19	0.04	48.70	0.40	0.54	2.39	2.56	−7.11
20	0.08	61.88	0.50	0.98	4.68	3.99	14.74
21	0.08	60.56	0.50	0.77	3.14	4.00	−27.38
22	0.16	100.85	0.50	1.49	3.11	2.90	6.75
23	0.04	26.48	0.40	0.37	2.03	2.41	−18.72
24	0.06	49.25	0.30	0.86	3.13	2.78	11.18
25	0.04	106.34	0.30	0.94	2.65	2.20	16.98
26	0.06	45.95	0.60	1.16	3.70	3.70	0.00
27	0.06	101.80	0.60	2.70	2.51	2.93	−16.73
28	0.16	45.99	0.30	1.15	5.87	5.54	5.62
29	0.16	104.20	0.30	2.23	3.36	3.89	−15.77
30	0.16	45.95	0.60	2.28	5.06	4.80	5.14
31	0.16	46.97	0.60	2.43	6.35	4.74	25.35
32	0.16	103.10	0.60	3.32	4.72	3.59	23.94

12.3 Fuzzy Set Theory

In the conventional crisp set theory, an element is either a member of the set or a non-member of the set. With each member, we can associate a number 1 or 0, depending on whether it is a member or non-member of the particular set. We may call this number the membership grade of the element in the set. In crisp set theory, the membership grades of the elements contained in the set are 1 and the membership grades of the elements not contained in the set are 0. In fuzzy set theory, the value of the membership grade of an element of the universe may have any value in the closed interval [0, 1]. A membership grade 1 indicates full membership and 0 indicates full non-membership in the set. Any other membership grade between 0 and 1 indicates partial membership of the element in the set.

Let us consider an example to understand the concept of crisp and fuzzy set. Assume that there are five machined shafts designated by a, b, c, d and e. The CLA surface roughness values of the machined shafts are 0.6, 0.8, 1.2, 3.0 and 4.1 µm, respectively. If we construct a set A as the set of shafts having surface roughness less than 1.5 µm, then clearly the shafts a, b and c are the members of the set and set A may be represented as

$$A = \{a,b,c\}. \tag{12.10}$$

Now, let us construct another set B of shafts having low surface roughness. There is a degree of subjectivity in deciding the definition of low surface roughness in a particular context. One possible fuzzy set B may be

$$B = \{\, 1/a, 0.8/b, 0.5/c, 0/d, 0/e\}, \tag{12.11}$$

In the above set, the shafts have membership grades of 1, 0.8, 0.5, 0 and 0 respectively, indicated before an oblique slash with each shaft. Here, shaft a is a full member of the set B, shafts b and c are partial members, and shafts d and e are non-members. Someone else may wish to form the set B as

$$B = \{\, 1/a, 0.9/b, 0.6/c, 0/d, 0/e\}, \tag{12.12}$$

Here, the membership grades of b and c have changed; nevertheless the membership grade of b is more than that of c. Thus, although the membership grades are subjective, they are not arbitrary. However, some skill is needed in the formation of a fuzzy set that properly represents the linguistic name assigned to the fuzzy set.

Fuzzy set theory may be called a generalization of conventional crisp set theory. Various set-theoretic operations commonly used in crisp set theory have been defined for fuzzy set theory as well. These operations reduce to their conventional forms for crisp sets. For example, the intersection of two fuzzy sets A and B, i.e., $A \cap B$ is defined as a set in which each element has a membership grade equal to the minimum of its membership grades in A and B. Similarly, the union of two fuzzy sets A and B, i.e., $A \cup B$ is defined as the set in which each element has a membership grade equal to the maximum of its membership grades in A and B.

The application of fuzzy sets extends to logic. In the classical binary logic, a statement is either true or false. Quantitatively, we can say that the truth value of a statement is either 1 or 0. In fuzzy logic, it is possible for a statement to have any truth value in the closed interval [0, 1]. For example, the statement "The CLA surface roughness value of 1.2 µm is a low surface roughness." may be assigned a truth value of 0.7 by some expert.

Another offshoot of fuzzy set theory is fuzzy arithmetic. Fuzzy arithmetic deals with operations on fuzzy numbers. The fuzzy numbers are generalization of interval numbers. An interval number is specified by an upper and a lower bound. For example, (3, 5) is an interval number with a lower bound of 3 and upper bound of 5. The interval numbers are used when the value of a variable is expected to lie in a range. A fuzzy number consists of the different interval numbers at different membership grades. A requirement of the fuzzy number is that the membership grade function of fuzzy number should be convex. Figure 12.8 shows examples of valid and invalid fuzzy numbers.

The fuzzy arithmetic operations are defined over each α-cut. An α-cut of a membership grade function $\mu(x)$ is the set of all x such that $\mu(x)$ is greater or

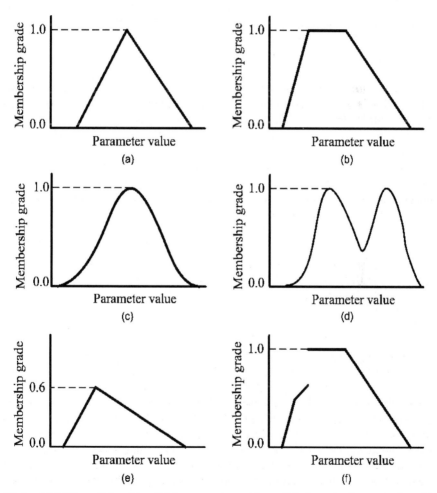

Figure 12.8. The valid and invalid fuzzy numbers: (a) a valid triangular fuzzy number, (b) a valid trapezoidal fuzzy number, (c) a valid bell-shaped fuzzy number, (d) a non-convex invalid fuzzy number (e), a non-normal invalid fuzzy number and (f) a discontinuous invalid fuzzy number

equal to α. Thus, at a particular α-cut, an interval number is obtained corresponding to the interval number at the membership grade of α. If two fuzzy numbers are represented by (a_1^α, a_2^α) and (b_1^α, b_2^α) at an α-cut, then four basic arithmetic operations are defined as follows:

$$\text{Addition: } \left(a_1^\alpha, a_2^\alpha\right) + \left(b_1^\alpha, b_2^\alpha\right) = \left(a_1^\alpha + b_1^\alpha, a_2^\alpha + b_2^\alpha\right), \qquad (12.13a)$$

$$\text{Subtraction: } \left(a_1^\alpha, a_2^\alpha\right) - \left(b_1^\alpha, b_2^\alpha\right) = \left(a_1^\alpha - b_2^\alpha, a_2^\alpha - b_1^\alpha\right), \qquad (12.13b)$$

$$\text{Multiplication: } \left(a_1^\alpha, a_2^\alpha\right) \times \left(b_1^\alpha, b_2^\alpha\right) = \left(a_1^\alpha \times b_1^\alpha, a_2^\alpha \times b_2^\alpha\right), \qquad (12.13c)$$

$$\text{Division: } \left(a_1^\alpha, a_2^\alpha\right) \div \left(b_1^\alpha, b_2^\alpha\right) = \left(a_1^\alpha \div b_2^\alpha, a_2^\alpha \div b_1^\alpha\right). \qquad (12.13d)$$

Other arithmetic operations can be derived from these four basic operations. The fuzzy arithmetic operations can be used to obtain the fuzzy value of the expression. Care must be taken when a variable occurs more than once in an expression. In this case, the blind application of fuzzy arithmetic will provide wider than realistic interval of the value of expression at an α-cut. One way to avoid this problem is to carry out one computation by taking the lower bound, and the other computation by taking the upper bound of the variable at each α-cut.

In many situations, a fuzzy parameter is estimated by a computer code rather than by a closed-form expression. In this case, at each α-cut, the upper bound of the parameter may be obtained by running the code for those combinations of lower and upper bounds of the independent variables that provide the lower and upper bounds of the parameter. In machining process, a number of imprecise variables are involved that can be represented by fuzzy numbers, and computation can be carried out to get an estimate of the dependent variables in the form of fuzzy numbers. The material parameters and the friction, for example, can be treated as fuzzy numbers in finite element code to predict the cutting forces in the form of the fuzzy numbers. This will provide more insight to a higher level decision-maker than the prediction in the form of the crisp numbers.

The fuzzy set theory can also be used to make prediction from the data. As an example, suppose that, in a finish turning process, the surface roughness of the turned job is to be predicted for a particular feed and cutting speed. The estimation of surface roughness consists of four steps. The first step is fuzzification, in which the crisp values of feed and cutting speed are fuzzified, *i.e.*, they are assigned to linguistic fuzzy sets. A variable may be associated with more than one fuzzy set with different membership grades for different sets. For example, the given feed value may be called "high feed" with a membership grade μ_1 and "low feed" with a membership grade μ_2. Similarly, the cutting speed may be called "low cutting speed" with a membership grade μ_3 and "high cutting speed" with a membership grade μ_4.

Once the independent variables are fuzzified, the next step is carried out. In this step, called the rule evaluation step, the strength of the various rules is evaluated.

Various rules are kept in a rule bank, which may be prepared by experts based on their experience or may be generated from data following systematic procedures [11]. For the present example, the four rules could be:

Rule 1: If the feed is low and the cutting speed is low, then the surface roughness is medium.

Rule 2: If the feed is low and the cutting speed is high, then the surface roughness is low.

Rule 3: If the feed is high and the cutting speed is low, then the surface roughness is high.

Rule 4: If the feed is high and the cutting speed is high, then the surface roughness is medium.

In each rule, the 'if' part is called the antecedent and the 'then' part is called the consequent. The strength of a rule is equal to the truth value of the antecedent. If the antecedent consists of the statements separated by 'and', which is equivalent to intersection operation, the truth value of the antecedent is equal to the minimum of the truth values of each of the statements. For example, consider the first rule. The membership of the given feed value in the fuzzy set 'low feed' is μ_2, as described before. Hence, the truth value of the statement 'the feed is low' is μ_2. Similarly, the membership of the given cutting speed in the fuzzy set 'low cutting speed' is μ_3. As both statements in the antecedent are separated by 'and', the truth value of rule 1 is min(μ_2, μ_3). This is the strength of the rule. If the antecedent consists of the statements separated by 'or', which is equivalent to the union operation, the truth value of the antecedent is equal to the maximum of the truth values of each of the statements. For a given feed and cutting speed, a number of rules are applicable with different strengths. Now, we pay attention to the consequent part of the rules. Let us assume that the strengths of the four rules in the above example are 0.06, 0.14, 0.24 and 0.56. Membership functions for the fuzzy sets "low surface roughness", "medium surface roughness" and "high surface roughness" are shown in Figure 12.9. These fuzzy sets are clipped at the membership grades corresponding to the rule strengths, as shown in Figure 12.10. The clipping is done by slicing off the portion of the membership function having membership grade more than the strength of the rule.

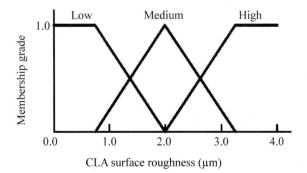

Figure 12.9. Fuzzification of CLA surface roughness

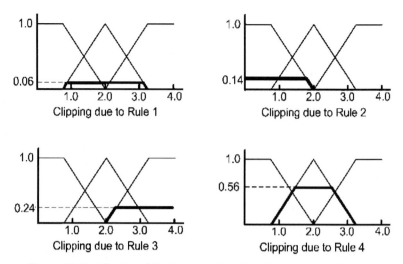

Figure 12.10. Clipping of the fuzzy sets based on the strengths of the rules

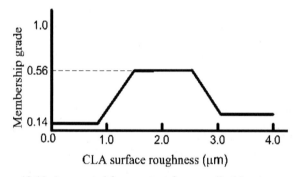

Figure 12.11. Aggregated fuzzy output for prescribed input parameters

The third step is rule aggregation. The clipped rules are aggregated by applying union operation as shown in Figure 12.11. This provides the output *viz.* surface roughness in this case, in the form of a fuzzy variable. This needs to be defuzzified in the fourth step. There are various methods of defuzzification. One method is finding out the centroid of the area covered by the membership function of the aggregated output. The defuzzified output corresponds to the horizontal coordinate of the centroid. Another simpler method is to take the output as the mean of the outputs at the maximum membership grade. In this way, the surface roughness can be predicted for a given set of input variables.

12.4 Neuro-fuzzy Modelling

Neural networks have a good learning capability. However, they act as black boxes. Fuzzy-set-based systems can incorporate expert knowledge and are generally transparent. Often, it is convenient to apply both these techniques together. There are two ways in which both these techniques can join hands. One way is to have two separate modules for the neural network and fuzzy set, and use these modules in the main task appropriately. For example, for predicting the surface roughness in a turning process, Abburi and Dixit [12] trained a neural network by using shop floor data. The trained network was used to generate a large number of predicted datasets. These datasets were used to feed a fuzzy-set-based rule-generation module. The generated rules were in the form of IF–THEN rules, providing transparency to a user. The developed rule base was used for predicting the surface roughness by using a fuzzy inference system. The second way is to have an entirely different technique which uses the features of both the fuzzy set and the neural network, but can be classified into neither category. One such technique is the adaptive neuro-fuzzy inference system (ANFIS) [13], which can be considered a child of neural network and fuzzy logic. In this section, we briefly describe the ANFIS.

A typical architecture of ANFIS having two inputs as feed f and cutting speed v is shown in Figure 12.12. The input data is fuzzified in the first layer. Each neuron in this layer represents a linguistic fuzzy set such as "small feed", "large depth", *etc.* Each neuron takes the value of an input and emits a membership grade corresponding to that input. Thus, each neuron houses a membership function. The membership function may be fixed based on the expert knowledge, or may contain some adjustable parameters called premise parameters.

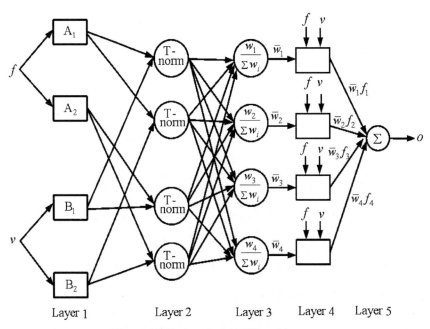

Figure 12.12. A typical ANFIS architecture

Layer 2 of the ANFIS contains neurons which basically emit the strength of various rules. The number of neurons in this layer will be equal to the number of fuzzy rules. The input to each neuron is the membership grades of feed and depth of cut. In the previous section, we mentioned that the strength of a rule may be evaluated by the "minimum" operator. Thus, if the feed value has a membership grade of 0.7 in the fuzzy set "large feed" and the depth of cut has a membership grade of 0.6 in the fuzzy set "large depth of cut", then the strength of the rule having the IF part of "If feed is large and depth of cut is large" can be considered as min(0.7, 0.6), $i.e.$, 0.6. Note that it is not the only way to calculate the strength of a rule containing two statements connected by an "and". In general any T-norm operator can be used to represent fuzzy intersection or "and" in the English language. Some of the most frequently used T-norm operators are:

$$Minimum: \; T_{min}(\mu_1,\mu_2) = min(\mu_1,\mu_2), \tag{12.14a}$$

$$Algebraic \; product: \; T_{ap}(\mu_1,\mu_2) = \mu_1 \mu_2, \tag{12.14b}$$

$$Bounded \; sum: \; T_{bs}(\mu_1,\mu_2) = max(0, \; \mu_1 + \mu_2 - 1), \tag{12.14c}$$

$$Drastic \; product: \; T_{dp}(\mu_1,\mu_2) = \begin{cases} \mu_1, \; if \; \mu_2 = 1, \\ \mu_2, \; if \; \mu_1 = 1, \\ 0 \quad otherwise. \end{cases} \tag{12.14d}$$

Layer 3 normalizes the strength of the rules. Each neuron in layer 4 emits the weighted output corresponding to a rule. The output of the ith fuzzy rule is taken as

$$f_i = p_i f + q_i v + r_i, \tag{12.15}$$

where p_i, q_i and r_i are the adjustable parameters called the consequent parameters. Note that the output of the fuzzy rule given by Equation (12.15) is in the form of a function. This type of rule is used in a Sugeno model. The types of fuzzy rules described in the previous section are used in a Mamdani model. Layer 5 consists of a single neuron, which just sums the inputs.

The network training is carried out by adjusting the premise and consequent parameters so as to minimize the error. In the forward pass, the consequent parameters p_i, q_i and r_i can be obtained by a least-squares procedure for fixed premise parameters of the membership functions. In the backward pass, the consequent parameters are fixed and the premise parameters can be obtained by a suitable optimization procedure to minimize the backpropagated error. Apart from the gradient descent algorithm, a number of optimization algorithms may be used for this purpose. The reader may also note that this is not the only possible topology for an ANFIS. For further details of soft computing techniques, the readers may refer to the textbooks [13–15].

12.5 A Note on FEM Modelling

We have discussed some of the soft computing techniques. These techniques require huge amount of data for proper modelling. It is always better to take advantage of the help of the physics of the process of making a model. The physics of the process needs to be expressed in the form of differential equations. The minite element method (FEM) is a tool to solve these differential equations. In this method, the domain is discretized into a number of small elements. The output variable to be determined is approximated by a combination of functions that are continuous inside the elements and possess at least some order of continuity at the interface of the two elements. Each element contains certain points, which are called nodes. The approximating function is expressed in the form of the nodal values. The nodal values of the output variable are obtained in such a manner that the differential equations are best satisfied with the given approximation. The details of the FEM can be found in a number of textbooks [16–18].

In machining, the FEM has been extensively used for the determination of cutting temperatures [19–22]. For determining the temperature distribution in the tool and the workpiece by a control-volume approach, the following heat conduction equation needs to be solved:

$$k\left[\frac{\partial^2 T}{\partial x^2}+\frac{\partial^2 T}{\partial y^2}+\frac{\partial^2 T}{\partial z^2}\right]+\dot{Q}=\rho c\left[\frac{\partial T}{\partial t}+u\frac{\partial T}{\partial x}+v\frac{\partial T}{\partial y}+z\frac{\partial T}{\partial z}\right], \qquad (12.16)$$

where k is the thermal conductivity, T is the temperature, $\dot{Q}$ is the rate of heat generation per unit volume, ρ is the density, c is the specific heat, t is the time and (u, v, w) are the velocity components of a particle at coordinates (x, y, z). The heat generation is due to plastic deformation of the work material and the friction at the tool surfaces.

For obtaining the heat generation due to plastic deformation and also for obtaining the cutting forces, the machining process may be simulated using the finite element method. There are two approaches for modelling: the updated Lagrangian [23] and Eulerian [24] methods. In the updated Lagrangian approach, the motion of each particle is followed, whereas in the Eulerian approach a control volume is chosen to find various quantities of interest at the spatial coordinates. In general, the finite element simulation of machining processes is computationally expensive due to the non-linear nature of the problem. The finite element simulation of machining process is carried out iteratively. In each iteration, a number of linear equations need to be solved. Among the Eulerian and Lagrangian formulation, the latter takes much more time than the former but is able to predict more detailed information such as the residual stresses in machining. The accuracy of the finite element simulation is dependent on the accuracy of the material parameters and friction characteristics. The machining process occurs at high strain rates and temperature. Therefore, the flow stress of the work material need to obtained from deformation tests at high strain rate and temperature. There is always some uncertainty in the determination of flow stress. The uncertainty is greater for the values of the friction at tool–job and tool–chip interface. It therefore becomes more meaningful to carry out finite element simulations with fuzzy parameters. This, however, further increases the computational time. One way to reduce the computational time is to train a neural network from the data obtained by FEM and use the neural network for prediction of the machining parameters.

Besides the simulation of machining process, the finite element model has been used for the predictions of tool wear and fracture of the cutting tool [25–28]. It has also been used for predicting the integrity (residual stresses, microhardness and microstructure) of machined surfaces [29, 30].

12.6 Machining Optimization

Optimization of machining processes is one of the most widely investigated problems in machining. The objective functions in the machining problems are: (1) minimization of cost of machining, (2) maximization of production rate and (3) maximization of profit rate. A weighted combination of these objectives may be taken, or the problem can be solved as a multi-objective optimization problem. In the machining optimization problem, there are constraints on tool life, surface finish, cutting force, machining power, *etc*. Usually the machining processes are performed in a number of passes, the last pass being the finishing pass and other passes being the rough passes. In a multipass machining process, the cutting speed, feed and depth of cut in each pass are the primary variables. We shall discuss the issues in machining optimization by taking the example of multipass turning.

12.6.1 Objective Functions and Constraints

Consider the multipass turning of a cylindrical work piece of length L, initial diameter D_0 and final diameter D_f. The production time per component is given by

$$T_p = T_{tR} + \frac{t_c T_{tR}}{T} + T_{tF} + \frac{t_c T_{tF}}{T} + T_l + t_{ts} ,\tag{12.17}$$

where T_{tR} is the total cutting time for rough machining, t_c is the time required for changing a tool, T is the tool life, T_{tF} is the total cutting time for finish machining, T_l is the job loading and unloading time, and t_{ts} is the total tool setting time for a job. The total cutting time for rough machining is obtained as the summation of the cutting time for m roughing passes. Thus,

$$T_{tR} = \sum_{i=1}^{m} \frac{\pi L D_{i-1}}{v_{R_i} f_{R_i}} ,\tag{12.18}$$

where v_{R_i} and f_{R_i} are the cutting speed and feed, respectively, at the ith roughing pass and D_{i-1} is the workpiece diameter at the beginning of that pass. Note that the cutting speed is defined as the surface speed of the work piece at the beginning of the pass. The total tool setting time is given as

$$t_{ts} = (m+1)t_s ,\tag{12.19}$$

where t_s is the setting time for each pass.

The tool life T is a critical parameter for the objective function. The tool life is a function of machining parameters, *viz.* feed (f), cutting velocity (v) and depth of cut (d) as well as the tool–job combination and the machine tool. For a given machine tool and tool–job combination, the tool life can be expressed as a function of f, v and d. A neural network can be used for online prediction of the tool life.

The cost of machining for a workpiece, C_p, can be expressed as

$$C_p = C_0 T_p + C_{t_1} \frac{T_{tR}}{T} + C_{t_2} \frac{T_{tF}}{T},$$

$$(12.20)$$

where C_0 is the operating cost per unit time, C_{t_1} is the tool change cost for a roughing tool and C_{t_2} is the tool change time for a finishing tool. If the same tool is used for roughing and finishing passes, then $C_{t_1} = C_{t_2}$. If the price of a work piece is fixed as P_p, then the profit rate P_R is expressed as

$$P_R = \frac{P_p - C_p}{T_p}.$$

$$(12.21)$$

Equations (12.17), (12.20) and (12.21) can be used as the objective functions.

The machining optimization problem may be subjected to various constraints. Some of the constraints may be as follows:

1. Constraint on the tool life: The tool life should not be too low or too high. If the tool life is too low, then it will fail without machining even one piece. While machining a workpiece, if the tool is changed in between, it will affect the surface finish and dimensional accuracy of the workpiece. As a thumb rule, the tool life should be at least such that 20 components can be produced without the need of replacing the tool. In other words, the deterioration in tool life should be limited to less than 5% in machining a component. At first sight, it appears that there is no need to have a constraint on the maximum permissible tool life. However, a close examination of shop floor practice reveals that there is no advantage in operating at parameters providing too high value of tool life. Most factories follow the policy of replacing tools after a certain interval of time. If the tool life is too high, tools may be underutilized. Also, if certain types of tools have been specifically procured for a particular batch production, often it may not be advantageous to preserve the tools once the entire batch has been produced. In that case also, too high a tool life will mean underutilization of tools. However, in most cases, the constraint on tool life may turn out to be an inactive constraint.

2. Constraint on the surface finish: Most of the time a constraint on the upper limit of surface roughness is prescribed. However, the possibility of having a constraint on the lower limit of the surface roughness cannot be ignored. Surface roughness affects the heat transfer characteristics and a rough surface is beneficial, for example, in order to have greater heat transfer in a pool boiling process. Similarly, a specific tribological characteristic may be obtained by controlling the surface roughness. For the given process pa-

rameters, the surface roughness may be predicted by a trained neural network as described in Section 12.2.

3. Constraints on the machining forces: The neural network can be used for the modelling of cutting forces in a machining process. These forces should be less than the prescribed forces to avoid tool breakage and excessive deflections of the job and the tool.

4. Constraint on the machining power: The machining power is obtained by dividing the product of the main cutting force and cutting speed by the efficiency of the machine tool. The machining power constraint requires that the machining power be less than the prescribed power.

5. Constraints on vibrations: The vibrations during a machining process should be low as they affect the accuracy and precision of the job.

6. Constraint on dimensional deviation: The dimensional deviation of the workpiece should be kept within the prescribed limit. If a model for predicting the dimensional deviation in machining is available, it can be used in this constraint. Otherwise the constraints on machining forces will indirectly impose the constraint on dimensional deviation.

7. Constraints on geometric relations: It is obvious that the number of passes m in a machining operation is an integer quantity. The following geometric relation has to be satisfied for the multi-pass turning process:

$$D_0 = D_f + 2\sum_{i=1}^{m} d_{R_i} + 2d_F \qquad (12.22)$$

where D_f is the diameter of the finished job, d_{R_i} is the depth of cut of the ith roughing pass and d_F is the depth of cut of the finish pass.

Apart from these constraints, there are variable bounds on the cutting speed, feed and depth of cut.

12.6.2 Optimization Techniques

The general form of a single-objective optimization problem is

$$Minimize\ f(x),$$

$$subject\ to\ certain\ constraints. \qquad (12.23)$$

where x is a vector containing the decision variables. The maximization of a function is equivalent to the minimization of the negative of the function. Thus, even a maximization problem can be expressed in the form of Equation (12.23). There are a plethora of optimization techniques available in the literature [31–33]. We shall provide a glimpse of some techniques.

12.6.2.1 Golden Section Search Method

For one-dimensional minimization, there is an efficient technique for finding the minimum of a unimodal function. A function which has got only one minimum in a certain interval is called a unimodal function in that interval. Figure 12.13(a) shows a unimodal function. Figure 12.13(b) shows a multi-modal function for comparison. The minimum of a unimodal function can be found by a number of region-elimination methods that require only the function values but not derivative information. The general procedure of a region-elimination method is as follows:

Step 1: Find an interval (a, b) in which the minimum is expected to lie. The interval can be obtained by physical consideration or by a systematic mathematical procedure. The length of the interval should be large enough to ensure the presence of a minimum in the range.

Step 2: Choose two points x_1 and x_2 in the interval (a, b) and evaluate the function $f(x)$ at these points.

Step 3: If $f(x_1) > f(x_2)$, then minimum does not lie in interval (a, x_1), requiring replacement of the interval (a, b) by the interval (x_1, b). Else, if $f(x_2) > f(x_1)$, then minimum does not lie in interval (x_2, b), requiring replacement of the interval (a, b) by the interval (a, x_2). Else, if $f(x_2) = f(x_1)$, the minimum lies in the interval (x_1, x_2), requiring the replacement of the interval (a, b) by the interval (x_1, x_2).

Step 4: If the current interval is not sufficiently small, go to step 2.

It is clear from Figure 12.13(b) that the region-elimination method will not be effective for a multi-modal objective function.

There are a number of region-elimination methods depending on how we choose the points x_1 and x_2. One of the efficient one-dimensional search methods in this category is the golden section search. In this method, for the interval (a, b), two points x_1 and x_2 are chosen as follows:

$$x_1 = \tau a + (1-\tau)b, \tag{12.24}$$

$$x_2 = \tau b + (1-\tau)a, \tag{12.25}$$

where τ is called a golden number and is equal to 0.618. This procedure ensures that in the subsequent iterations (when the interval gets changed), only one new point has to be chosen. The other point becomes common with a point of the pre-

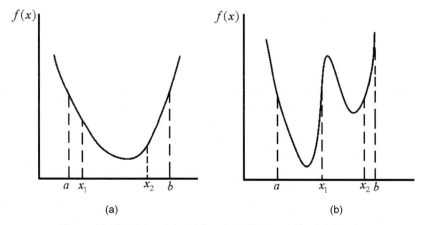

Figure 12.13. (a) A unimodal function. (b) A multimodal function

vious iteration. Thus, at each iteration only one function evaluation is needed. The interval reduces by a factor of $(0.618)^{n-1}$ after n function evaluations. Once the interval has become sufficiently small, the minimum can be taken as the middle point of the interval. For further accuracy, one may take three points in the reduced interval and fit a quadratic function to find the minimum.

12.6.2.2 Sequential Quadratic Programming (SQP) Method

For constrained optimzation of multivariable problems, there are a number of methods, which do not require the gradient information, *viz.* complex search method, random search method, *etc*. These methods are called direct search methods. However, we discuss here a powerful method requiring gradient information. The method is called sequential quadratic programming (SQP) and is quite efficient in handling constrained optimization problems. The principle of this algorithm is explained here.

Consider the following optimization problem with l equality and m inequality constraints:

$$\text{Minimize } f(x),$$
$$\text{subject to } h_i(x)=0, \ i=1 \text{ to } l; \ g_i(x)\le 0, \ i=1 \text{ to } m \tag{12.26}$$

At a particular guess point, Equation (12.26) is converted to the following quadratic form:

$$\text{Minimize } (\nabla f(x))^T d + \frac{1}{2}d^T Hd,$$
$$\text{subject to } h_i(x)+(\nabla h_i(x))^T d = 0, \ i=1 \text{ to } l; \tag{12.27}$$
$$g_i(x)+(\nabla g_i(x))^T d \le 0, \ i=1 \text{ to } m.$$

In the above equation, $d = (d_1, d_2, \ldots, d_n)^T$ is the vector of decision variables of the problem, ∇ is the gradient operator and H is the Hessian matrix. In particular,

$$(\nabla f(x))^T = [\partial f / \partial x_1, \partial f / \partial x_2, \ldots, \partial f / \partial x_n], \tag{12.28}$$

and

$$H = \begin{bmatrix} \dfrac{\partial^2 f}{\partial x_1^2} & \dfrac{\partial^2 f}{\partial x_1 \partial x_2} & \cdots & \dfrac{\partial^2 f}{\partial x_1 \partial x_n} \\[2mm] \dfrac{\partial^2 f}{\partial x_2 \partial x_1} & \dfrac{\partial^2 f}{\partial x_2^2} & \cdots & \dfrac{\partial^2 f}{\partial x_2 \partial x_n} \\[2mm] \cdots & \cdots & \cdots & \cdots \\[2mm] \cdots & \cdots & \cdots & \cdots \\[2mm] \dfrac{\partial^2 f}{\partial x_n \partial x_1} & \dfrac{\partial^2 f}{\partial x_n^2} & \cdots & \dfrac{\partial^2 f}{\partial x_n^2} \end{bmatrix}. \tag{12.29}$$

The gradients and the Hessian matrix are evaluated at the guess point.

The variable d is actually a search direction. It indicates that, moving in this direction, one will get the minimum. The question arises how much we should move. For this, we assume that the minimum point x_m is obtained by the following expression:

$$x_m = x + \lambda d ,$$

(12.30)

where λ is a step length chosen to reduce the value of a suitable merit function. Once the value of x_m is substituted in the merit function, the merit function becomes a function of the scalar λ. The optimum value of λ can be found by a one-dimensional optimization method such as the golden section search method, described in Section 12.6.2.1. The merit function can be a penalty function of the following form:

$$M = f(x) + R\left(\sum_{i=1}^{l} |h_i(x)| + \sum_{i=1}^{m} max(g_i(x),0) \right),$$

(12.31)

where R is a very large value called the penalty parameter.

At the obtained optimum point x_m, the objective function is again converted to the form of Equation (12.27), *i.e.*, a quadratic approximation for the function and a linear approximation for the constraints is used. After this, the procedure is repeated. Thus, the optimization process is iterative and the iterations are continued until convergence is obtained.

12.6.2.3 Genetic Algorithms

Conventional optimization methods are not suitable for optimization problems with multiple optimal solutions. There are newer methods, which start the search process with a population of solutions. These methods are inspired by natural processes and a sound mathematical basis for these methods needs to be developed. These methods can find multiple optimum solutions and globally optimum solutions and are suitable for multi-objective optimization problems. One category of such methods is genetic algorithms (GAs). GAs are inspired by the mechanics of natural genetics and natural selection. Although there are a lot of variants of genetic algorithms, we describe here a simple binary-coded genetic algorithm with reference to the optimization of the finish pass turning considering the feed and the cutting speed as decision variables.

In the binary-coded genetic algorithm, the variables are represented in binary form. For example, if feed varies from 0 to 0.15 mm/rev in a turning operation, it can be represented with a 4-bit binary number with 0000 representing zero feed and 1111 representing a feed of 0.15 mm/rev. A finer resolution in feed values may be obtained by choosing a higher bit number. Similarly let us consider that a cutting speed in the range of 0 to 310 m/min is represented by a 5-bit number with 11111 representing the cutting speed of 310 m/min in the usual way. A typical combination of feed and cutting speed can be represented by putting these two binary numbers together to make it a 9-bit string. For example, the reader may observe that 101010000 is a string representing a feed of 0.1 mm/rev and cutting speed of 160 m/min.

The search for the optima in GA is started by taking a number of random strings forming the population. Choosing an appropriate population size is crucial. The population of the strings is successively evolved by three operators: reproduction, crossover and mutation.

In reproduction, the good strings in the population are probabilistically assigned to a large number of copies. The fitness of the string decides how good the string is. In minimization problem, the fitness function can be taken as

$$F(x) = 1/(1 + f(x)), \tag{12.32}$$

For infeasible solution, a very low value of fitness may be assigned. There are a number of ways to do reproduction. One popular and easy-to-implement method is the tournament selection. In this, each string plays duel tournaments with two other randomly chosen strings. In a tournament, the fitnesses of the two strings is compared and the winning string is retained. It is clear that the best string will always remain and the worst string will be eliminated from the population. The fate of the other strings is dependent on chance. Finally, it is expected that, as a result of this operation, the number of good strings will be more than the bad ones in the population.

The second operation is crossover. In this, two new strings are created by exchanging the bits between two strings. For example, consider two strings chosen at random from a population, called parent strings. A single-point crossover operation is performed by randomly choosing a crossing site along the string and by exchanging all the bits on the right-hand side of the crossing site as shown;

$$\begin{matrix} 010001|111 \\ 100110|011 \end{matrix} \xrightarrow{} \begin{matrix} 010001|011 \\ 100110|111 \end{matrix}$$

The new strings formed are called the children strings. The crossover operation is performed with certain a crossover probability, usually lying between 0.75 to 0.95.

In the last operation, mutation, one or more bit of a string may be changed randomly. However, the probability of mutation is kept quite low, between around 0.01 and 0.05. The mutation operation serves the crucial role of preventing the algorithm from becoming stuck in the local optimum.

After the application of three GA operators in succession, a new generation is formed and an iteration is said to be complete. The iterations are carried out until the average fitness in successive generations more or less becomes constant. The entire methodology is illustrated by a flowchart in Figure 12.14. The application of GA to machining process optimization has been carried out by a number of researchers [34–36].

There are other newer population-based techniques that are proving to match the capabilities of GAs and on occasions outperform GAs. Two such techniques are particle swarm optimization (PSO) that was developed by Eberhert and Kennedy [37] and ant colony optimization (ACO) that was first proposed by Dorigo *et al.* [38]. PSO is a population-based algorithm that simulates the social behaviour of a flock of birds for the search of approximate solutions to optimization problems. The ACO algorithm is a population-based algorithm that simulates the social behaviour of a colony of ants for the search of approximate solutions to optimization

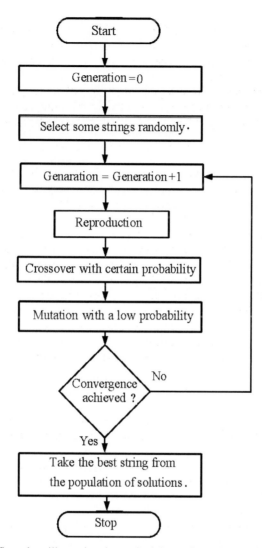

Figure 12.14. A flow chart illustrating the methodology of genetic algorithm

problems. ACO algorithm has been used for the determination of optimal machining parameters such as cutting speed, feed, depth of cut and number of cuts in order to minimize the production cost subject to various machining constraints such as tool life, surface finish, cutting forces and power, and temperatures [39].

12.7 Future Challenges

Earlier in this chapter, we compared the intelligent system of a machine tool with a skilled human operator. The performance of a skilled operator is dependent to a great extent on the sensory organs and the brain. Therefore, to develop an intelligent system that is able to function like a skilled operator, it is necessary to develop good sensors and efficient computing tools. A lot of research needs to be done to develop effective, compact and low-cost sensors. At the same time, computational methods and optimization algorithms need to be further improved. The compuational methods that model based on the available data need to be made more robust. They should be able to eliminate noise from the data and make the best out of the available information. There should be simultaneous efforts to understand the physics of the process and use that knowledge to complement the data modelling task. The proper formulation of optimization problems for different processes and choosing of an efficient algorithm is another area that needs to be explored further. Lastly, the intelligent system must exploit the developments in Internet technologies, which requires research in the Internet-based machining area.

References

[1]   Szecsi T (1999) Cutting force modelling using artificial neural networks. J Mater Process Technol 92:344–349.

[2]   Hao, W, Zhu, X, Li, X and Turyagyenda, G (2006), Prediction of cutting force for self-propelled rotary tool using artificial neural networks, Journal of Materials Processing Technology, Vol. 180, pp. 23-29.

[3]   Azouzi R, Guillot M (1997) On-line prediction of surface finish and dimensional deviation in turning using neural network based sensor fusion. Int J Mach Tools Manuf 37:1201–1217.

[4]   Risbood KA, Dixit US, Sahasrabudhe AD (2003) Prediction of surface roughness and dimensional deviation by measuring cutting forces and vibrations in turning process. J Mater Process Technol 132:203–214.

[5]   Kohli A, Dixit US (2004) A neural-network-based methodology for the prediction of surface roughness in a turning process. Int J Adv Manuf Technol 25: 118–129.

[6]   Özel T, Karpat Y (2005) Predictive modelling of surface roughness and tool wear in hard turning using regression and neural networks, Int J Mach Tools Manuf 45:467–479.

[7]   Ezugwu EO, Arthur SJ, Hines EL (1995) Tool-wear prediction using artificial neural networks. J Mater Process Technol 49: 255–264.

[8]   Tosun N, Ozler L (2002) A study of tool life in hot machining using artificial neural networks and regression analysis method. J Mater Process Technol 124:99–104.

[9]   Ojha DK, Dixit US (2005) An economic and reliable tool life estimation procedure for turning. Int J Adv Manuf Technol 26: 726–732.

[10]   Sonar DK, Dixit US, Ojha DK (2006) Application of radial basis function neural network for predicting the surface roughness in turning process. Int J Adv Manuf Technol 27:661–666.

[11]   Chen JC, Black JT (1997), A fuzzy-nets in-process (FNIP) systems for tool-breakage monitoring in end-milling operations. Int J Mach Tools Manuf 37: 783–800.

[12]   Abburi NR, US Dixit (2006) A knowledge-based system for the prediction of surface roughness in turning process. Robot Comput-Integr Manuf 22: 363–372.

[13]   Jang JSR, Sun CT, Mizutani E (2002) Neuro-Fuzzy and Soft Computing – A computational Approach to Learning and Machine Intelligence. Prentice-Hall of India, New Delhi.

[14] Kosko B (1992) Neural Networks and Fuzzy Systems: a Dynamical Systems Approach to Machine Intelligence. Prentice-Hall, NJ.

[15] Tsoukalas LH, Uhrig RE (1996) Fuzzy and Neural Approaches in Engineering. Wiley, New York, NY.

[16] Reddy JN (1993) An Introduction to the Finite Element Method, second edition. McGraw-Hill, New York, NY.

[17] Cook RD, Malkus DS, Plesha ME (1989) Concepts and Applications of Finite Element Analysis, 3rd edn. Wiley, New York, NY.

[18] Bathe KJ (1996) Finite Element Procedures. Prentice-Hall of India, New Delhi.

[19] Grzesik W (2006), Determination of temperature distribution in the cutting zone using hybrid analytical-FEM technique. Int J Mach Tools Manuf 46: 651–658.

[20] Fang G, Zeng P (2005) Three-dimensional thermo-elastic-plastic coupled FEM simulations for metal oblique cutting processes. J Mater Process Technol 168: 42–48.

[21] Abukhshim NA, Mativenga PT, Sheikh MA (2006) Heat generation and temperature prediction in metal cutting: A review and implications for high speed machining. Int J Mach Tools Manuf 46:782–800.

[22] Tay AOO, Lee KH (1992) Calculation of temperature distribution in machining using a hybrid finite-element-boundary-element method. J Mater Process Technol 29:47–62.

[23] Ozel T (2006) The influence of friction models on finite element simulations of machining. Int J Mach Tools Manuf 46:518–530.

[24] Joshi VS, Dixit PM, Jain VK (1994) Viscoplastic analysis of metal cutting by finite element method. Int J Mach Tools Manuf 34:553–571.

[25] Filice L, Micari F, Settineri L, Umbrello D (2007) Wear modelling in mild steel orthogonal cutting when using uncoated carbide tools. Wear 262: 545–554.

[26] Xie L-J, Schmidt J, Schmidt C, Biesinger F (2005) 2D FEM estimate of tool wear in turning operation. Wear 258: 1479–1490.

[27] Ahmad MM, Draper WA, Derricott RT (1989) An application of the finite element method to the prediction of cutting tool performance. Int J Mach Tools Manuf 29:197–206.

[28] Cakir MC, Sik YI (2005) Finite element analysis of cutting tools prior to fracture in hard turning operations. Mater Des 26:105–112.

[29] Monaghan J, Brazil D (1997) Modelling the sub-surface damage associated with the machining of particle reinforced MMCs. J Comput Mater Sci 9:99–107.

[30] Wen Q, Guo YB, Todd BA (2006) An adaptive FEA method to predict surface quality in hard machining. J Mater Process Technol 173:21–28.

[31] Rao SS (1996) Engineering Optimization – Theory and Practice. Wiley and New Age International (P), New Delhi.

[32] Deb K (2005) Optimization for Engineering Design – Algorithms and Examples. Prentice-Hall of India, New Delhi.

[33] Deb K (2002) Multi-Objective Optimization Using Evolutionary Algorithms. Wiley, Singapore.

[34] Wang X, Da ZJ, Balaji AK, Jawahir IS (2002) Performance-based optimal selection of cutting conditions and cutting tools in multi-pass turning operations using genetic algorithms. Int J Prod Res 40:2053–2065.

[35] Sarvanan R, Asokan P, Vijyakumar K (2003) Machining parameter optimization for turning cylindrical stock into a continuous finished profile using genetic algorithm (GA) and simulated annealing (SA). Int J Adv Manuf Technol 21:1–9.

[36] Abburi NR, Dixit US (2007), Multi-objective optimization of multipass turning processes. Int J Adv Manuf Technol 32:902–910.

[37] Kennedy J, Eberhart R (1995) Particle swarm optimization, Proceedings of the IEEE international conference on neural networks (Perth, Australia), pp. 1942–1948. Piscataway, NJ.

[38] Dorigo M, Maniezzo V, Colorni A (1996) Ant system: optimization by a colony of cooperating agents. IEEE Trans Syst Man Cybern 26:29–41.

[39] Vijayakumar K, Prabhaharan G, Asokan P, Saravanan R (2003) Optimization of multi-pass turning operations using ant colony system. Int J Mach Tools Manuf 43:1633–1639.

Index